Indian Institute of Metals Series

More information about this series at http://www.springer.com/series/15453

Bikramjit Basu · Sourabh Ghosh

Biomaterials for Musculoskeletal Regeneration

Applications

Springer

Bikramjit Basu
Indian Institute of Science
Bangalore, Karnataka
India

Sourabh Ghosh
Indian Institute of Technology Delhi
New Delhi, Delhi
India

ISSN 2509-6400 ISSN 2509-6419 (electronic)
Indian Institute of Metals Series
ISBN 978-981-10-3016-1 ISBN 978-981-10-3017-8 (eBook)
DOI 10.1007/978-981-10-3017-8

Library of Congress Control Number: 2016955926

Printed on acid-free paper

This Springer imprint is published by Springer Nature
The registered company is Springer Nature Singapore Pte Ltd.
The registered company address is: 152 Beach Road, #22-06/08 Gateway East, Singapore 189721, Singapore

Bikramjit Basu would like to dedicate this book to his father, Manoj Mohan Basu & mother, Chitra Basu

Foreword

Nature has provided us with tissue interfaces with remarkable complexity and functional excellence. It is therefore clear that restoring damaged or diseased musculoskeletal tissues to healthy states represents an immense challenge. There is also a growing clinical need for such repairs for patients around the world, and this need drives continued research and development efforts to elucidate, understand and exploit the underlying mechanisms involved in these complex tissue interfaces. These efforts will lead to the next generation of repairs and medical devices to help patients with failing tissues and organs. This textbook provides an important view and teaching tool into the state of this process, from both research and development/regulatory perspectives. The work includes the rationale, the progress and the challenges ahead, along with the opportunities available to guide students and professionals into the field of musculoskeletal tissue regeneration. This textbook is timely in serving as a useful guide to those getting started in the field, covering key aspects of biomaterials and composite systems, mechanics, case studies, biocompatibility, clinical trials and the need for integrated approaches in terms of theory and experiments in materials science and biology focused on a clinical context.

David L. Kaplan
Department of Biomedical Engineering
Tufts University
Medford, MA USA

Series Editors' Preface

The Indian Institute of Metals Series is an institutional partnership series focusing on metallurgy and materials sciences.

About the Indian Institute of Metals

The Indian Institute of Metals (IIM) is a premier professional body (since 1947) representing an eminent and dynamic group of metallurgists and materials scientists from R&D institutions, academia and industry mostly from India. It is a registered professional institute with the primary objective of promoting and advancing the study and practice of the science and technology of metals, alloys and novel materials. The institute is actively engaged in promoting academia–research and institute–industry interactions.

Genesis and History of the Series

The study of metallurgy and materials science is vital for developing advanced materials for diverse applications. In the last decade, the progress in this field has been rapid and extensive, giving us a new array of materials, with a wide range of applications and a variety of possibilities for processing and characterizing the materials. In order to make this growing volume of knowledge available, an initiative to publish a series of books in metallurgy and materials science was taken during the Diamond Jubilee year of the Indian Institute of Metals (IIM) in the year 2006. IIM entered into a partnership with Universities Press, Hyderabad, and as part of the IIM book series, 11 books were published, and a number of these have been co-published by CRC Press, USA. The books were authored by eminent professionals in academia, industry and R&D with outstanding background in their respective domains, thus generating unique resources of validated expertise of interest in metallurgy. The international character of the authors' and editors has enabled the books to command national and global readership. This book series includes different categories of publications: textbooks to satisfy the requirements of undergraduates and beginners in the field, monographs on select topics by experts in the field and proceedings of select international conferences organized by IIM after mandatory peer review. An eminent panel of international and national experts constitute the advisory body in overseeing the selection of topics, important areas to be covered, in the books and the selection of contributing authors.

Current Series Information

To increase the readership and to ensure wide dissemination among global readers, this new chapter of the series has been initiated with Springer. The goal is to continue publishing high-value content on metallurgy and materials science, focusing on current trends and applications. Readers interested in writing for the series may contact the undersigned series editor or the Springer publishing editor, Swati Meherishi.

About This Book

This book has two volumes, one on "Biomaterials for Musculoskeletal regeneration: Concepts" by Prof. B. Basu and another on "Biomaterials for Musculoskeletal Regeneration: Applications" by Prof. B. Basu and Prof. S. Ghosh. The volume on "Biomaterials for Musculoskeletal regeneration: Concepts" has comprehensive coverage on biological and material concepts in bone tissue engineering. It comprises of 12 chapters covering the basics of biological sciences relevant to biocompatibility; an overview of biomaterials and concepts of bone tissue engineering; desired mechanical properties of biomaterials and osteogenic tissues; aspects of ceramic processing; scope of additive manufacturing; concepts of fracture toughness; toughening mechanisms, friction and wear behaviour; fundamental aspects of experimental techniques to evaluate degradation, toxicity, and *in vivo* biocompatibility of biomaterials; properties of 3D porous scaffolds and simulator studies; new design concepts for developing multifunctional biomaterial; and newer approaches for enhancement of biocompatibility properties. The next volume on "Biomaterials for Musculoskeletal Regeneration: Applications" jointly authored by Prof. B. Basu and Prof. S. Ghosh provides important features demonstrating the opportunity of biomaterial development using various conventional and advanced fabrication techniques, prototype development and clinical trials This book comprises of 12 chapters with 10 chapters exclusively presenting case studies of illustrative examples of biomaterials development. Toughness enhancement of hydroxyapatite-based bioceramic implants without compromising on cytocompatibility; new design concepts and newer approaches for the development of multifunctional biomaterials and with multiscale porosity; osteointegration and implant stability of strontium-containing glass ceramics; injection moulding route; and 3D print cartilage-based prototype development are all described in detail. As a whole, this textbook with both volumes is providing a new horizon for the readers to develop the basic and advanced understanding of the bone tissue engineering, and applications, without any inhibitions of a non-biology background. The author and the editors are of conviction that this book will motivate students and researchers to develop new bioengineering strategies that trigger novel biomaterial development. We wish all readers enrichment in knowledge and motivation. Also, we await the feedback for improving the book when it goes to the second edition.

Baldev Raj
Editor-in-chief, and
Director
National Institute of Advanced Studies, Bangalore

U. Kamachi Mudali
Co-editor-in-Chief
Outstanding Scientist and Associate Director
Indira Gandhi Centre for Atomic Research, Kalpakkam

Preface

The impact of the field of biomaterials for musculoskeletal regeneration has been truly reflected in high impact publications in leading science journals and research funding for large interdisciplinary research programs around the world. A common thread of motivation for such intense research activities globally has been the immense potential of the biomaterials in treating various musculoskeletal diseases. For example, the bone/joint degenerative and inflammatory problems affect millions of people worldwide. Because of the growing geriatric population, these problems account for half of all chronic diseases in people over 50 years of age. Osteoporosis is another prevalent bone disorder, being one of the most common health problems in the middle aged populations and post-menopausal women, worldwide. The prosthetic infection has been another cause for revision surgery in many patients undergoing orthopedic surgeries. A number of chapters in this book will discuss the case studies related to some of these human diseases.

The ultimate impact on human healthcare has attracted a large number of materials engineers (more than other engineering disciplines) to pursue research on biomaterials over the last few decades around the world. In fact, many materials science and engineering departments in those universities offer the compulsory course on this subject at undergraduate and graduate levels. Different books treat this subject in different manners. The sustained progress in this field demands a textbook highlighting some of the major applications of the biomaterials for musculoskeletal applications. One of the authors, Dr. Bikramjit Basu, being a metallurgist by formal undergraduate and postgraduate training and a ceramicist by virtue of doctoral training, and the other author, Dr. Sourabh Ghosh with a background of textile technology and surgical oncology, together have conceptualized this book in a different manner than the existing books on this subject. In particular, this book has the following important features: (a) coverage of the various conventional and advanced manufacturing of materials as well as mechanical and biocompatibility properties of newly developed biomaterials through a number of case studies demonstrating the opportunity of biomaterial development and (b) a discussion on how to take lab scale research to a patient's bedside through technology readiness level assessment of biomedical device prototype development and clinical trials with examples. In view of such important features, this book will be extremely useful for the non-specialist and the beginners in the field of biomaterials.

It is expected that this text book will help the young researchers to appreciate the fact that one does not need to know the entire spectrum of biological sciences to conduct cutting edge research on biomaterials.

This text book entitled "*Biomaterials for Musculoskeletal Regeneration: Application*" contains altogether 12 chapters with 10 chapters on case studies illustrative examples of biomaterials development. The introductory chapter emphasizes the clinical perspective of this field and this is followed by several case studies, essentially illustrating various aspects of development of biomaterials for different biomedical applications. In the first of such chapters, the biocompatibility, cell adhesion and proliferation of hydroxyapatite–titanium (HA-Ti) bulk nanocomposites are presented. The toughness enhancement of bioceramic implants, particularly for hydroxyapatite-based materials, has been a major concern. This aspect has been discussed in reference to the HA-Ti-based composites with moderately high fracture toughness properties. The results of *in vitro* studies are also summarized to demonstrate that cytocompatibility is not compromised while achieving better toughness properties. This is followed by a chapter on micro- and macroporous scaffolds, primarily based on HA. Some of the new design concepts in developing implants and the perspective of the development of multifunctional biomaterials with multiscale porosity are emphasized. This chapter closes with the brief discussion on some of the newer approaches for the enhancement of biocompatibility properties.

The following chapters thereafter discuss strontium-containing glass ceramics and mica glass ceramic (Chap. 5). The results of both in short-term and long-term osseointegration together and implant stability are majorly discussed in Chap. 4, and to this end, the results of histology and microcomputer tomography are summarized. The possibility of developing some fascinating microstructures and composition-dependent *in vitro* properties are discussed in Chap. 5. The next chapter, i.e. Chapter 6, presents the case study on HDPE-based hybrid composites using injection moulding route, and their biocompatibility properties are also summarized. One important aspect is the biomaterial-based prototype development. This aspect is discussed in Chap. 7 in reference to HDPE hybrid composite-based acetabular socket development for total hip surgery. In Chap. 8, strategies and unsolved challenges to develop printed cartilage has been addressed.

The clinical trials are important aspect in translational research on biomaterials. The basic premise as well as a thorough discussion on randomized control trials is provided in Chap. 9. This is followed by a chapter on clinical trials for biomaterial constructs for maxillofacial reconstruction (Chap. 10). The technology and manufacturing readiness level to take laboratory-scale research to prototype development is discussed in Chap. 11. The last chapter provided author's perspective on the present status and future growth of this societally important field of research. In summary, the various chapters globally present the authors' perspective on various stages to be followed while developing a new biomaterial—from materials processing to physical properties/mechanical properties measurement and to *in vitro/in vivo* biocompatibility assessment. Such an integrated approach requires one to acquire a healthy theoretical foundation of both materials science and biological sciences, and these aspects are broadly discussed in a large number of chapters in this book.

This book is an outcome of several years of teaching undergraduate- and postgraduate-level courses in the area of biomaterials, being offered to students of Indian Institute of Technology Kanpur, Indian Institute of Technology Delhi, and Indian Institute of Science, Bangalore, India. Several chapters of this will also reflect on the extensive research from the author's research group, both at IIT Kanpur and at IISc, Bangalore, as well as IIT Delhi in last one decade, which is being supported by Council of Scientific and Industrial Research (CSIR), Department of Biotechnology (DBT), Department of Science and Technology (DST), Indo-US Science and Technology Forum (IUSSTF), and UK-India Education and Research Initiative (UKIERI).

One of the authors, Dr. Bikramjit Basu, would like to acknowledge some of his present and past group members, who deserve special mention, including Greeshma T., B. Sunil Kumar, Ravikumar K., Yashoda Chandorkar, Sharmistha Naskar, Anupam Purwar, Ragini Mukherjee, Gopinath N.K., Atasi Dan, B.V. Manoj Kumar, Amartya Mukhopadhyay, G.B. Raju, Indu Bajpai, Shekhar Nath, Subhodip Bodhak, D. Sarkar, Atiar R. Molla, Naresh Saha, Shouriya Dutta Gupta, Garima Tripathi, Alok Kumar, Shilpee Jain, Ashutosh K. Dubey, Shibayan Roy, Ravi Kumar, Prafulla Mallik, R. Tripathy, U. Raghunandan, Divya Jain, Nitish Kumar and Sushma Kalmodia. The author also acknowledges the past and present research collaborations with a number of researchers and academicians, including Drs. Omer Van Der Biest, Jozef Vleugels, R.K. Bordia, Dileep Singh, M. Singh, T. Goto, Sanjay Mathur, T.J. Webster, Amar S. Bhalla, Ruyan Guo, Mauli Agrawal, Artemis Stamboulis, G. Sundararajan, K. Chattopadhyay, K. Balani, V. Verma, K. Biswas, N.K. Mukhopadhyay, M. Banerjee, R. Gupta, Mira Mohanty, P.V. Mohanan, Ender Suvaci, Hasan Mondal, Ferhat Kara, Nurcan Kalis Ackibas, S. J. Cho, Doh-Yeon Kim, J.H. Lee, Alok Pandey, Arvind Sinha and Animesh Bose.

One of the authors, Dr. Bikramjit Basu, is grateful for the suggestions from several colleagues, including Prof. David Williams, Prof. G. Padmanaban, Prof. Indranil Manna and Prof. N.K. Mukhopadhayay. The author also would like to acknowledge the long-term association with a few of his colleagues, including Profs. Seeram Ramakrishna, Mauli Agrawal, Raman Singh, Anish Upadhyaya, Kantesh Balani and Krishanu Biswas, who have provided some suggestions to me during the course of writing this book. A few chapters of this book are critically reviewed by Profs. M.S. Valiathan, Abhay Pandit, Brian Derby, P. Balaram, H.S. Maiti, Dieter Scharnweber, Alok Dhawan, Ranjna C. Dutta, Deepak Saini, Aditya Murty and S. Bose. We are grateful to them. The author also acknowledges the support from the colleagues of SCTIMST, Trivandrum, including Drs. C. P. Sharma, H.K. Varma, C.V. Muraleedharan, A. Sabareeswaran and Sahin J. Shenoy during the course of writing this book. The clinical and commercial perspective on biomaterials development, as summarized briefly in this book, is largely credited to the author's continuous discussion with his colleagues, Dr. D.C. Sundaresh (honorary director, Sri Sathya Sai Institute of Higher Medical Sciences, Bangalore), Mr. Ravi Sarangapani (vice president, Smith and Nephew, Pune), Dr. Tanvir Momen (orthopaedic surgeon, Woodlands Hospital, Kolkata), Dr. K.H. Sancheti (Sancheti Hospital, Pune) and Dr. T.R. Rajesh (cardiothoracic surgeon, Sparsh Hospital, Bangalore).

One of the authors, Dr Bikramjit Basu, would also like to thank the Centre for Continuing Education, IISc, Bangalore, and the Indian Institute of Metals (in particular the office bearers of the IIM Headquarters, Kolkata, and other colleagues involved in the Springer-IIM book series, Dr. K. Bhanu Shankar Rao and Dr. K. Mudali) for extending necessary support and financial help during the writing of this book. The author likes to express his gratitude to his long-time friend and collaborator, Dr. Jaydeep Sarkar, for his constant inspiration during the writing of this book. Finally, we would like to acknowledge the moral support extended by our parents and family members during the course of writing this book. Last but not least, the authors are extremely grateful to Dr. Baldev Raj, director, National Institute of Advanced Studies, Bangalore, for his constant inspiration to motivate us to take up this important and satisfying assignment to write this book. The author also expresses a great pleasure to publish this book as the first one under the umbrella of "Springer-IIM book series". The great help rendered by Prerana S., Shubham Jain, Sherine Alex, Srimanta Barui, Sourav Mandal, Subhadip Basu, Subhomoy Chatterjee and Rahul Upadhyaya and Shikha Chawla during the writing of this book is gratefully acknowledged.

Bangalore, India
New Delhi, India
May 2016

Bikramjit Basu
Sourabh Ghosh

Contents

About the Authors

Dr. Bikramjit Basu is currently the professor at the Materials Research Center and holds the associate faculty position at the Center for Biosystems Science and Engineering, Indian Institute of Science (IISc), Bangalore. He is currently an adjunct faculty at the Indian Institute of Technology Kanpur (IITK). After his undergraduate and graduate degrees in metallurgical engineering, he earned his Ph.D. in ceramics at Katholieke Universiteit Leuven, Belgium, in March 2001. After a brief postdoctoral stint at University of California, Santa Barbara, he joined IITK in November 2001 as an assistant professor and was promoted to professor in March 2012.

Over close to last two decades, Bikramjit Basu has made outstanding contributions to the field of biomaterials, wherein he has brought in a strong materials engineering perspective to address biologically and clinically relevant problems. An ardent believer of "bedside to bench to bedside" concept, he is currently leading, as principal investigator, the largest Center of Excellence on Biomaterials in India with an interdisciplinary team of 15 co-investigators and 20 young researchers (of seven institutions) from diverse background of materials science, histopathology, polymer science, biomechanics, ceramics engineering, prosthodontics, orthopaedics and biomedical entrepreneur to translate biomaterial innovations to orthopaedic and dental applications. As principal investigator of multi-institutional research programmes on biomaterials funded by Indo-US Science and Technology Forum (2008–2012) and UK-India Education and Research Initiative (UKIERI) (2009–2012), he has played a pivotal role towards the success of multiple bilateral projects in the area of orthopaedics and cardiovascular tissue engineering. His collaborative research with Brown University on developing PLGA-carbon nanofibre-based cardiac patches has received media attention worldwide. One of the landmark outcomes of his UKEIRI project has been to establish long-term implant stability and osseointegration of a new generation of strontium-substituted glass ceramic implants for osteoporosis treatment. Prof. Basu's international standing and impact on the field are illustrated by his prolific publication record (more than 225 peer-reviewed journal articles, including 25 papers in journals with high impact factor (>5.0), more than 20 invited review papers/chapters) and citation record (total citation: ~5300, H-index: 41). He currently serves on editorial board of 12 SCI journals. A critical citation analysis by Elsevier Scopus has placed him as the topmost biomaterial scientist of India, based on the publications during 2010–2015. In order to promote biomaterials research, he has been instrumental in initiating a new international conference series, BIND (Design of Biomaterials) and organized it at IITK (2006) and IISc (2012). In addition, he organized 20 thematic sessions at major conferences organized by American Ceramic Society and Society for Biomaterials. He has delivered 20 plenary/keynote lectures and 140 invited seminars in USA, UK, France, Germany and Japan. Dr. Basu has served as a research adviser to 16 Ph.D. students and mentored 10 young academic colleagues.

Professor Basu's contributions in engineering science have been widely recognized. He is one of the youngest recipients from the metallurgy/materials science community as well as the only biomaterial scientist so far to receive India's most coveted science award, Shanti Swarup Bhatnagar award (2013). He is an elected fellow of the Indian National Academy of Engineering (2015), West Bengal Academy of Science and Technology (2014), Society for Biomaterials and Artificial Organs (2014) and National Academy of Sciences, India (2013). In 2015, he received National Bioscience award from the government of India for unique contributions in the frontier areas of biological sciences. He remains the only Indian from India to receive the prestigious "Coble Award for Young Scholars" (2008) from the American Ceramic Society.

Dr. Sourabh Ghosh has been working as an associate professor at IIT Delhi since 2008. After completing his Ph.D. in surgical oncology from University of Basel, Switzerland (2003–2006), he received Fellowship for Prospective Researchers ("Stipendien für Angehende Forschende"), a highly prestigious grant for young scientists, who are at the point of starting their independent research career, from Swiss National Science Foundation, for pursuing research at the Tufts University, Boston (2006–2007).

His research highlights include designing bioinks with tailored composition and rheology to develop complex 3D scaffolds. Using 3D-printed hydroxyapatite, he conducted clinical trials for oral cancer patients at the Lady Hardinge Hospital, New Delhi. Results demonstrated successful healing in severe maxillofacial osteotomies, for which he received the Gandhian Young Technological Innovation award. In parallel, preclinical investigations of such 3D-printed scaffolds are ongoing to assess their osteogenic and angiogenic potential at AIIMS. He has developed injectable silk hydrogel to deliver stem cells and/or glucosamine in degenerated disc, which activated pathways of organogenesis for back pain relief. This work was featured in Rajya Sabha TV. Some other *in vitro* models successfully established include *in vitro* osteoarthritis model, healthy as well as scarred human skin models and micropatterned thermoresponsive matrices for engineering corneal cell sheets.

He has been honoured with the MAHE award by the Society of Biomaterials and Artificial Organs and received Outstanding Young Faculty Fellowship by Kusuma Trust, IIT Delhi. So far, his team has published more than 40 research papers in peer-reviewed journals with 4 chapters and 2 patents, with 2 more currently in process.

Abstracts and Keywords

Chapter 1

Abstract: In this introductory chapter, the clinical perspective highlighting the relevance of biomaterials towards human healthcare is discussed. The major driving force behind the development of new generation of biomaterials is the implant failure, longer healing time post-surgical implantation or leading to clinically unacceptable host response. These issues are emphasized after defining a few concepts of biomaterials science. It is envisaged that the brief discussion in this chapter will set the platform for the readers to realise the significant thrust to develop implantable biomaterials with better host response for musculoskeletal applications.

Keywords: Bioceramics, Biocomposites, Biomaterials, Biopolymers, Bone, Ceramics, Clinical perspective, Implant failure, Implants, Knee replacement, Metals, Polymers, Revision surgery, Scaffolds, Tissue engineering, Total hip replacement.

Chapter 2

Abstract: The research on bulk hydroxyapatite (HA)-based composites is driven by the need to develop biomaterials with better mechanical properties without compromising biocompatibility properties. Despite several years of research, the mechanical properties of the HA-based composites still need to be enhanced to match the properties of natural cortical bone. In this regard, the scope of the present chapter is limited to discuss the processing and the mechanical as well as biocompatibility properties in the context of bone tissue engineering applications of a model system i.e. HA–Ti. It will be discussed as how hydroxyapatite-titanium (HA–Ti) based bulk composites can be processed to have better fracture toughness and strength together with uncompromised biocompatibility. On the materials fabrication aspect, the recent results are discussed to demonstrate that advanced manufacturing technique like, spark plasma sintering can be adopted as an advanced processing route to restrict the sintering reactions, while enhancing the mechanical properties. Various toughening mechanisms are discussed with an emphasis to synergize multiple toughening mechanisms, which requires careful tailoring of microstructure. The *in vitro* cytocompatibilty, as well as *in vivo* biocompatibility results are also reviewed.

Keywords: Ag, Al_2O_3, $BaTiO_3$, Biocompatibility, Brittleness, $CaTi_4(PO_4)_6$, $CaTiO_3$, Compressive strength, Conventional sintering, Crack bridging, Crack deflection, Crack growth resistance, Densification, FGM, Field Activated Sintering Technology (FAST), Fluorescence microscope, Fracture toughness, HA, HA–ZrO_2, HA–Ag, HA–Al_2O_3, HA–$BaTiO_3$, HA–$CaTiO_3$, HA–Ti, *in vitro*, *in vivo*, K_{IC}, Phase stability, SEM, SEVNB, Sintering, Spark plasma sintering, SPS, TCP, TEM, Ti, Toughening mechanisms, Toughness, ZrO_2, α-TCP, β-TCP.

Chapter 3

Abstract: This chapter will present some of the unique processing approaches to develop porous scaffolds with porosity scaling in the range of either 1–50 μm or in the range of 100–300 μm. In the first part of this chapter, the results will be summarized to illustrate how hydroxyapatite scaffolds with micro/mesoscale porosity in the range of 1–50 μm can be produced using the polymer blend method using PMMA (poly methyl methacrylate) as porogenous template. The cytocompatibility assessment using human osteoblast cells (Saos2) confirm that the adopted processing approach to produce porous hydroxyapatite scaffolds can stimulate significant cell adhesion and osteoblast differentiation. In the second part of this chapter, the efficacy of polymer sponge replication method to prepare the macroporous hydroxyapatite scaffolds with interconnected oval shaped pores of 100–300 μm with pore wall thickness of ~50 μm will be demonstrated. The enhanced cellular functionality and the ability to support osteoblast differentiation for porous scaffolds in comparison to dense HA has been explained in terms of higher protein absorption on porous scaffold. The last part of the chapter will present the results on the protein adsorption and release kinetics as well as *in vitro* biodegradability of cryogenically cured hydroxyapatite-gelatin based micro/macroporous scaffolds (CHAMPS). The adsorption and release of bovine serum albumin (BSA) protein exhibits steady state behavior over the incubation period up to 10 days. The extensive micro-computed tomography (micro-CT) analysis establishes cancellous bone-like highly interconnected and complex porous architecture of CHAMPS scaffold. Importantly, excellent adsorption (up to 50 %) and release (up to 60 % of adsorbed protein) of BSA has been uniquely attributed to the inherent porous microstructure of the CHAMPS scaffold.

Keywords: ALP assay, Cell adhesion, Cell differentiation, Cell growth, Cell spreading, Cell viability, Freeze drying, Macroporous scaffold, Micro-CT, Microporous scaffold, MTT assay, Osteoblast cells, Polymer replication method, Pore interconnectivity, Porogen, Porosity, Protein adsorption, Scaffold, Sintering, TEM, HA.

Chapter 4

Abstract: The most important property of bone cement or a bone substitute in load bearing orthopaedic implants is good integration with host bone with reduced bone resorption and increased bone regeneration at the implant interface. Long term implantation of metal-based joint replacements often results in corrosion and particle release, initiating chronic inflammation leading onto osteoporosis of host bone. An alternative solution is the coating of metal implants with hydroxyapatite (HA) or bioglass or the use of bulk bioglass or HA-based composites. One of the desired properties for any new biomaterial composition is its long term stability in a suitable animal model and such property cannot be appropriately assessed by performing short term implantation studies. While hydroxyapatite or bioglass coated metallic biomaterials are being investigated for *in vivo* biocompatibility properties, such study is not extensively being pursued for bulk glass ceramics. In view of their inherent brittle nature, the implant stability as well as impact of long term release of metallic ions on bone regeneration have been a major concern. In the above perspective, the present study reports the *in vivo* biocompatibility and bone healing of the strontium (Sr)-stabilized bulk glass ceramics with the nominal composition of $4.5SiO_2$–$3Al_2O_3$–$1.5P_2O_5$–$3SrO$–$2SrF_2$ during short term implantation of up to 12 weeks in rabbit animal model followed by long term implantation for 26 weeks in cylindrical bone defects in rabbit model. The progression of healing and bone regeneration was qualitatively and quantitatively assessed using fluorescence microscopy, histological analysis and micro-computed tomography. The overall assessment of the present study establishes that the investigated glass-ceramic is biocompatible *in vivo* with regards to local effects after short term implantation in rabbit animal model. Excellent healing was observed, which is comparable to that seen in response to a commercially available implant of HA-based bioglass.

Keywords: Biocompatibility, Bone implant interface, Bone labeling, Bone morphometric analysis, BV/TV ratio, Explants, Femoral defect model, Fluorescence image, Fluorochrome dosage, Glass-ceramics, Histology, *in vivo*, Long term biocompatibility, Micro-CT, Neo-bone, Osseointegration, Rabbit model, Short term biocompatibility, Sr, X-ray radiography.

Chapter 5

Abstract: The design and development of glass ceramic materials provide us the unique opportunity to study the microstructure development with changes in either base glass composition or heat treatment conditions and thereby developing an understanding of processing-microstructure-property (mechanical/biological) relationship. Among various brittle materials, the mica based glass ceramics with crystalline ceramic embedded in a glass matrix are of greater scientific interest, because of their machinability. Considering the potential of these materials as dental implants, this chapter summaries the published results on K_2O–B_2O_3–Al_2O_3–SiO_2–MgO–F glass ceramics to demonstrate the microstructure dependent mechanical, tribological and cytocompatibility properties. Among the high hardness of around 8 GPa together with 3-point flexural strength and elastic modulus of 80 MPa and 69 GPa, respectively were obtained in glass ceramics with maximum amount of crystals. While analyzing influence of environment on the friction and wear behavior systematic decrease in wear rate with test duration was recorded with a minimum wear rate of 10^{-5} mm^3/Nm after 100,000 fretting cycles in artificial saliva. The *in vitro* results illustrate how small variation in fluorine and boron in base glass composition influences significantly the cytocompatibility and antimicrobial bactericidal property, as evaluated using a range of biochemical assays. Overall, the mechanical, tribological property, *in vitro* cytocompatibility study, when taken together clearly reveals that microstructure and base glass composition play an important role in enhancing the cellular functionality and antimicrobial property.

Keywords: ALP assay, Antibacterial, Artificial saliva, Bacteria colony, Bactericidal, Biomineralisation, Cell adhesion, Cell differentiation, Cell proliferation, Cell spreading, Ceramisation, Crystallisation, Cytocompatibility, Décor, Dental restoration, Dissolution, Fluorophlogopite, Fretting damage, Glass, Glass ceramics, Glass melting, Hardness, Heat treatment, *in vitro*, Machinability, Mica, MTT assay, Nucleation, Osteoblast cells, *S. epidermidis*, Tooth, Wear rate.

Chapter 6

Abstract: In this chapter, the published research results of author's own group are summarized to establish the HDPE–HA–Al_2O_3 based hybrid composites with enhanced mechanical properties and good biocompatibility properties. The processing related concerns in injection molding route are discussed in reference to addition. The tensile and flexural fracture properties are analyzed. The cytocompatibility with osteoblast-like cells and *in vitro* mineralization are also discussed. More importantly, the osseointegration in rabbit model is qualitatively and quantitatively analyzed.

Keywords: Al_2o_3, Bone growth, Bone remodeling, Ceramic filler, Fluorescence microscopy, Fracture toughness, HA, HDPE, Histology, Histomorphometry, Host bone, Hybrid composite, Injection molding, K_{IC}, MTT assay, Osseointegration, Osteogenic cells, Protein adsorption, SEM, Shear ate, Tensile deformation, Torque, Viscosity, *in vitro*, *in vivo*, Implant stability, Surface energy, Mineralisation, Cell functionality, Flexural test, Viscoelastic deformation.

Chapter 7

Abstract: In some of the preceding chapters, the processing and biocompatibility property are discussed with a focus on 'lab-scale' research of designing new biomaterials. The translation of bench to bedside requires the fabrication of biomedical device prototype based on the lab-scale tested biomaterials. While addressing this aspect, this chapter and ZrO_2—toughened Al_2O_3 reports design and development of the compression molded high density polyethylene (HDPE)-based biocompatible acetabular socket with 20 wt% hydroxyapatite (HA) and 20 wt% alumina (Al_2O_3) ceramic fillers for the total hip joint replacement applications. This new implant material can be used either for non-cemented socket or as a liner for a metal back porous coated cup. This is more relevant as the total hip joint replacement (THR) arthroplasty has reduced pain clinically and tremendously improved the quality of life for millions. In the context of a growing need to develop patient-specific biomedical devices, this chapter describes some physical properties and more importantly 3D microstructural characterization using micro-computed tomography.

Keywords: Acetabular socket, Al_2O_3, Bone implantation, CAD, Ceramic filler, COF, Compression molding, Cytocompatibility, Femoral defect, Fretting wear, HA, HDPE, HDPE–HA–Al_2O_3 composite, Hip prosthesis, *in vitro*, *in vivo*, Osseointegration, Osteolysis, Polymer composite, Prototype, Segmental defect, THR, Total hip joint replacement, Wear debris, Wear rate, Wear resistance, X-ray radiography, ZrO_2.

Chapter 8

Abstract: In the past decade, cartilage tissue engineering research envisaged on the development of engineered constructs to repair cartilage defects, which could be inflicted due to degenerative disease or traumatic injury. However, despite significant efforts, development of load bearing functional cartilage remains elusive. 3D bioprinting offers a fascinating approach to replicate the complex anatomical cartilaginous tissue architecture by precise delivery of encapsulated cells and morphogens at pre-determined location. Silk fibroin protein can be used for cartilage 3D bioprinting, as it possesses unique features such as shear thinning behaviour, self-supporting filamentous extrusion, instant cytocompatible sol-to-gel transition and tailorable mechanical strength. But systematic optimization of chemistry and rheology of bioink, topographical, physico-chemical and biomechanical functionality of printed cartilage constructs should be done to achieve this target. In this chapter we tried to summarize how chondrogenic differentiation is supported in 3D printed construct and signaling mechanisms minimizing hypertrophic differentiation of progenitor cells towards development of phenotypically stable engineered cartilage constructs.

Keywords: 3D Printing, Alveolar ridge, Hydroxyapatite, Maxillofacial, Osteocytes, Textile braids.

Chapter 9

Abstract: An important concept that has been increasingly recognised in the biomedical research community is the 'bedside-bench-bedside' concept. Once a new biomaterial is found to exhibit a pre-clinically acceptable biocompatibility, the final and the most important stage of research is the clinical trial using ethically-approved protocols. The research on biomaterials should evolve around a specific disease model and, therefore, *in vitro* as well as *in vivo* biocompatibility assessments should involve the use of relevant cell lines or animal models, respectively. This chapter introduces the importance of clinical trials in biomaterials' research, which has received increasing attention, but is yet at a lower intensity than those in the case of new drug designs and development. Further more, various possible avenues to conduct clinical trials are also briefly mentioned. Subsequently, this chapter discusses the concept of Randomised Control Trials (RCTs). Some illustrative examples are also provided to substantiate how clinical trials can establish the clinical efficacy of some of *in vivo* biocompatible implants. The various stages of the clinical trials are also highlighted. One specific example is discussed to describe the rationale for the study of design as well as the outcomes of clinical trials, particularly in the context of dental implants in prosthodontics. Subsequently, the chapter closes with the ethical issues/approval stages to be critically considered prior to conducting any clinical trials. The content and discussion in this chapter should be used only for academic purposes as the ethical approval and associated legal issues involved in clinical trials can vary from country to country. In order to illustrate some published clinical study reports in the field of bone tissue engineering applications are described for the readers to develop an understanding of how to conduct such experiments in clinical settings as well as how to evaluate the outcomes of such important studies. As a cautionary note to the readers, this chapter should neither be used to design any clinical trial study nor does this document have any legal binding.

Keywords: Allocation concealment, Blindness, Clinical trial, CONSORT, Cranioplasty, Good clinical practice, Good laboratory practice, Osteoporosis, Phase I trial, Phase II trial, Phase III trial, Prosthodontic, Randomization, Randomized control study.

Chapter 10

Abstract: Increasing number of clinical incidences of maxillofacial disorders has developed the quest for the fabrication of improved synthetic materials to aid in complete craniofacial restoration. Replicating the complex 3D architecture and functional dynamics of maxillofacial bone tissue is a challenging proposition which aggravates the need for a custom-made, on demand tissue replacement strategy for rendering patient specificity which could not be achieved till date. Textile technology offers versatility to develop 3D spatial structures with tailor-made mechanical properties in the order of micro- and macro meters. 3D printed structures have fascinating potential for reconstruction of maxillofacial deformations due to the ability to fabricate patient-specific, defect site-specific structural features in the order of several nanometers along with the flexibility of being tailored into any desired shape or size. These case studies highlight clinical trials to evaluate the key properties of high performance textile braided structures for preservation of dimension of alveolar ridge as well as 3D printed Hydroxyapatite Direct-write scaffolds for maxillofacial reconstruction and how tailoring their architecture could enhance patient-specificity and defect-specificity in situ.

Keywords: Biocompatibility, Bone implant interface, Bone labeling, Bone morphometric analysis, BV/TV ratio, Explants, Femoral defect model, Fluorescence image, Fluorochrome dosage, Glass-ceramics, Histology, *in vivo*, Long term biocompatibility, Micro-CT, Neo-bone, Osseointegration, Rabbit model, Short term biocompatibility, Sr, X-ray radiography.

Chapter 11

Abstract: Any attempt to pursue translational research requires adopting a truly interdisciplinary approach by integrating the ideas drawn from multiple disciplines such as Mechanical engineering, Materials Science, Biological sciences and Biomedical engineering. While emphasizing the need to develop a scalable and commercially viable strategy to fabricate biomedical implants, this chapter will discuss the concepts of Technology Readiness Levels (TRLs) and Manufacturing Readiness Levels (MRLs). In discussing various TRLs, the different aspects of property measurements or process control are emphasized. The maturation of technology can be realized once one travels across different TRLs. Two illustrative examples are provided so that one can judge how to assign various TRLs at different levels of technology development in research on bone tissue engineering.

Keywords: GLP, GMP, Good laboratory practice, Good manufacturing practice, Manufacturing readiness level, MRL, Technology readiness level, Translational research, TRL.

Chapter 12

Abstract: The last one decade has witnessed a significant impetus towards patient-specific solutions of biomedical implant prototypes. The discussion in preceding chapters emphasize that multidisciplinary efforts are required to establish such patient-specific implants. Finite element (FE) modelling is used to predict the site-specific mechanical properties, which in-turn requires 3D reconstructed models of macroscopic biological entities based on CT/MRI scan data. The importance of low temperature additive manufacturing processes in fabrication of the patient-specific implants have been particularly highlighted. Also the major challenges related to the design, development and performance limiting properties of the scaffolds are discussed in this chapter. It has been emphasized that the optimization of the shape and size of the pores in the scaffolds is a challenging task in order to obtain the desired *in vivo* cytocompatibility property. The chapter closes with the author's perspective on developing biomedical research programs leading to device/implant fabrication.

Keywords: 3D plotting, 3D printing, Additive manufacturing, Binder, Biocompatibility, Biomedical device, Bone implant, CAD, Clinical trial, CT scan, ECM, Extracellular matrix, FDM, Finite element (FE), Hemocompatibility, Histopathology, Implant, *in vivo*, Layer-by-layer manufacturing, Manufacturing, Micro-CT, Patient-specific implant, Pore shape, Pore size, Porous architecture, Prototype, Regulatory approval, Scaffold, Split-Hopkinson pressure bar test, Toxicology, Weibull modulus.

Chapter 1
Introduction

1.1 Clinical Perspective

Musculoskeletal health problems are considered as prevalent healthcare disorder, particularly in human population for many developing nations, including India, China, etc. Though this was recognised by the United Nations and the WHO through the Bone and Joint Decade (2000–2010), it can be said that significantly more steps are required towards affordable healthcare. The human population in many nations is gradually tending towards older average age and concomitantly reports higher incidences of osteoporosis, osteoarthritis, and other musculoskeletal diseases. This is reflected in the orthopaedic devices market, exhibiting a rapid growth to be valued at up to $41.2 billion by 2019. Several technological innovations in diverse spheres of science are driving growth in this sector; these include biodegradable implants and internal fixation devices, design of patient-specific implants, real time sensor embedded implants etc. It is imperative that manufacturing capabilities be strengthened alongside translating research knowledge into clinical practice.

Presently, a large proportion of healthcare-related treatment involves the use of medical devices in diagnosis, monitoring and implantable applications. The quality of healthcare in the developing nations largely depends on imported medical devices, and there exists an immediate unmet need to engage the manufacturing sector in making medical devices. Many of these devices, particularly those manufactured in the US and Europe, are not designed to take account of the characteristics of patients in various developing nations. Alongside the increased availability of minimally invasive procedures, it can be anticipated that many nations will urgently require large-scale but precise engineering and manufacturing of patient-specific orthopaedic products. Specifically, medical implant manufacturing is a sector which requires patient customization for best results, as well as extreme precision and strict quality control. Moreover, a highly regulated pre-marketing approval process coupled with the cutting edge nature of medical

B. Basu and S. Ghosh, *Biomaterials for Musculoskeletal Regeneration*,
Indian Institute of Metals Series, DOI 10.1007/978-981-10-3017-8_1

research also implies short product life cycles. Medical device manufacturing thus needs to be highly versatile and responsive to changes.

One of the important bottlenecks in any medical device development is the fact that most manufacturing modalities are developed by mechanical/ material scientists using standard production materials with little consideration of suitability of such manufacturing processes to be adopted for medical devices.

In India, approx. 17 % of the overall population suffers from ailments related to musculoskeletal disorders (MSD), amongst which osteoarthritis is the most common (14 %) followed by lower back pain (8 %). Traditionally, customized implants were introduced to treat cases when the available 'off-the-shelf' implant did not fit into the anatomical site, as in the case of patients with juvenile rheumatoid arthritis (JRA), or with unusual bone geometry, or for revision procedures with previous implant failures. However, given the increase in efficiency of technology, physicians are hoping to extend the benefits of patient-specific implants to more common procedures like hip replacements, knee arthroplasty and treatment of recurring tumours.

The orthopaedic device market comprises sales of reconstructive joint implants, trauma devices, spinal implants, bone graft substitutes, etc. The present market for joint replacement alone in one of the developing nations, India, hovers above Rs. 2000 million and most commercial market analysts, e.g. Frost and Sullivan, have forecast a robust growth of this segment. Such anticipated growth is primarily powered by an increasingly aging demographic profile of the country, who require the use of orthopaedic devices in view of their sedentary lifestyle and the recent spurt in medical tourism. There are about 200,000 knee and hip replacements performed in India annually as of 2014, including perhaps 100,000 medical tourism procedures, while the overall growth rate in orthopaedic implants is estimated to be more than 25 % per annum for the next 5–6 years.

The basic limitations, however, are economic disparity, inaccessibility of medical technology in remote areas and a general sense of fear and dissatisfaction amongst patients undergoing interventions. It may here be noted that, along with governmental policies which promote improvements in economic conditions and accessibility and affordability for patients, technological interventions which can overcome the limitations can phenomenally add to market growth apart from improving clinical outcomes. The use of additive manufacturing (AM) technologies to produce patient-specific medical implants is undoubtedly one such technological innovation, which can radically change orthopaedic surgical practice and make it more accessible and economically acceptable to the people. Precise advantages will be obtained through the transmission of data over the internet to manufacturing sites directly from hospitals, which could be located in remote areas. In addition, there could be reduced costs of materials required for production, and reduction in surgical times and hospitalization duration due to better fixation along with removal of revision procedures. The major global players among orthopaedic device manufacturing companies include ArthroCare, B. Braun Melsungen, Biomet, Maxx Orthopedics, Inc., ConMed, Johnson and Johnson, DePuySynthes, Medtronic, Otto Bock HealthCare, Smith and Nephew, Stryker, and Zimmer. This apart, some start-up companies are also venturing into orthopaedic manufacturing. However, the requirement of a large infrastructure for manufacturing and testing of orthopaedic implants can often be an obstacle to the start-ups, as opposed to their opportunities in the tissue engineering field.

1.2 Failure of Implants

The excitement at the fact that a synthetic material can replace a diseased organ/tissue has been a major catalyst for the growth of research on biomaterials. The field of biomaterials is extensive, covering all divisions of materials science and many other scientific disciplines. For practical use, a biomaterial should be amenable to being formed or machined into different shapes, have relatively low cost, and be readily available. Figure 1.5 presents a schematic of the various human body parts, which can be potentially replaced by synthetic biomaterials.

In the context of orthopaedic application, biomaterials are used in different parts of the body for a specific function, e.g. femoral stem, femoral ball head and acetabular cup in THR (Total Hip Replacement), dental implant etc. (see Fig. 1.5) [1–3]. The failure of an implant necessitates that the patient undergo revision surgery [4]. The reasons for the failure of an implant include (a) mismatch in the elastic modulus between implant and the host bone; (b) poor mechanical properties (e.g. fracture toughness); (c) generation of corrosion product; (d) dissolution of material; (e) wear debris; and (f) bacterial infection etc. Therefore, these issues are to be considered during the designing of a biocompatible material for an orthopaedic application.

As discussed in the preceding section, the failure of orthopaedic implants limits durability/lifetime and this has been an important aspect that has driven research on developing new implants. The implant failure can be attributed to a multitude of factors, which are summarised in Fig. 1.1. Material removal in biomechanically loaded articulating orthopaedic joints is caused by wear at interacting surfaces. Continuous sliding, often with low relative displacement between mating surfaces, causes friction particularly in older patients and such friction leads to considerable pain, if physiological fluid contained at osteochondral sites is dried up. The related disease or patients' discomfort is widely known as osteoarthritis (OA), which is found mostly in knee joints. Depending on the severity of osteoarthritic pains, the injection of hyaluronic acid based FDA approved drugs can be one of the clinical solutions for severe OA cases. However, knee replacement is the only clinical solution for fourth degree OA, when OA related pain extends from the knee region to the lower part of the legs, i.e. ankles.

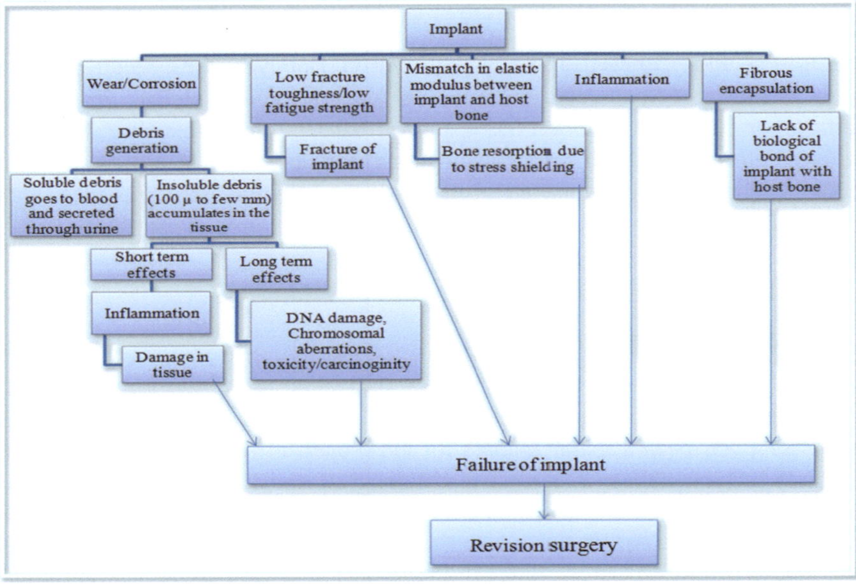

Fig. 1.1 Schematic illustration of various factors responsible for revision surgery (redrawn) [1]

In the context of the biomedical applications of metallic implants, the corrosion and wear have been widely perceived to limit the long-term performance at articulating joints.

Another aspect of wear has much more biological ramifications. The finer sized wear debris particles, generated at articulating joints, can be transported through blood and subsequently through urine. However, insoluble and finer debris particles are often accumulated in the host tissue and can cause inflammation as a short term host response. Over a longer timescale, the wear particles can cause genotoxicity, i.e. DNA damage or carcinogenicity. The debris particles are not only generated through friction and wear, but also through corrosion of metallic implants/degradation of polymeric scaffolds. The dynamic change in pH or other physiological parameters in terms of ionic environment around an implant leads to material degradation. All these aspects drive the development of the new generation metallic implants.

For ceramic implants, the low fracture toughness or strength properties (fatigue/bending/compression/tensile) often lead to failure under the existing biomechanical loading conditions, *in vivo*.

The lack of our predictability for implant failure is the absence of long term device level testing. Depending on targeted applications, specific joint simulator tests are adopted. This is preceded by the lab-scale development of any new biomaterial. It needs to be emphasized here that the lab scale testing involves a simplistic uniaxial or monotonic loading condition. In actual clinical conditions, an implant is subjected to loading in an oblique manner as well as to a combination of various modes (e.g. compression and bending together in the case of hip implants). Moreover, an implant is tested for strength properties in isolation. However, an orthopaedic device often contains an assembly of bearing materials and the biomechanical load transfer at the contacting interface can additionally contribute to the failure of implants. An illustrative example of stem failure in total hip replacement is due to two contrasting cases of mismatch in elastic modulus shown in Fig. 1.2. The failure of an implant is analysed using X-ray radiography, and such failure causes significant pain, which necessitates revision surgery.

The mismatch in elastic modulus between the implant and host bone leads to the differential loading of the implant, which causes stress shielding and bone resorption.

The above aspect can be addressed using either (i) implants with matching elastic modulus without compromising the strength/toughness properties or (ii) innovative design of gradient in porosity on the outer surface of implant or (iii) porous biocompatible coatings. It is worthwhile to note that a given material

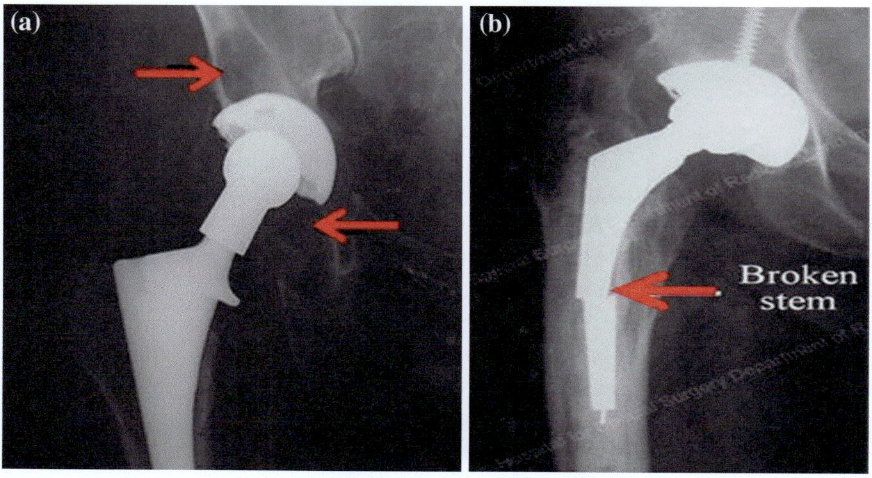

Fig. 1.2 Representative X-ray radiographs revealing two clinical cases caused by a significant mismatch between the host bone tissue and the synthetic implant in total hip replacement surgery: **a** aseptic loosening ($E_{implant} > E_{tissue}$) and **b** mechanical failure leading to stem fracture ($E_{implant} < E_{tissue}$), as indicated by arrow

can have matching modulus, but may have poor load bearing property (inferior strength) or biocompatibility property. Therefore, options (ii) or (iii) can provide clinically relevant solutions. FEM simulation results also proved that an implant with gradient porosity can be a better solution to address stress shielding compared to dense implants, particularly in the case of the stem in THR.

The inflammation due to the lack of structural or mechanical compatibility can cause implant failure. In the case of a bioinert implant, the fibrous encapsulation shields an implant from its surrounding osseous structure and such a scenario in the long term causes implant failure. Summarising, the discussion in this section emphasises the growing need to develop new implantable materials with tailored mechanical properties for bone tissue engineering applications. Such a discussion therefore establishes the basis for subsequent discussion on defining biomaterials and biocompatibility as well as briefly on different material classes.

1.3 Defining Biomaterials and Related Concepts

Biomaterials can be defined as synthetic materials which have been designed to induce a specific biological activity. In order to perform a particular function, it is necessary for the implanted material to be compatible with its environment and not to produce any toxic effect in the body. Williams defines biomaterials as follows.

> A biomaterial is a substance that has been engineered to take a form which, alone or as part of a complex system, is used to direct, by control of interactions with components of living systems, the course of any therapeutic or diagnostic procedure, in human or veterinary medicine.

The major difference between biomaterials and other classes of materials is their ability to remain in a biological environment without damaging the surrounding tissues/bones and without getting damaged in that process [5]. Therefore, biomaterials require both biological and materials properties to fit in with a specific biomedical application. After fabricating a biomaterial, detailed structural and functional characterization need to be conducted prior to implantation, as depicted by the biomaterials science tetrahedron in Fig. 1.3. For instance, a material for nerve regeneration must be compatible with the surrounding environment, and at the same time it should be flexible and stable during nerve regeneration. It must be emphasized here that the biological properties/response of a material in a physiological environment are by far the most important consideration, as opposed to superior mechanical properties, for selecting/defining biomaterials. From health care perspective, it is desirable that a biocompatible material interrupts normal body functions as little as possible. The most important aspect of a biomaterial is,

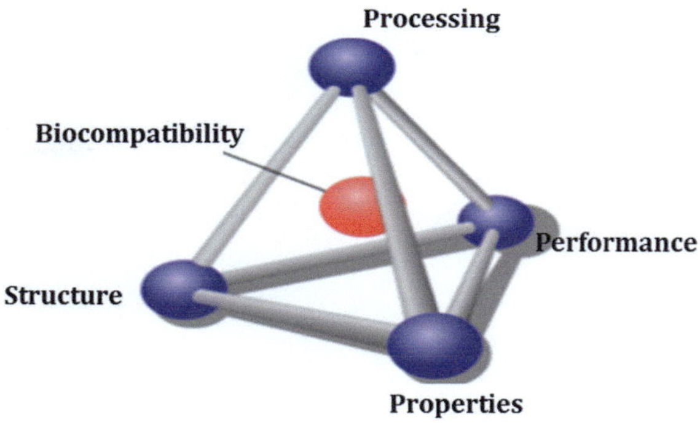

Fig. 1.3 The materials science tetrahedron relating the four cornerstones of materials science and engineering: microstructure, properties, processing and performance with characterization playing an important role

therefore, how a material interacts with living systems, when it is implanted in the human/animal body.

It is instructive to examine the implications of such important terms in the introductory section of this book.

> Apart from its formal definition, biocompatibility, in the author's opinion, should ideally be considered as a concept, rather than a simple material property.

This concept has been described in Fig. 1.3 with the help of the materials science tetrahedron. A change of process parameters is bound to influence the microstructure, which in turn will have an effect on physical property (hardness, strength, toughness). In addition, the biomedical application specific *in vitro* and *in vivo* properties are the most important aspects of biomaterials research.

> The final determinant for any implant to see the light of day is the performance assessment, first in appropriate animal models to assess host response (tissue-level compatibility), followed by clinical trials in human subjects.

The fact that 'biocompatibility' sits at the center of the tetrahedron in Fig. 1.3 makes the point that this concept is to be understood at the crossroads of process-microstructure-property-performance. In fact, any single property measurement, e.g. *in vitro* results, can be used as one of the components of the 'biocompatibility' concept, but an understanding of the above-mentioned range of properties is required to achieve the ideal of biocompatibility.

Extending the discussion further, Fig. 1.4 summarizes various aspects to be considered when developing bone analogue materials. It is interesting to note that 50 % of the aspects mentioned in Fig. 1.4 relate mainly to the biological properties, which include both *in vitro* and *in vivo* property assessment. Moreover, many of the

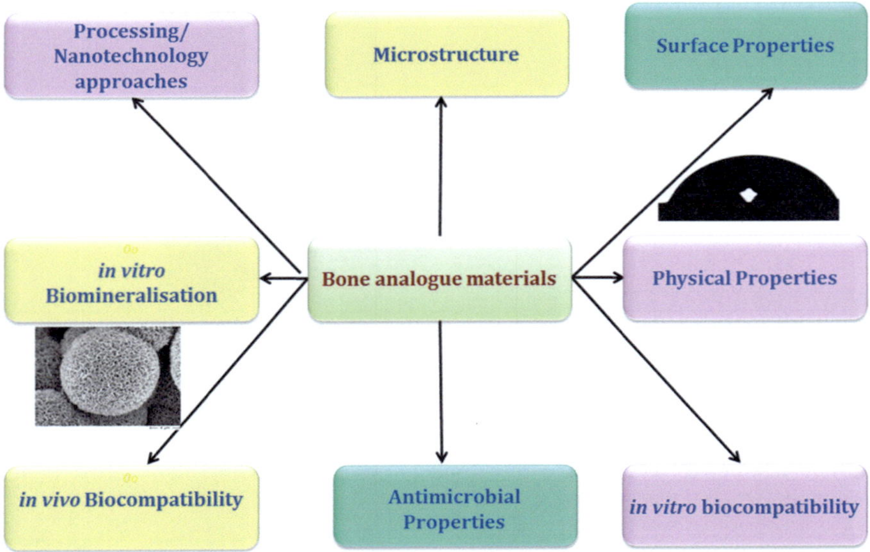

Fig. 1.4 Various facets involved in research on the development of bone replacement materials

biological properties can be correlated with the microstructure and surface properties, while processing conditions influence both microstructure and properties. Among the biological properties, *in vitro* biomineralisation is more specific to bone replacement materials, while antimicrobial properties, cytocompatibility and tissue compatibility are common to biomaterials for diverse biomedical applications. Biomineralisation is related to the ability of a material (which may not contain any CaP phase in bulk) to support the CaP-rich layer formation in simulated body fluid. All the *in vitro* properties (biomineralisation, antimicrobial and cytocompatibility) are used to screen the material prior to in vivo osseointegration study on a limited number of samples in view of the animal ethical considerations. The endpoint results after *in vivo* test guide the researchers to plan for human clinical trials, which require the strictest ethical/regulatory committee approval is required. The above discussion also reflects the interdisciplinary nature of the field of bone tissue engineering.

In the context of pre-clinical study, the host response is an important aspect of scientific relevance. This specific term implies the biological acceptance or integration of non-living biomaterial in a living system. The signature of such response is reflected in the appropriate tissue response around the implanted biomaterial.

Depending on the host response, biomaterials are classified as follows.

(a) Bioinert/biotolerant (absence of any significant biological bond between implants and bone),

(b) Bioactive (implant and bone attach directly to each other, post-implantation),
(c) Bioresorbable (implant gradually resorbed and substantially replaced by new bone ingrowth).

It is important to note that no material implanted in living tissues can be considered as inert. In reality, all materials elicit a response from the host tissue. It is critical that any implant material must avoid a toxic response, which kills cells in the surrounding tissues or releases chemicals that can migrate within tissue fluids and cause systemic damage to the patient (Fig. 1.5).

A biomaterial should be non-cytotoxic, biocompatible, and for many applications it should also show osteoconductive properties. **Osteoconductivity** is defined as the ability of the implant material to serve as a scaffold for new bone formation

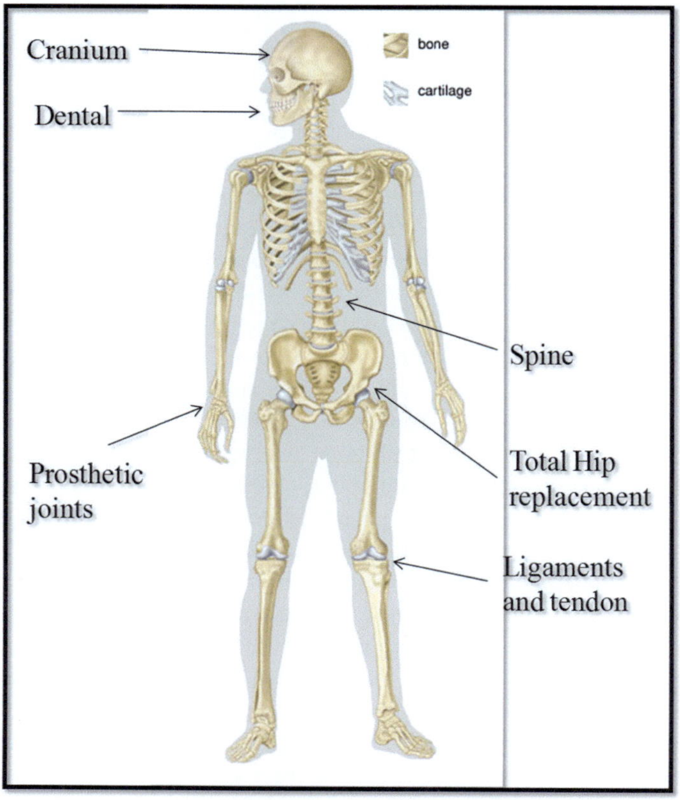

Fig. 1.5 The anatomical areas in the human skeletal system, wherein synthetic biomaterials can be used for repair/replacement of injured part

and therefore promote osteoblast adhesion, proliferation and differentiation [6, 7]. **Osteoinductivity** is the ability of a biomaterial implanted at a non-bone-forming site *in vivo* to induce new bone formation.

In summary, a biomaterial should not damage the surrounding tissue under such specific conditions (Fig. 1.1). Therefore, on the basis of the above-mentioned properties, biomaterials can be differentiated from the other classes of materials. It should be noted that the choice of biomaterial solely depends upon the anatomical location of use. In this book, significant emphasis is placed on bioinert and bioactive materials primarily for hard tissue replacement applications.

1.4 Conclusion

The procedure for complete replacement of a hip joint started its journey with implant and bone cement and provided more or less tolerable results [8]. The large number of operations performed each year shows that total hip arthroplasty is a well-established procedure for the management of hip joint anomalies, with an average of 500,000 cases in the USA alone [9]. This huge statistic indicates the demand for functional development of the implants. The 3D FE model was introduced in 1983 by Rohlmann et al. [6] to show the stress effect in implant and bone. Previous researchers showed that the prostheses with collar in contact with femur bone may generate physiological stresses [7, 10]. Some scientists commented that highly stiffened prostheses generate higher stress and lower cement interface stress [6, 7, 11–13]. Combined with the rapid development of advanced manufacturing techniques, FEM helped us to think about the concept of the customized prosthesis in the mid-1990s. It was observed that cement-less press-fitted implants were mostly showing better performance over cemented ones. However, tendencies to intra-operative cracks and fractures, early stem loosening and secondary proximal bone loss were seen, possibly due to stress shielding and femoral osteolysis. Most of these problems were thought to be related to a geometric mismatch between the stem and the endosteal bone due to variability of upper femoral geometry [14]. These problems prompted several research groups to become involved in the development of customized femoral stems to optimize the fit of the stem to the femur and to optimize the biomechanics of the joint [15–19].

Despite such attempts, the following issues remain to be addressed, (a) the position-wise optimum quality of bone (according to Hounsfield unit) that should remain after implantation for better bone growth, (b) the best possible interfacial gap between implant and bone, (c) the optimum pore size of the porous implant material for proper bone growth and (d) the arrangement of functionally graded materials to achieve optimum stiffness. The aspect that is still not considered in most of FE-based studies is the condition of the existing bone, which would actually be exposed to the stresses from the implant. The difference between actual situation and simulated situation from CT scan data is also a matter of concern. In view of the difference in the elastic properties of the bone and the implant material, difference in strain conditions can be perceived, when subjected to physiological loads. This, in turn, creates highly uneven stress distribution with few regions of stress concentration, along with other regions of stress shielding. This leads to damage of the bone, and bone tends to breakdown at the regions of stress concentration with cyclic loading during gait activities of daily living, while it suffers resorption at the regions of stress shielding. Both of these lead to misfit and loosening. The proper straining of the implant is anticipated to be very important for even distribution of stress and promotion of bone growth and bone remodelling at the interface which would, in turn, improve the grip of the implant.

Besides the design problem, the manufacturing aspect is also important here for the intricate 3D shape of many implants/devices. Laser Engineered Net Shaping (LENS), Rapid Prototyping (RP), electron beam machining, 3D CNC milling machining etc. are the most acceptable techniques to develop those implants. Some attempts are made to evaluate the mechanical and biomechanical performance of novel custom-made dental implants fabricated by the selective laser melting technique with simulation and *in vitro* experimental studies. It was concluded that this method can provide an efficient means of printing fully dense customized implants with high strength and sufficient dimensional accuracy. Adding the threaded characteristic to the customized root-analog threaded implant design is reported to enable the approximate geometry of the natural root to be maintained with better stress distribution and primary stability [20]. The research results reveal that many materials have CT numbers very near to human tissues, while further work and process improvements are necessary to obtain homogenous internal structures. In order to reduce the risk of a mal-aligned component and notching with subsequent femoral neck fracture, a computer navigation system for setting the femoral component stem shaft angle during hip resurfacing surgery can be adopted. Using this imageless method, the surgeons identified anatomical landmarks on the shaft, and the computer created a PSM for setting the shaft angle. In advanced clinics, the surgeons are able to accurately and consistently place the femoral component at the planned target angle using the computer navigation system. The system was validated using a relatively high number of human patients undergoing hip resurfacing surgery [21]. Finally, Pahr and Zysset [22] published a study aimed at creating numerically efficient FE models of femoral and vertebral bone. The authors detail a modelling technique that includes a new self-correcting cortical shell thickness evaluation algorithm for generating bone iso-surface meshes. The importance of developing reliable softwares for the AM process was also emphasized, as significant inaccuracies and differences in patient skull data caused by DICOM to STL conversion for additively manufactured medical skull models were observed. However, improved processing techniques undertaken along with manufacturing can be employed to solve this problem. An alternative method for manufacturing a customized implant could involve obtaining a plurality of medical images of the injury site, converting the images into 3D data. This can be followed by designing a customized mould for the defect area and fabricating a customized implant from a biocompatible plate using the customized mould via an additive manufacturing technique. The above discussion therefore emphasizes the need to adopt an appropriate manufacturing strategy while developing patient—specific implants.

References

1. David J. Apple and John Sims. Harold Ridley and the invention of the intraocular lens. Surv Ophthalmol. 1996;40(4):279–92.
2. Hench Larry L. The story of bioglass®. J Mater Sci Mater Med. 2006;17(11):967–78.
3. Hench LL. Biomaterials. Science. 1980;208(826).
4. Habal MB. The biologic basis for the clinical application of the silicones: a correlate to their biocompatibility. Arch Surg. 1984;119(7):843–8.
5. Ratner BD, Hoffman AS, Schoen FJ, Lemons JE. Biomaterials science: an introduction to materials in medicine. Academic press; 2004.
6. Rohlmann A, Mossner U, Bergmann G, Kölbel R. Finite-element-analysis and experimental investigation in a femur with hip endoprosthesis. J Biomech. 1983;16:727–42.
7. Fagan M. Material selection in the design of the femoral component of cemented total hip replacements. Clin Mater. 1986;1:151–67.
8. Huiskes R. A survey of finite element analysis in orthopaedic biomechanics: the first decade. J Biomech. 1983;16:385–409.
9. Latham B, Goswami T. Effect of geometric parameters in the design of hip implants paper IV. Mater Des. 2004;25:715–22.
10. Prendergast PJ. Finite element models in tissue mechanics and orthopaedic implant design. Clin Biomech. 1997;12(6):343–66.
11. Christel P, Meunier A, Leclercq S, Bouquet P, Buttazzoni B. Development of a carbon-carbon hip prosthesis. J Biomed Mater Res. 1987;21:191–218.
12. Lewis J, Askew MJ, Wixson RL, Kramer GM, Tarr RR. The influence of prosthetic stem stiffness and of a calcar collar on stresses in the proximal end of the femur with a cemented femoral component. J Bone Jt Surg (Am). 1984;66A:280–6.
13. Maher SA, Prendergast PJ. Discriminating the loosening behavior of cemented hip prostheses using measurements of migration and inducible displacements. J Biomech. 2002;35:257–65.
14. Bargar WL. Shape the implant to the patient. A rationale for the use of custom fit cementless total hip implants. Clin Orthop Relat Res. 1989;249:73–8.
15. Aldinger G, De Pellegrin M, Küsswetter W. The personalized hip prosthesis. Ital J Orthop Traumatol. 1988;14(4):429–33.
16. Mulier JC, Mulier M, Brady LP, Steenhoudt H, Cauwe Y, Goossens M, Elloy M. A new system to produce intraoperatively custom femoral prosthesis from measurements taken during the surgical procedure. Clin Orthop Relat Res. 1989;249:97–112.
17. Robertson DD, Walker PS, Granholm JW, Nelson PC, Weiss PJ, Fishman EK, Magid D. Design of custom hip stems prostheses using three-dimensional CT modeling. J Comput Assist Tomogr. 1987;11(5):804–9.
18. Rubin PJ, Leyvraz PF, Aubaniac JM, Argenson JN, Esteve P, de Roguin B. The morphology of the proximal femur. A three-dimensional radiographic analysis. 2409, J Bone Jt Surg (Br). 1992;74(1): 28-32.
19. Stuhlberg SD, Stuhlberg BN, Wixson RL. The rationale, design characteristics, and preliminary results of a primary custom total hip prosthesis. Clin Orthop Relat Res. 1989;249:79–96.

20. Jianyu C, Zhiguang Z, Xianshuai C, Chunyu Z, Gong Z, Zhewu X. Design and manufacture of customized dental implants by using reverse engineering and selective laser melting technology. J Prosthet Dent. 2014;112(5):1088–95.
21. Bailey C, Gul R, Falworth M, Zadow S, Oakeshott R. Component alignment in hip resurfacing using computer navigation. Clin Orthop Relat Res. 2009;467(4):917–22.
22. Pahr DH, Zysset PK. A comparison of enhanced continuum FE with micro-FE models of human vertebral bodies. J Biomech. 2009;42:455–62.

Chapter 2
Case Study: Hydroxyapatite–Titanium Bulk Composites for Bone Tissue Engineering Applications

2.1 Background: Brittleness of Hydroxyapatite

Healthy bone and joints (e.g. articulating, knee, and hip joints) are necessary for the structural stability and pain free movements. However, these healthy bones and joints can get damaged in some unfortunate cases due to accident/diseases and fail to perform their normal functions. The damaged bones can be replaced or repaired with allogeneic bones or allografts [1–5]. However, the associated risk of transmission of life threatening infections with allogeneic bones is significant particularly due to inferior osteoconductivity as well as poor mechanical properties of the freeze dried [1–3, 6–13]. In contrast, the designed biomaterials with bone-mimicking properties can be used as an alternative to the allografts and autografts.

Biomaterials development has a long history. In 1940, Jean and Robert Judet of Paris performed the replacement arthoplasty to replace the femoral head with acrylic resin, which opens up the possibility of joint replacement [14]. However, after an initial success, this implant failed due to its poor mechanical properties as well as inapt design. More than 10 decades later, when most of the issues related to joint replacement have been settled down, the interest in new biomaterials remain as fierce ever. A biomaterial is a synthetic biocompatible material/device that has been intently designed to persuade a definite activity in the biological system, which can simulate the desired biological function, without generating a short and/or long term damage to the nearby or distant tissue as well as without being getting damaged in this procedure [15–17].

For many biomedical applications, a biomaterial should show the bioactivity, osteoconductivity, and osteoinductivity. Bioactivity is defined as the property of implant material to make direct bonding with living bone [18–21]. The bonding is mediated by an intervening (apatite) layer formation between the implant and the bone [22]. Osteoconductivity is defined as the ability of implant material that can serve as a scaffold for new bone formation and therefore, promotes the osteoblast adhesion, proliferation, and differentiation [23, 24]. Osteoinductivity is the ability

B. Basu and S. Ghosh, *Biomaterials for Musculoskeletal Regeneration*,
Indian Institute of Metals Series, DOI 10.1007/978-981-10-3017-8_2

of a biomaterial to induce the new bone formation, implanted at non bone forming site, *in vivo* [25].

It is now well recognized that in case of bone applications the success of an implant material depends on biocompatibility, osseointegration property, physical/mechanical properties as well as antibacterial property etc. [26–29]. Metals have an advantage of good mechanical properties like high fracture toughness and flexural strength. But the effects such as corrosion as well as wear (results in leaching of metal ions), and stress shielding results in irritation, inflammation, and loosening of implants [30, 31]. Additionally, bioinert nature of metal implants result in weak bonding with host bone, *in vivo* [32, 33]. Although ceramics based biomaterials are found suitable for hard tissue applications due to its biocompatibility, corrosion resistance, and compressive strength properties [34, 35], inherent brittleness of ceramics may result in the early stage failure of implants during load bearing applications [36]. A classic example of a brittle bioceramic is hydroxyapatite, which widely researched due to its structural similarity with major inorganic component of bone composition [36]. In the context of load bearing implants, high elastic modulus of the implant with respect to neighbouring bone causes stress shielding, which is responsible for bone resorption near to the implant [37–41]. Therefore, the elastic modulus of an implant ideally needs to be matched with bone, while efforts are to be made to enhance both fracture toughness and strength.

In the above backdrops, the hydroxyapatite (HA)-based metal or ceramic reinforced composites can be used as a potential bulk implant material, where second phase contributes specifically towards better mechanical properties. The choice of second phase is limited as it needs to have good mechanical property as well as biocompatibility. As compared to Al_2O_3, ZrO_2 and stainless steels, titanium (Ti) is found to be more promising second phase material for orthopedic applications due to its excellent corrosion resistance property under physiological environment, low density (4.540 g/cc) as well as lower elastic modulus ($\sim$110 GPa) than Co–Cr alloy ($\sim$230 GPa)/stainless steel ($\sim$205 GPa) [26, 41] and some data are shown in Table 2.1. The design of HA–Ti composites underlies the fact that the Ti addition significantly improves the mechanical properties of the composites without degrading the cytocompatibility properties [49–51].

Table 2.1 A comparison of physical and mechanical properties of bone with some metallic and ceramic biomaterials [26, 42–48]

Properties	Cortical bone	Cancellous bone	HA	Ti	Al_2O_3	ZrO_2
Young modulus (GPa)	14–20	0.05–0.5	80–110	117	390	205
Compressive strength (MPa)	170–193	7–10	400–900	250–600	3900	3000
Fracture toughness (MPa $m^{1/2}$)	2–12	0.1	0.7–1.2	60	5.2	10
Apparent density (g/cm^3)	1.8–2.0	0.1–1.0	3.16	4.54	3.9	6

HA–Ti composites can be prepared by different sintering routes, such as pressureless sintering, pulse electric current sintering (PECS), spark plasma sintering (SPS) etc. It is difficult to retain the Ti in the composite by conventional pressureless sintering techniques. For example, Nath et al. conducted a number of pressureless sintering experiments on the HA–Ti system and reported the extensive formation of TCP and $CaTiO_3$ in the composites [52]. On the other hand the SPS as well as PECS techniques with suitably chosen sintering parameters will allow the sintering of HA–Ti composites with retention of almost all Ti in the sintered compacts [52–55]. Nakahira et al. developed the HA–Ti composites with Ti content of 20 and 25 wt%, using PECS technique and compared the results with hot pressed HA–Ti composites of similar compositions [53]. Extensive study by our group indicates that it is possible to retain the Ti in the composite after sintering. However, in limited studies on HA–Ti compositions [52–55], no comprehensive attempt has been made to understand the influence of Ti content on the mechanical and biocompatibility property, when both HA and Ti are retained in the sintered composites.

In developing a bone replacement material, the choice of the second phase depends on the application as and the desired functionality. For example, $BaTiO_3$, $CaTiO_3$, Fe_3O_4, and ZnO or Ag can be added to HA as a second phase to introduce the piezoelectric, electrical conductivity, magnetic, and bactericidal properties, respectively [56–60]. However, brittle nature and the load bearing incapability of the ceramics may limit wider application of such types of ceramic–ceramic composites and some examples includes HA–20$HA_{whisker}$, HA–20ZrO_2, and HA–(Al_2O_3 coated) ZrO_2 with fracture toughness of 1.4, 1.5, and 3 MPa $m^{1/2}$, respectively [61–63]. Interestingly, a glaring example of HA-based ceramic composite with fracture toughness higher or equal to that of the cortical bone is still unavailable (see Fig. 2.1). In contrast, the incorporation of metallic reinforcement as second phase may improve the fracture toughness of the HA-based composites. However, the sintering reactions between HA and metal phase are of major concern and can affect the biocompatibility properties *in vitro* and *in vivo*. Although stainless steel and Co–Cr alloys are competitively better than ceramics counterpart in terms of fracture toughness, the high elastic modulus of metals however may lead to stress shielding during *in vivo* performance [29]. In contrast, Ti was found to be most suitable for biomedical applications due to excellent biocompatibility, corrosion resistance property and stability in physiological conditions [64]. Thus, in order to incorporate both load bearing capability without fracture under peak loading as well as osseointegration property, the HA–Ti composites can be used as a bulk biomaterials. Over the various composites (HA–$BaTiO_3$, HA–$CaTiO_3$, HA–Al_2O_3, HA–ZrO_2, HA–ZnO, HA–Ag), the selection of HA–Ti composites as a model system in the present chapter can be justified on the basis of following aspects: (a) low density of HA (3.16 g/cc) [65] and Ti (4.54 g/cc); (b) high fracture toughness of metallic Ti (60 MPa $m^{1/2}$) [26]; (c) bioactivity of HA and bio-inertness of Ti; (d) corrosion resistant properties of Ti; (e) limited dissolution of HA under physiological condition [66].

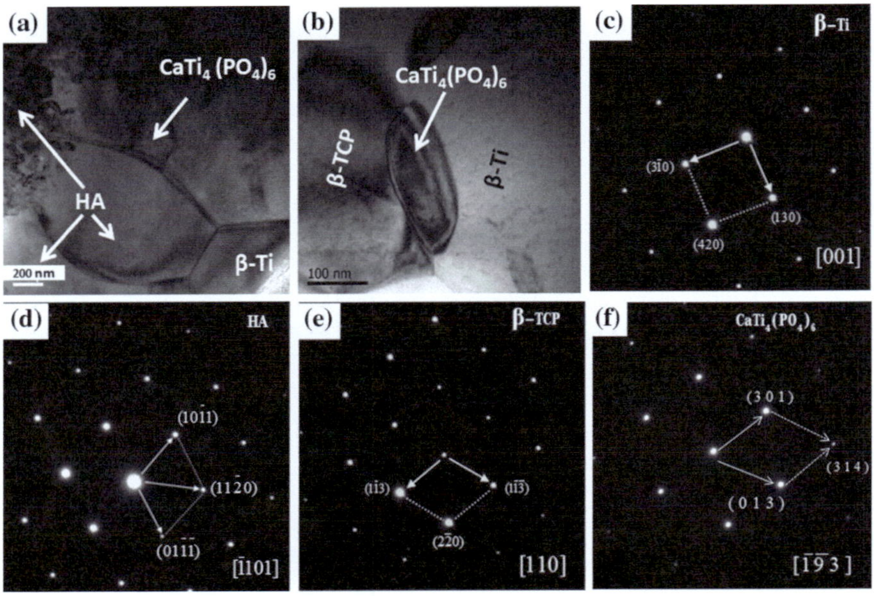

Fig. 2.1 Representative bright field TEM micrographs (**a**, **b**) of HA–20Ti composite, showing the different phases, formed during sintering. SAD patterns confirm the presence of β-Ti (**c**), HA (**d**), β-TCP (**e**), and $CaTi_4(PO_4)_6$ grains (**f**) [67]

2.2 Processing of HA–Ti Composites

HA–Ti composites can be prepared either by conventional pressureless sintering or advanced sintering, such as spark plasma sintering (SPS). SPS has the advantage of higher heating rate and shorter soaking time over conventional sintering. The advanced sintering techniques have many other advantages as compared to the conventional sintering.

Traditionally, HA–Ti composites were prepared by the homogeneous mixing of the HA and Ti powders using a ball-mill, followed by green compaction and then pressureless sintering at relatively higher temperature (≥1100 °C) and soaking time (≥1 h) [52]. Recently, the FAST (field assisted sintering technique) has been utilized to prepare dense compact from powder mixture within a short time [53, 67, 68].

2.2.1 Conventional Sintering

In this subsection, the literature reports on conventional sintering of HA–Ti are briefly discussed. Fahami et al. [69] used $CaHPO_4$, CaO, and Ti powders as precursors to prepare the HA–Ti composites by the mechanochemical synthesis of mixture of HA and Ti according to the following reaction:

$$6CaHPO_4 + 4CaO + Ti \rightarrow Ca_{10}(PO_4)_6(OH)_2 + Ti + 2H_2O \quad (2.1)$$

It is important to note that the lower crystallinity of the HA powder results in the increased bioresorption and solubility under physiological conditions [70]. Therefore, further attempt has been made by Fahami et al. [69] to increase the crystallinity from 13 to 69 % by annealing the ball-milled powder at 650 °C for 2 h, synthesized after reaction (2.1). However, this results in the oxidation of Ti and decomposition of HA to β-TCP (beta-tri-calcium phosphate) phase. In a different approach, Chu et al. [71] prepared the HA–20 wt% Ti composite by mixing HA and Ti powders in a ball mill, followed by hot pressing at 900 and 1000 °C. Importantly, the sintering at 1000 °C leads to the formation of α-TCP and $Ca_4O(PO_4)_2$ phases. The difference in the linear thermal expansion coefficient (α) of HA (17.3×10^{-6} °C^{-1}) [72] and Ti (8.4×10^{-6} °C^{-1}) [73] phases results in inferior sinterability in case of HA–20 wt% Ti composite with ~67 % relative density, achieved when sintered at 1200 °C [71]. However, the same sintering conditions result in ~88 % relative density in monolithic HA. Importantly, the addition of Ti to HA leads to an increase in fracture toughness from 0.7 to 1.0 MPa m$^{1/2}$. Although Ti addition to HA was intended to increase the fracture toughness significantly, the marginal improvement in the fracture toughness can be attributed to poor sinter density of the composite. In an earlier work, Nath et al. [52] prepared the HA–Ti composites with varying amount of Ti (10–40 wt%) using pressureless sintering. It has been reported that the sintering of different compositions at 1000–1400 °C for 2 h leads to the extensive sintering reactions and therefore the formation of TiO_2, CaO, TCP and $CaTiO_3$ (Table 2.2). The thermodynamic calculations show the higher probability of oxidation of Ti during conventional sintering due to higher sintering temperature and longer holding time [52]. Such calculations also indicate the possibility of reaction between HA and TiO_2 above 750 °C.

Considering the thermodynamical aspect, Yang et al. [55] prepared the HA–10 wt% Ti and HA–20 wt% Ti composities by sintering at 1100 °C in vacuum to minimize the oxidation of Ti. The FT-IR and XRD analysis of the HA–Ti samples show the decomposition of HA to α-TCP, TTCP (tetracalcium phosphate) as well as $Ca_2Ti_2O_5$. As the sintering temperature further increases, HA is reported to decompose to HA with deficiency of hydroxyl group in following way [74, 75]:

$$Ca_{10}(PO_4)_6(OH)_2 \xrightarrow{vacuum,\, T > 800\,^{\circ}C} Ca_{10}(PO_4)_6(OH)_{2-2x}O_x + xH_2O \uparrow \quad (2.2)$$

The water vapor produced from reaction (2.2) reacts with Ti to form TiO_2 [see Eq. (2.3)], which can finally react with HA to form β-TCP and $CaTi_2O_5$ [see Eq. (2.3)].

$$Ti + 2H_2O \longrightarrow TiO_2 + 2H_2 \quad (2.3)$$

$$Ca_{10}(PO_4)_6(OH)_2 + 2TiO_2 \longrightarrow 3Ca_3(PO_4)_2 + CaTi_2O_5 + H_2O \quad (2.4)$$

Table 2.2 XRD results, showing the effect of sintering parameters (pressure, sintering temperature, and soaking time) on the sintering reaction of HA–Ti composites [52, 53, 67]

	Conventional sintering (soaking time ≥1 h)				Advanced sintering (soaking time ≥5 min)		
	Pressureless sintering in air at temperature				Pulse electric current sintering in argon		Spark plasma sintering in argon
	1000 °C	1200 °C	1300 °C	1400 °C	900 °C	1000 °C	950 °C
HA	HA	HA	HA	HA, β-TCP, α-TCP, CaO	–	–	HA
HA–10Ti	HA, β-TCP, α-TCP, CaO, TiO_2, $CaTiO_3$	HA, β-TCP, CaO, TiO_2, $CaTiO_3$	HA, β-TCP, α-TCP, CaO, TiO_2, $CaTiO_3$	HA, β-TCP, α-TCP, CaO, TiO_2, $CaTiO_3$	–	–	HA, Ti, TiO_2, Ti_2O, β-TCP, CaO, $CaTi_4(PO4)_6$
HA–20Ti	HA, CaO, TiO_2, $CaTiO_3$	HA, β-TCP, CaO, TiO_2, $CaTiO_3$	HA, β-TCP, CaO, TiO_2, $CaTiO_3$	β-TCP, CaO, TiO_2, $CaTiO_3$	HA, Ti_3O	HA, Ti_3O, Ti_2O	HA, Ti, TiO_2, Ti_2O, β-TCP, CaO, $CaTi_4(PO4)_6$

Therefore, it is not impossible but extremely difficult to retain Ti in the sintered product using conventional sintering routes. The phase stability aspect in summarized in Table 2.2.

The sintering under a protective atmosphere (e.g. argon) is another promising approach towards retaining the metallic Ti in the sintered product. In order to see the effect of protective environment on the sintering reaction in the HA–Ti composites, Ning et al. [76] conventionally sintered the HA–50Ti composite at 1200 °C for 30 min in argon atmosphere. However, the sintering leads to the formation of Ti_2O, $CaTiO_3$, CaO and TiP-like phases along with α-Ti.

In summary, conventional sintering of HA–Ti composites even at very low temperature of 1000 °C leads to the extensive dehydroxylation of HA as well as reaction between HA and Ti. The formation of TCP phases will result in higher dissolution of the implant materials and therefore, decreased mechanical strength during service period, *in vivo*. As mentioned earlier, it is extremely difficult to retain the metallic Ti in the composite during the conventional sintering. The sintering in protective atmosphere of Ar was not found to be helpful in retarding the oxidation reactions in conventional sintering due to the longer sintering duration (≥1 h) and higher holding temperature (≥1000 °C). In the following, we shall discuss the applicability of advanced sintering techniques (e.g. SPS) to prepare the composites with predominant retention of metallic Ti.

2.2.2 *Advanced Sintering*

Among various advanced sintering techniques, field activated sintering (FAST) is widely investigated to resolve the issues related to the conventional sintering, discussed earlier. For example, Nakahira et al. [53] reported the pulse electric current sintering (PECS) of HA–20 vol.% Ti and HA–25 vol.% Ti composites at 800 and 1000 °C under 30 MPa in a protective argon atmosphere. Although PECS suppresses the oxidation of Ti at 900 °C, the formation of Ti_2O was found in samples sintered at 1000 °C. Mondal et al. [77] used the spark plasma sintering technique to sinter β-TCP and Ti powders at 1200 °C temperature under 50 MPa pressure. Despite the retention of β-TCP and Ti, this results in the formation of $CaTiO_3$.

In a recent work, the author's research group has reported the results of extensive study on spark plasma sintering and phase evolution of HA–xTi (x = 5, 10, 20 wt%) composites [67]. The processing conditions were intelligently tailored to achieve high toughness with minimum sintering reaction between the HA and Ti. The XRD analysis confirms the predominant presence of HA and Ti.

In order to observe the finer scale microstructure of the sintered HA and HA–Ti samples as well as to identify the reaction product at HA/Ti interface, transmission electron microscope (TEM, FEI, UTWIN T-20) with an accelerating voltage of 200 kV was used. Figure 2.1a shows the representative bright filed TEM micrograph of HA–20Ti composite, revealing the presence of different phases. The presence of these phases has also been confirmed at different locations in the same sample. Figure 2.1b represents the bright field micrograph of such a location. The detailed analysis of the selected area diffraction pattern (Fig. 2.1c–f) indicate the presence of β-Ti, HA, β-TCP, as well as the formation of $CaTi_4(PO_4)_6$ phase at the interface between HA and Ti. It is to be noted that β-TCP is a phase formed by the dehydroxylation of HA.

Overall, the critical analysis of SAD patterns reveals the formation of $CaTi_4(PO_4)_6$ as grain boundary phase, which is likely to be formed either due to limited sintering reaction between β-TCP and TiO_2 or between HA and TiO_2. The presence of β-Ti also indicates the suppression of $\beta \rightarrow \alpha$ transformation of Ti phase during cooling through 882 °C (transformation temperature), perhaps due to faster cooling.

We shall now discuss the formation and morphology of finer scale sintering reaction product. It has been reported that dehydroxylation of HA does not occur at a sintering temperature below 1300 °C [52, 62]. In the presence of Ti and

Ti-oxides, HA starts to dissociate at a much lower temperature [55]. The detailed microstructural investigations of HA–Ti composites using TEM indicates the possibility of formation of reaction product between HA and Ti during sintering. Based on extensive TEM-SADP analysis, it has conclusively been shown that the reaction product to be $CaTi_4(PO_4)_6$, which is deemed to be the result of following reactions:

$$Ca_{10}(PO_4)_6(OH)_2 \longrightarrow 3Ca_3(PO_4)_2 + CaO + H_2O\uparrow \tag{2.5}$$

$$3Ca_3(PO_4)_2 + 4TiO_2 \longrightarrow CaTi_4(PO_4)_6 + 8CaO \tag{2.6}$$

$$Ca_{10}(PO_4)_6(OH)_2 + 4TiO_2 \longrightarrow CaTi_4(PO_4)_6 + 9CaO + H_2O\uparrow \tag{2.7}$$

The reaction (2.10) is known as dehydroxylation of HA and leads to the formation of β-TCP. The reaction product $CaTi_4(PO_4)_6$ can form due to reaction (2.6) and/or (2.7). In view of the lack of data on Gibbs free energy of formation of related phases, any analysis on the thermodynamic feasibility of reaction (2.10–2.12) could not be made. Although TEM investigation does not reveal the presence of CaO phase, detailed analysis of XRD data shows the peaks corresponding to CaO to a detectable extent. Also, Ti_2O or TiO_2 could not be detected during TEM analysis of any of the HA–xTi samples.

2.3 Mechanical Properties

In the following, the mechanical properties of HA and HA–Ti composites are analysed. The hardness values shows a decreasing trend with addition of Ti to HA (HA $\approx$ 7.1 $\pm$ 0.5 GPa, HA–10Ti $\approx$ 6.1 $\pm$ 0.6 GPa). One can measure the plastic work of indentation by measuring the area under P–h curve (= area under loading curve − area under unloading curve) [78]. Such measurements indicate that the plastic work of indentation is higher than elastic work of indentation for both HA and HA–10Ti.

The elastic moduli, measured using nanoindentation are 94 $\pm$ 1.8 GPa for HA and 80.6 $\pm$ 0.9 GPa for HA–10Ti composite. The elastic modulus has also been measured using impulse excitation method. The measured values show a similar trend of the decrease of elastic modulus with Ti addition. The measured elastic moduli for different materials are as follows, HA = 61.4 $\pm$ 1.4 GPa, HA–5Ti = 58.1 $\pm$ 0.2 GPa, HA–10Ti = 54.4 $\pm$ 0.3 GPa and HA–20Ti = 50.2 $\pm$ 0.8 GPa [79]. Since the resonance frequency method measures the elastic stiffness properties of a bulk sample, such results are more realistic and representative of the true modulus, rather than the nanoindentation results. The latter is based on the estimation from the slope of unloading response from an indented region of length scale of less than 1 μm.

In further characterization of mechanical properties, the flexural strength of the sintered composites has been measured using 3-point flexural test with 12 mm gauge length. The flexural strength systematically increases with increase in Ti-content in the composite. The flexural strength of pure HA is 70.7 ± 2.3 MPa. However, the maximum measured value of the flexural strength is measured as 99.7 ± 1.2 MPa in case of HA–20Ti composite. Therefore, the addition of Ti up to 20 wt% leads to about 41 % increase in the flexural strength. This shows a marked improvement in the flexural strength in HA–Ti biocomposite. It is to be noted here that the flexural strength of cortical bone is 150–170 MPa [80].

The fracture toughness of the sintered products have been measured using SEVNB test in 4-point flexural mode and a test sample is shown in Fig. 2.2b. As mentioned earlier, this technique is proved to be more acceptable than the indentation fracture toughness measurement technique in terms of providing reliable estimate of long crack fracture toughness. The SEVNB toughness results are plotted in Fig. 2.8a against amount of Ti addition to HA. For pure HA, the measured K_{IC} value is found to be 3.4 ± 0.2 MPa $m^{1/2}$. As compared to the monolithic HA, HA–10Ti exhibits about 35 % increase in the K_{IC} value (4.7 ± 0.1 MPa $m^{1/2}$). However, the addition of 5 wt% Ti does not show any significant improvement. On the other hand, 20 wt% Ti addition leads to about 26 % increase in K_{IC} value as compared to that of monolithic HA. K_{IC} value decreases in case of HA–20Ti sample as compared to HA–10Ti. Therefore, 10 wt% Ti has been found to be optimum Ti addition to HA to achieve substantial improvement in fracture toughness (see Fig. 2.2a). It is to be noted that the fracture toughness of the cortical bone is reported be in the range of 2–12 MPa $m^{1/258}$. Therefore, the present measurements are higher than the lower limit of fracture toughness of the cortical bone.

In order to investigate the statistical significance in terms of the property difference among the samples were used and results are summarized in Figs. 2.2, 2.3, and 2.4. Figure 2.8 shows the effect of Ti addition to HA on enhancement in fracture toughness (K_{IC}) and work of fracture (G_{IC}). Dunnett's t test shows the significant difference, when HA was compared with HA–10Ti and HA–20Ti samples for both K_{IC} and G_{IC}. However, no significant difference between HA and HA–5Ti samples was noted. Dunnett's C test found the significant difference between HA and HA–10Ti as well as between HA–5Ti and HA–10Ti in terms of both K_{IC} and G_{IC}.

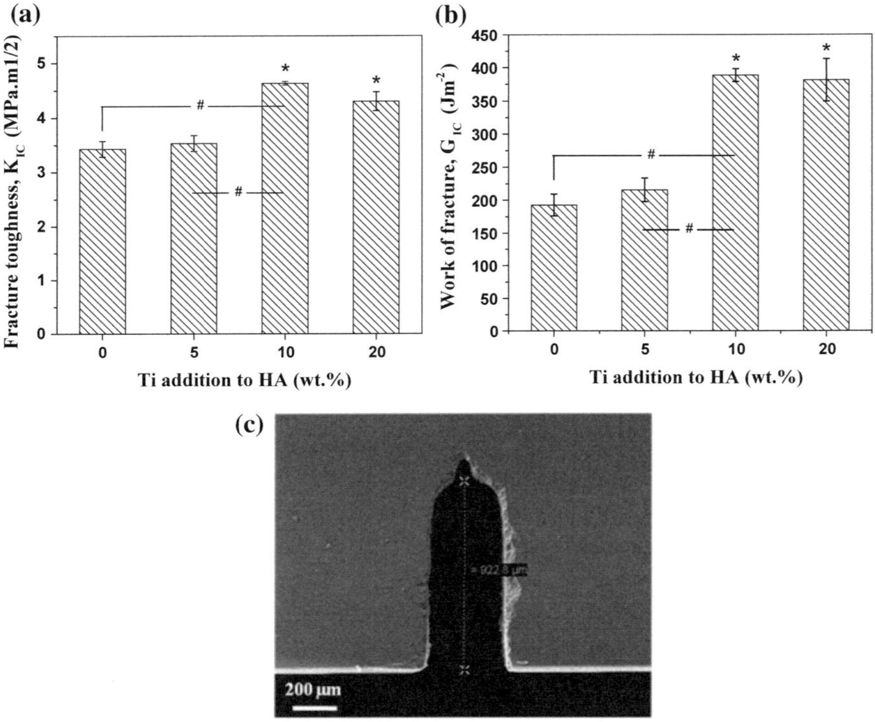

Fig. 2.2 Variation in fracture toughness (**a**) and work of fracture with wt% Ti in HA (**b**). Typical profile of V notch in one of the SEVNB test sample, as seen using SEM (**c**). Data in **a** and **b** has been represented as mean ± standard error. Dunnett's t (2 side) were used to compare the HA with HA–Ti samples (marked with *); while Dunnett's C test used to know the significant difference among all samples (marked with #). Statistical analysis shows the significant difference at 0.05 level for n = 5 [67]

Summarizing, the mechanical properties (hardness, flexural strength and fracture toughness) of spark plasma sintered HA–Ti composites show substantial improvement as compared to the properties obtained with earlier developed HA-based composites (see Fig. 2.1). Figure 2.4 compares the fracture toughness results of present investigation with other HA-based composites. We can analyse the results in terms of four categories: I (HA-0x), II (HA-5x), III (HA-10x), and IV (HA-20x), where x is the amount of the second phase. In all categories, Dunnett's t test (2 side) shows the significant difference, when K_{IC} from the present work is compared with earlier reported results. Also, various symbols, like *, **, #, ## indicate the

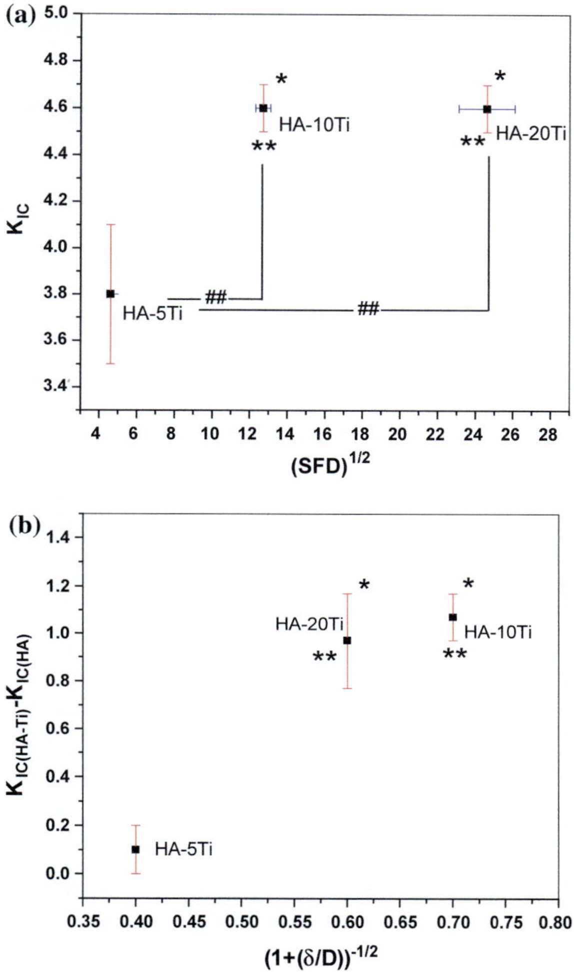

Fig. 2.3 Plot showing the variation in fracture toughness of HA–Ti composites versus fraction of particles nearby to the crack tip 'S', Ti particle size 'D' and volume fraction 'F'. **b** Plot showing dependence of the toughness increment on Ti particle size 'D' and nearest neighbor distance for Ti, 'δ'. In plot **a**, * and ** symbols show the significant difference between HA–5Ti and other HA–Ti samples for K_{IC} and $(SFD)^{1/2}$, respectively, when Dunnett's t test was applied. The symbol, ## is showing the significant difference among the samples, measured by Dunnett's C test. In plot **b**, * and ** showing the significant difference between HA–5Ti and other HA–Ti samples for $K_{IC(HA-Ti)} - K_{IC(HA)}$ and $(1 + (\delta/D))^{-1/2}$, respectively, when Dunnett's t test was applied [67]

significant difference at 0.05 level. In the following section, the reason for such an increase by correlating the microstructure of the composite processed by novel SPS technique will be discussed with an effort to establish microstructure-property correlation.

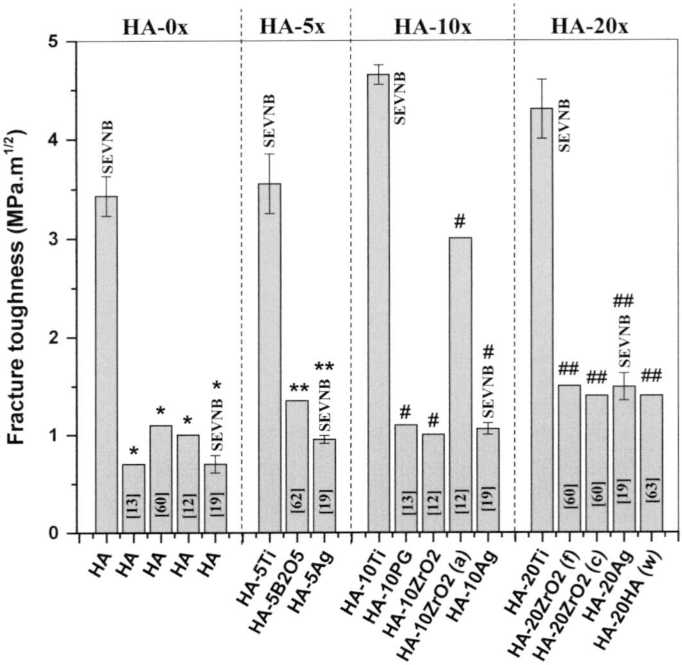

Fig. 2.4 A comparison of fracture toughness of various HA-based biocomposites. For comparison, plot has been divided into four categories: HA-0x, HA-5x, HA-10x, and HA-20x, where x is the wt% of second phase. HA (w) indicates the use of HA whisker reinforcement and HA–10ZrO_2 (a) indicates the Al_2O_3 coated ZrO_2 particles as second phase. ZrO_2 (f) and ZrO_2 (c) stands for fine grained and coarse grained ZrO_2 particles. The toughness values measured with SEVNB are mentioned and otherwise, those are estimated on the basis of indentation cracking. References related to the respective work are mentioned on individual bar and otherwise, results are from present work. The symbols *, **, #, and ## show the significant difference (measured by Dunnett's t (2 side) test) between reference sample (from present investigation) and other samples in each category [67]

2.4 Toughening Mechanisms and Toughness Properties

Fracture in brittle solids can be correlated with propagation of crack under tensile loading. Importantly, the crack propagates if and only if the stress at crack front is higher than a critical value. In brittle solids, the resistance against the crack propagation can be increased by introducing the mechanisms that reduce the driving force for the crack propagation by dissipating the additional energy associated with crack tip and therefore leads to the blunting of crack tip [81]. In polycrystalline ceramic materials, the microstructure dependent fracture toughness can be correlated with the dissipation of the crack tip energy and thereby with the reduction in the driving force for the crack propagation [82]. It is generally recognized that weak interfaces, such as grain boundaries and/or matrix-reinforcement boundary are the preferred crack deflection paths. Moreover, the presence of second phase disturbs the crack propagation by forcing the crack either to go around it or through it. The extent, to which such phenomenon is able to dissipate the energy, depends on the morphology/distribution and mechanical properties of the second phase in ceramic matrix composite [83].

Although the presence of second phase may improve the fracture toughness of composites, the chemical reaction at the interface during processing (e.g. sintering) always offers the challenge for the materials scientists to optimize the toughness. On the basis of microstructural characteristics of the second phase; i.e., nature, shape and size, the toughened composite materials can be categorized as [84]: (a) particulate reinforced composite, (b) whisker reinforced composite, and (c) fiber reinforced composite. In the context of the current review paper, the toughening in metallic particulate reinforced ceramic composites is attributed to the shielding effect to the crack front propagation due to ductile metal reinforcements. In such case, the crack resistance energy of the matrix can be given by [82],

$$R_o = (1 - V_f)2\gamma_B \tag{2.8}$$

where, γ_B is the intrinsic cohesive energy of the matrix material and V_f is the volume fraction of second (reinforcement) phase. Thus, both the matrix as well as the reinforcement material contributes towards the fracture toughness. Therefore, enhanced fracture toughness can be achieved by intelligently designing the composites having mixed shielding mechanisms.

In case of the polycrystalline monolithic ceramic materials, the frontal zone shielding results due to the interaction of the crack front with dislocation cloud and/or microcrack cloud. The activation of the dislocations results in generation of dislocation cloud and this restricts the crack front propagation by applying a stress field on the crack tip [82]. The stress field generated around the dislocation is proportional to the shear modulus (G) and can be given as [85],

$$\sigma \propto \frac{Gb}{r} \tag{2.9}$$

where, G, b, and r are the shear modulus of the material, Burger vector, and distance from the dislocation, respectively. On the other hand, the microcrack cloud can be activated at the stress field of primary cracks at the relatively weaker zone like, grain boundaries and interphase flaws.

Unlike the combination of dislocation as well as microcrack clouds towards the improvement in fracture toughness in brittle ceramics, the phase transformation in zirconia (ZrO_2) can also contribute to the fracture toughness improvement in ZrO_2 reinforced ceramic matrix [82]. In such case, the partially stabilized ZrO_2 in metastable tetragonal state transforms to stable monoclinic phase at the crack tip stress field [81]. This transformation can be controlled by using additives like, MgO, CaO, Y_2O_3 and CeO_2 [81]. In addition, the grain bridging of crack in monolithic polycrystalline materials results in the toughness improvement due to the dissipation of crack energy as a result of grain sliding and/or fracture. For example, Swanson et al. [86] reported the evolution of grain bridging in pure polycrystalline alumina. In such case, the transgranular fracture may occur in large grains along with primary crack due to high frictional stresses at sliding grain boundary surfaces.

In 1983, Faber and Evans studied the effect of geometrical aspect of second phase on the crack deflection and therefore, effect on fracture toughness [87, 88]. The strength of the interfacial boundary plays an important role in deciding whether a crack will pass though second phase or get deflected. In a case when interface is strong enough to deflect the crack, it penetrates through the second phase. This may result in pinning the advancing crack front at the interaction site. In case of ductile metal phase, this results in dissipation of crack tip energy due to crack bridging.

In the context of biomaterials, the benefits of second phase addition to a ceramic matrix are twofold: (1) second phase improves the fracture toughness dramatically due to frontal wake and/or bridge interface shielding [67] and (2) second phase leads to the modification in surface energy [89], which can used as a tool to manipulate the cell-material interaction.

Although the reinforcement of ceramic matrix with fiber or whisker results in improvement in fracture toughness, the release of nanosized and needle shaped whisker/fibrous can lead to the severe toxicity to the cells caused by lipid layer damage and genotoxicity [90, 91]. Therefore, the fiber/whisker toughened ceramic composites are not considered for clinical applications, although related toughening mechanisms can be very well adapted to develop HA-based high toughness composites. In parallel, the reinforcement of ceramic matrix with ductile metal particles results in dramatic improvement in mechanical properties due to additional advantage of both ceramic matrix and metal particles. The crack bridging at metal particles results in dissipation of energy associated with crack tip [67, 82]. The fracture energies of pulsed electric current sintered (PECSed) HA–Ti composites, measured by Nakahira et al. [53] using single edge double notched beam method, show a substantial increment with the Ti content. The fracture energies of 9 and 20 J/m^2 were reported for HA–20 vol.% Ti and HA–25 vol.% Ti samples, respectively. The reason for the improvement in fracture toughness was crack deflection and crack bridging by Ti particles, which are retained in the HA matrix.

Similarly, the significant improvement in the fracture toughness of the spark plasma sintered HA–Ti composites was measured by the present authors (see Fig. 2.8) [67]. An addition of 10 wt% Ti to HA increases the fracture toughness (K_{IC}) from $\sim$3.4 to 4.7 MPa $m^{1/2}$. Further addition of Ti (20 wt%) to HA does not improve the fracture toughness (4.3 MPa $m^{1/2}$). The increased fracture toughness of HA and HA–Ti composites can be explained on the basis of crack deflection and crack bridging. In the context of toughness enhancement, the synergetic interaction of multiple toughening mechanisms can be explored. The presence of Ti particles provides the significant contribution toward the fracture toughness by crack bridging. Fracture toughness can also be further enhanced with the addition of ZrO_2 particles. The transformation toughening due to ZrO_2 phase along with crack bridging due to Ti phase can therefore provide the synergetic effect in order to improve the fracture toughness beyond the 4.7 MPa $m^{1/2}$ (fracture toughness of HA–10 wt% Ti composite). Also, the linear thermal expansion coefficient of ZrO_2 (10.5×10^{-6} °C^{-1}) is found to be in between HA (17.3×10^{-6} °C^{-1}) and Ti (8.4×10^{-6} °C^{-1}). This helps in minimizing the difference in thermal expansion coefficient between HA and Ti and therefore, does not lead to the interfacial debonding in sintered composites.

Several aspects can be discussed to explain the significant increase in the fracture toughness of the HA–Ti composites. It is worthwhile to mention that the reliable measure of toughness properties of HA–Ti has been obtained only using long crack toughness measurement techniques, like SEVNB etc. The most important aspect is that the crack bridging due to lath shaped Ti phase as well as crack wake debonding at HA and Ti interface are responsible for the fracture toughness enhancement ($\sim$4.7 MPa $m^{1/2}$) in HA–10 wt% Ti. However, no further improvement in fracture toughness of HA–20Ti composites was noted due to non-uniform distribution as well as lower volume fraction of Ti, resulted due to unification of Ti during ball-milling. This can be confirmed using theoretical models proposed by Ashby [92] as well as Evans [93]. The detailed analysis of fracture toughness of particulate (Ti) reinforced ceramic (HA) has been reported elsewhere [67].

According to these models, the particle size and the distribution of the metallic phase near the crack tip (which is directly proportional to the volume fraction of metal phase in the HA matrix) has a strong influence on the fracture toughness . The details are reported elsewhere [92, 93]. Although the shape of metal particles has an effect on fracture toughness, the average number of particles near the crack tip is the deciding factor in the ceramic-metal (e.g. HA–Ti) composites (see Fig. 2.3). The unification of Ti in HA–20Ti composite results in the reduction in volume fraction Ti in HA matrix and therefore reduces the distribution of Ti particles near the crack tip. Therefore, fracture toughness of HA–20 wt% Ti is found to be lower than HA–10 wt% Ti composite.

In future, cryo-milling of the HA–Ti composition may be found as a good alternative to conventional ball-milling in this particular case. The cryo-milling is expected to produce well distributed Ti particles in the starting powder mixture. This will allow well dispersed Ti in HA–Ti composites with concentration of Ti more than 20 wt% causing further improvement in fracture toughness. It may be possible to reduce the initial particle size considerably by cryo-milling. This will lead to the HA–Ti composites with even higher fracture toughness. Therefore, designing the composites with different shape and size of Ti particles may be useful in improving the fracture toughness.

The increased Ti particle distribution by restricting the unification of Ti during ball-milling may provide an opportunity to further increase the fracture toughness. Therefore, cryo-milling can be used in place of conventional high energy ball-milling to restrict the extensive deformation and therefore, unification of Ti particles. The uniform dispersion of metallic reinforcement would also increase the strength properties. In future, cryo-milling parameters need to be optimized and similar SPS experiments can be performed to prove this point.

In last two decades, the fractal modeling approach has been used to a modest extent to predict the fracture toughness of the ceramic composites [94, 95]. Xu and co-workers successfully used the Mandelbrot's modified fractal model to calculate the fracture toughness (K_{IC}) of alumina with equiaxed grains as well as reinforced with elongated grains [96]. According to this model, the microscopic (characterized by irregular crack propagation path) fracture toughness can be given as,

$$K_{IC} = (2E\gamma_s)^{1/2}\left(\frac{1}{r}\right)^{(d_f)/2} \quad (2.10)$$

where, E is the elastic modulus in tension; γ_s is the interfacial energy equals to (Eb_o/π^2), where, E and b_o are the elastic modulus and atomic dimension, respectively [82]. In Eq. 2.8, d_f is the fractal dimension, which defined as,

$$d_f = \frac{\log(N)}{\log\left(\frac{1}{r}\right)} \quad (2.11)$$

where, r is the ratio that describe the segments generated in self-similar object. A self-similar object (fractal curve) can be broken into arbitrary small elements with each small element being a replica of the entire object; N is calculated from fractal curve and equal to the number of patterns generated for a ratio 'r' [97, 98].

Recently, Rishabh et al. [99] have also used the Mandelbrot's modified model to estimate the fracture toughness of monolithic alumina with equiaxed grains. They also calculated the fracture toughness of alumina reinforced with elongated alumina grains as well as with single wall carbon nanotubes. It has been found that the Mandelbrot's modified model can be used to estimate the fracture toughness of ceramic composites instead of using Vickers indentation method. The fractal modeling approach in case of bioceramics is not being extensively utilized to predict the toughness properties and this is one area, which requires attention in future.

In fact, the fracture toughness of monolithic HA (3.4 ± 0.2 MPa $m^{1/2}$) and HA–10Ti (4.7 ± 0.1 MPa $m^{1/2}$) composite reflect on significant improvement compared to earlier literature reports (see Fig. 2.4). In the present case, K_{IC} values are measured by SEVNB method and the toughness values are reproducibly measured on multiple samples.

In Fig. 2.4, the fracture toughness values for HA based biocomposites with 20 % of less reinforcement content are compared with presently investigated materials. Such a comparison would be more effective if we look at the toughness data measured at a constant reinforcement content (e.g. 5, 10, 20 %). Another point to be noted is that indentation cracking method has largely been used to measure the toughness of HA based composites with the exception of the present work and the study by Zhang and co-workers [100]. In view of the dependence of toughness on volume fraction of reinforcement, we only compared our results with the toughness data for those with systematic variation in reinforcement content (≤20 %).

As far as monolithic HA is concerned, the present work demonstrates at least three to four times higher toughness than earlier reported values. In case of HA reinforced with 5 and 20 % toughening phases, the present study once again confirms the superiority of spark plasma sintered HA–Ti in terms of exhibiting three to four times higher toughness compared to HA with similar reinforcement content. While similar trend is also observed in case of HA with 10 % reinforcement content, the study by Kong et al. [61] demonstrates that HA, when reinforced with Al_2O_3-coated ZrO_2 (10 %) can exhibit 70 % of the toughness, that we have achieved with the currently developed HA–10 % Ti. Interestingly, the qualitative trend of the toughness increase through a maxima at 10 % reinforcement addition is observed in case of both Ti and ZrO_2 (see Fig. 2.1). A drop in toughness value in both the cases with the addition of larger than 20 % reinforcement therefore rationalizes our effort of not to develop HA composites with 30 % Ti or larger. In the following, high toughness property will now be critically analyzed in the light of the available crack bridging model due to the presence of Ti, as any toughness increment in ceramic composite containing metallic particles must be due to the bridging of crack faces because of the yielding of ductile metallic phase (Ti in the present case) as well as crack deflection.

First of all, the spark plasma sintered monolithic HA exhibits $K_{IC} \approx 3.4 \pm 0.2$ MPa $m^{1/2}$, which is 3 times higher than the reported values in the literature [62, 63, 100–102]. Such an increase in K_{IC} for HA can be related to unique microstructure developed. It is important to mention that the microstructure of HA is characterized by a mixture of submicron equiaxed grains and elongated grains with an average width 1 μm (see Fig. 2.6a). While the major fraction of the grains are equiaxed, a significant fraction is rod shaped grains with aspect ratio of 5 or larger. It is to be noted here that the initial HA particles are of rod shape in nature. This allows the orientation and stacking of HA particles in such a way that results in the increase in the relative intensity corresponding to the (300) plane as compared to the (211) plane. Chaudhry et al. also reported this kind of features in spark plasma sintered nanograined monolithic HA [103].

In order to understand the interaction of a crack path with microstructure, representative SEM images of indented regions on monolithic HA and HA–10Ti composite are shown in Fig. 2.5. Mostly, the montage images are shown to illustrate the crack path over a wider microstructural region. In Fig. 2.5a, the crack emanating from a Vickers indent, taken at 500 gm indent load, is shown and a wavy nature of crack path is visible, as marked by a series of white arrows. Clearly, the crack goes along the grain boundary or deflected/bridged by elongated grains on its path of propagation.

Now, it is worthy to study the effect of Ti addition along with the effect of reaction products (formed at the HA/Ti interface) on the interface debonding as well as mechanical properties. The BSE-SEM images confirm the homogenous distribution of Ti phase in HA matrix (see Fig. 2.5). However, the lath-like Ti phase at higher Ti content results in higher contact area at HA/Ti interface as compared to spherical shaped Ti particles. The observation of irregular or agglomerate like Ti

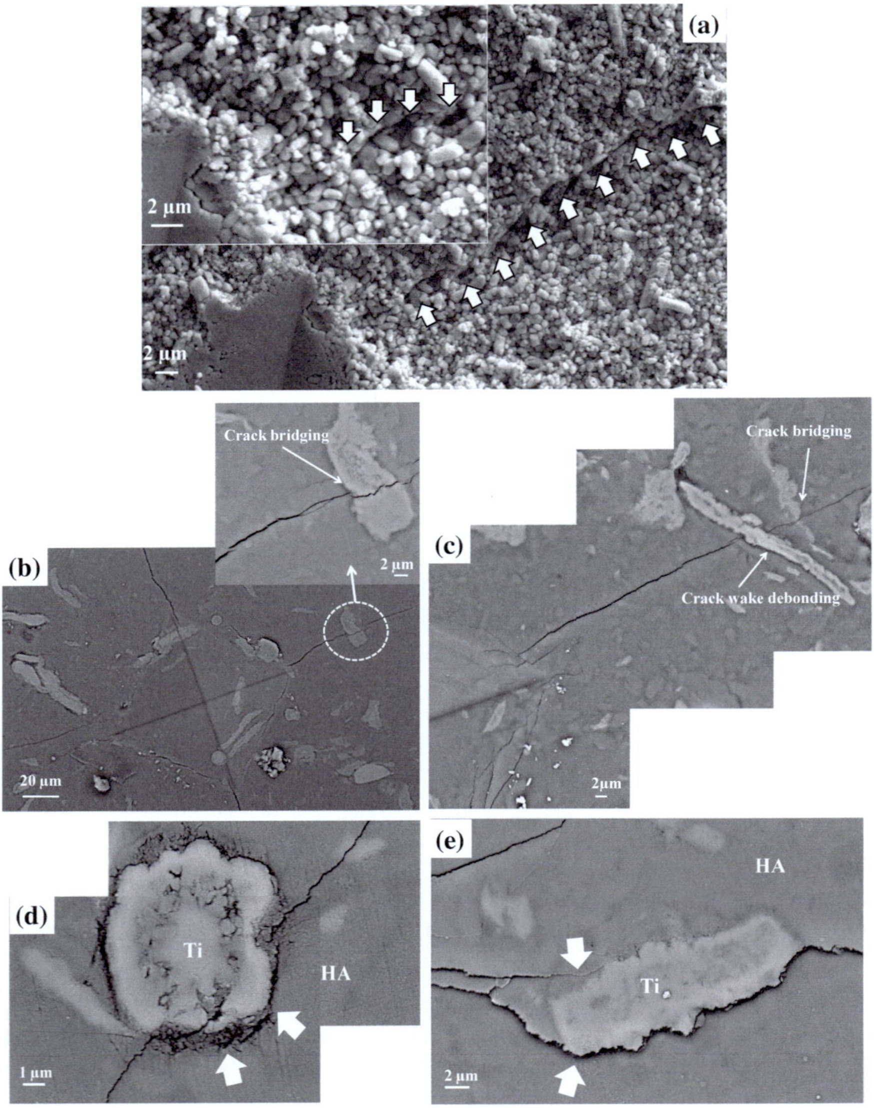

Fig. 2.5 Representative secondary electron SEM micrographs of chemically etched HA (**a**), showing the indent (500 g load) region and crack path (marked with white arrow). **b** and **c** represent the back scattered electron SEM images of HA–10Ti, showing the indent (at 1000 g load) region and crack path. The crack bridging (marked on '**b**' and '**c**') by lath shaped/elongated Ti particles and crack wake debonding (marked on '**c**') is visible. The crack wake debonding at HA/Ti interface with spherical and lath shaped particles is clearly visible and marked with white arrow on '**d**' and '**e**', respectively [67]

(bright contrast) in Fig. 2.5b–e indicate either extensive deformation or unification of finer Ti particles during SPS.

BSE-SEM images, shown in Fig. 2.5b, c, provide evidences for passage of crack through Ti-agglomerates as well as crack bridging around lath-shaped Ti phase. Interestingly, these toughening microstructural elements, i.e. Ti phase have an interesting/useful shape with width of 5–10 μm and aspect ratio of 10 or larger. It is worthwhile to mention that such shape of Ti phase is akin to elongated grains in Si_3N_4-based materials or short whiskers, both of which impart useful toughness. A closer inspection of the inset image of Fig. 2.5b reveals the crack, while going through Ti phase, is being bridged on its original path with apparent decrease in crack face width. In contrast, another crack causes debonding at HA/Ti interface and bridges the crack on the interface (Fig. 2.5c). Significant debond length of over 10 μm could be noticed. More evidences for crack wake debonding at the matrix/particulate interfaces with spherical or lath shaped Ti phase are provided respectively in Fig. 2.5d, e. The crack wake debonding can perhaps be attributed to the presence of Ti and/or $CaTi_4(PO_4)_6$ phase and the presence of $CaTi_4(PO_4)_6$ was earlier confirmed earlier using TEM analysis. From the above, it can be conclusively stated that both crack deflection and crack bridging contribute towards the toughness enhancement in HA–Ti composites. It can be mentioned that such crack bridging/deflection has also been reported in several ceramic-matrix composites reinforced with whiskers/fibers as well as in case of Ag-reinforced HA [100, 104].

2.5 Correlation with Existing Theoretical Models

In this sub-section, an attempt is being made to correlate the experimentally measured toughness properties with the theoretical model predictions. A theoretical model proposed by Becher et al. [105] was used to estimate the fracture toughness in the present case. It can be noted that Becher's model can successfully explain the toughness of Si_3N_4-based ceramics with large anisotropic tabular/elongated grains. In such case, fracture toughness can be correlated with microstructure using the following equations [105],

$$K = C\left(\sum_i V_i W_i (AR_i)^2\right)^{1/2} \tag{2.12}$$

where, V, W and AR represents to the volume fraction of elongated grains, width of elongated grain and aspect ratio of elongated grains, respectively and the suffix denotes the ith types of elongated grains.

In the above equation,

$$C = \left(\frac{E\tau}{6(1-\nu^2)}\right)^{1/2} \tag{2.13}$$

where, E (94 GPa), τ (≈15 MPa) [106] and ν (≈0.3) [107] are the elastic modulus, interfacial shear stress and Poisson's ratio of the elongated grains of HA, respectively. Using above equations and microstructural parameter, the fracture toughness of HA is found to be around ~3.2 MPa $m^{1/2}$, which is very close to the experimentally measured value (~3.4 MPa $m^{1/2}$). In estimating the fracture toughness from the above equation, elastic modulus measured by nanoindentation has been used. This is due to the fact that the indented zone dimension would scale with the interaction length due to crack bridging by elongated grain in monolithic HA.

In the next step, in order to explain the influence of ductile second phase (here Ti) in further improvement of fracture toughness, two different models has been adopted. The first model, proposed by Ashby et al. [92] highlights the effect of volume fraction (F) and particle size (D) of Ti phase on the fracture toughness of HA–Ti composites. The second model, proposed by Evans et al. [93], considers the effect of matrix phase on the fracture toughness of composite. In this model, the fracture toughness of composite strongly depends upon the particle size (D) and distance between near neighboring particles (δ).

Adopting Ashby's model, the increment in the fracture toughness (ΔK_{IC}) in a ceramic matrix composite reinforced with ductile metal particles can be given by the following expression [92],

$$\Delta K_{IC} = \left(CSFE\sigma_y D\right)^{1/2} \tag{2.14}$$

where, C is a constant which depends upon the interfacial strength; S is the fraction of particles nearby to the crack tip; F is the volume fraction of the ductile metal particles which equals to the area fraction on the crack plane; E is the elastic modulus of the metallic (Ti) particles (103 GPa) [108]; σ_y is the yield strength of the metallic Ti particles (325 MPa) [109]; D is the size of the Ti particles. The values of F and D were estimated using image processing software (ImageJ) from SEM images captured in BSE mode and some representative images are shown in Fig. 2.5.

Therefore, ΔK_{IC} can be considered to be directly proportional to the $(SFD)^{1/2}$,

$$\Delta K_{IC} \propto (SFD)^{1/2} \tag{2.15a}$$

$$\Delta K_{IC} = C^{*}(SFD)^{1/2} \tag{2.15b}$$

where, C^{*} is a constant.

In present case, the volume fraction of Ti in HA–5Ti, HA–10Ti, and HA–20Ti were 0.095, 0.205, and 0.292, as estimated using ImageJ software. The value of S (the fraction of particles nearby the crack tip) is directly related to the volume fraction of Ti in HA matrix (F). Therefore, S will increase with an increase in F. If we consider the number fraction of particles near the crack tip in case of HA–5Ti i.e. S = n, such assumption will lead to, S ≈ 2n and S ≈ 3n for HA–10Ti and HA–20Ti, respectively.

Based on Eq. (2.16) and above consideration, the increment in fracture toughness varies according to the volume fraction of reinforcement particles and size of the reinforcement particles. In Fig. 2.3a, ΔK_{IC} versus $(SFD)^{1/2}$ for investigated materials is plotted. The results obtained from theoretical model (Fig. 2.2a) shows statistically significant difference between HA–5Ti and HA–10Ti as well as HA–5Ti and HA–20Ti samples, when Dunnett's t (marked with *) test was applied. The Dunnett' C test does not show any significant difference in this case. However, a significant difference between HA and HA–10Ti as well as between HA–5Ti and HA–20Ti samples was found, when Dunnett's t (marked with **) and Dunnett' C (marked with ##) test were used for $(SFD)^{1/2}$. In Fig. 2.2b, the results reveal the significant difference for both $K_{IC(HA-Ti)} - K_{IC(HA)}$ and $(1 + (\delta/D))^{-1/2}$, when Dunnett's t test was applied. However, Dunnett's C test fails to reveal any significant difference in case of $K_{IC(HA-Ti)} - K_{IC(HA)}$ and $(1 + (\delta/D))^{-1/2}$. The strength of HA/Ti interface may lead to the deviation of ΔK_{IC} from its linear pattern. The positive deviation is possibly due to the result of high interfacial strength. In contrast, the negative deviation may be a result of the weak interface.

In order to validate the obtained results from the Ashby's model, we have used another model to estimate the fracture toughness and an attempt has been made to correlate the result obtained with the experimental values (see Fig. 2.3b). The yield strength of the ductile metal particle as well as size and interparticle spacing can also affect the toughness of a ceramic matrix-metal composite, as per the following equation [93],

$$K_{IC} = K_{ICm} + f(u^*, \sigma_y)\frac{1}{\sqrt{1+\frac{\delta}{D}}} \tag{2.16}$$

where, K_{IC} is the fracture toughness of composite; K_{ICm} is the fracture toughness of ceramic matrix (HA); $f(u^*, \sigma_y)$ is a constant depends upon the metal (Ti) strain to failure (u^*) and the yield strength of the Ti particles (σ_y); δ is the particles near-neighbor distance; D is the size of the Ti particles. In the present case, the values of δ and D were estimated using an image processing software (ImageJ) from SEM images captured in BSE mode (see Fig. 2.5).

For a given ceramic matrix and metal reinforcement, the fracture toughness of composite largely depends on δ and D. For a particular volume fraction of reinforcement, the ratio $(\frac{\delta}{D})$ is independent of particle size and therefore, $(\frac{\delta}{D})$ values do not affect the fracture toughness. However, in case of composite with varying particle size and amount of metal reinforcement, the ratio $(\frac{\delta}{D})$ is highly significant. In the present case, the microstructural analysis shows a lower value of $(\frac{\delta}{D})$ for

HA–20Ti as compared to HA–10Ti. This is due to the unification of Ti particles during ball milling/SPS, which leads to an increase in final particle size.

2.6 Biocompatibility Properties of HA–Ti Composites

The improvement in mechanical properties must not lead to the degradation of biological properties. Although HA and Ti (and Ti alloys) were investigated independently for their good biocompatibility property, only few reports are available on the study of biocompatibility of the HA–Ti composites. In few forthcoming sections, we will discuss briefly the bioactivity, cytocompatibility and biocompatibility of HA–Ti composites processed by advanced sintering route.

The cytotoxicity of a biomaterial can be investigated on the basis of cell viability and proliferation, *in vitro*. Nath et al. [52] reported the results of L929 mouse fibroblast cell adhesion on the conventionally sintered HA–10Ti, HA–20Ti, and HA–30Ti composites. The cultured L929 cells show more flattened morphology on HA–Ti samples (see Fig. 2.6) [52]. In a follow-up work [50], SEM studies reveal the extensive cell-to-cell contact as well as spreading of the Saos-2 cells followed by ECM (extracellular matrix) formation (Fig. 2.7). The well defined morphology of Saos-2 cells on the HA–Ti substrates reveals their viability and establishes the cytocompatibilty of the HA–Ti composites. The fluorescence microscopic investigation supports the finding of the SEM study. Extensive actin filament reorganization (green color) and unfragmented nucleus (blue color) confirm the cytocompatibility of the HA–Ti composites (see Fig. 2.7).

2.6.1 in vivo Biocompatibility

Although *in vitro* experiments can be used for initial screening of the biomaterials, the absence of actual physiological conditions during these experiments limit the applicability of such experiments. In fact, the material is to be tested for *in vivo* biocompatibility property in order to understand the effect of more diverse environment enriched with enzymes, proteins and variable pH as well as internal and/or external stimuli like stress/electric field. Therefore, *in vivo* experiments are the endpoint study of the biocompatibility of a biomaterial. In order to evaluate the biocompatibility of HA–Ti composites, Chu et al. [71] performed the implantation of HA–20 vol.% Ti in the skull of New Zealand white rabbits. Histological observations after 3 weeks of implantation revealed the partial integration of implant material with host bone. At the beginning of *in vivo* experiment (3 weeks), it has been noticed that the presence of Ti (bioinert) results in a decrease in the

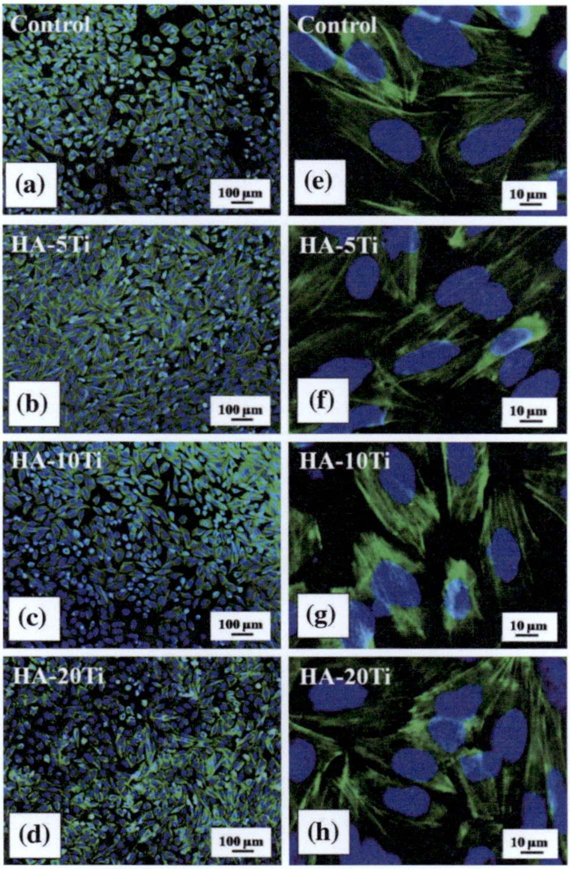

Fig. 2.6 Representative fluorescence microscopy images of $SaOS_2$ cells on control and different HA–Ti composites [50]

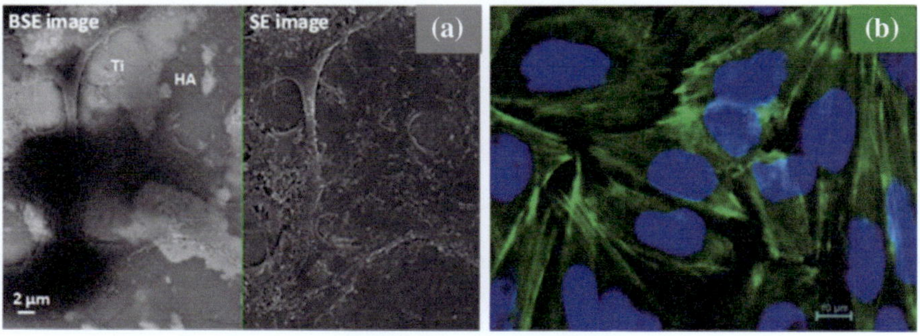

Fig. 2.7 **a** The SEM images showing the cell-to-cell contact and spreading of Saos-2 cells over HA as well as Ti surface of HA–Ti composites. **b** The fluorescence microscopy image of Saos-2 cells, grown on HA–Ti samples shows the extensive reorganization of the actin filaments (*green color*) and the healthy nuclei (*blue color*) [50]

osseointegration of the composite. However, after 12 weeks of implantation period, a complete integration of the composite material with host bone was noticed. Therefore, HA–20Ti composite is found to behave similar to HA in terms of long term implantation. Importantly, the decomposition products of HA in HA–20Ti composite such as α-TCP dissolve faster than monolithic HA in the physiological environment and therefore, supply the Ca^{2+} and PO_4^{3-} ions, results in faster healing at the defect site [83].

In a different work, Chu et al. [51] investigated the biocompatibility of HA–Ti FGM (Ti/Ti-20 vol.% HA/Ti-40 vol.% HA), *in vivo*. For this, FGM samples (3.3 mm × 3.3 mm × 6 mm) were implanted into the skull of New Zealand white rabbits. The bonding strength of FGM was found to be higher than pure Ti at any instant (see Fig. 2.8). It was found that the presence of TiO_2 (rutile) layer on the Ti metal surface provides the nucleation site for the apatite and therefore, enhances the bonding with host tissue [110].

In summary, the presence of HA in the HA–Ti FGM helps in formation of intervening apatite layer and therefore, good integration with host bone. The presence of oxide layer on Ti surface synergistically improves the bioactivity of FGM. Importantly, the addition of Ti to HA does not impair the biocompatibility of the HA–Ti composites.

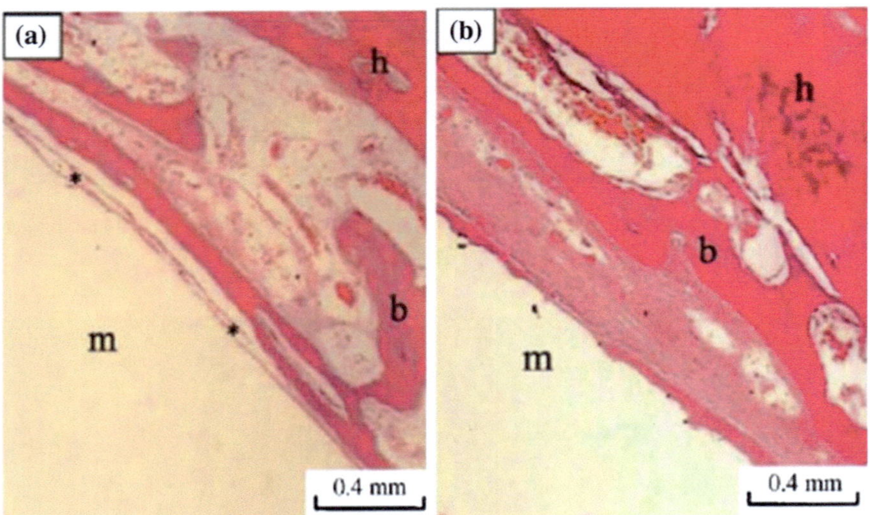

Fig. 2.8 Representative images showing the neobone formation at the interface of implant material **a** Pure Ti and **b** Ti–40 vol.% HA graded layer in the FGM) and host bone, after 4 weeks. Acronyms used: *m* implant, *b* newborn bone, *h* host bone, * fibrous tissue [51]

2.7 Future Perspective

This review summarizes the processing, characterization and properties of HA–Ti based bulk composites in the perspective of bone tissue engineering applications. The processing related challenges in context of phase stability in the HA-based systems are mainly discussed and the shortcomings of the conventional pressureless sintering are highlighted. Such discussion also places the benefits of using advanced processing techniques such as spark plasma sintering (SPS) to consolidate hydroxyapatite–titanium (HA–Ti) composites with maximum retention of HA and Ti in the composite by intelligently tailoring the sintering conditions.

The results of the present study is very significant as with several years of research, the fracture toughness value could not be increased beyond 2 MPa $m^{1/2}$ even when conventional indentation technique is used and many reinforcements (Al_2O_3 particles, HA whiskers, ZrO_2 particles, silver particles) were attempted [61, 100, 102, 111]. This is the first study showing the promise that the toughness of HA-based materials can be enhanced to 5 MPa $m^{1/2}$ or beyond. The present study also establishes that Ti addition can be intelligently tailored to obtain a desired combination of fracture toughness, strength and elastic modules. It is possible that synergistic toughening approach with incorporation of ZrO_2 (transformation toughening) can be attempted to further enhance the fracture toughness beyond 5 MPa $m^{1/2}$. However, this should require a careful control over sintering reactions. To this end, multistage SPS route with two or more intermediate holding at intermediate temperatures can be very effective.

At the closure, the development of high toughness HA biocomposite should have a significant impact on the potential biomedical applications. There has been a great need to develop mechanically reliable HA materials for heavy load bearing implants such as artificial teeth roots [102]. Earlier research reported the *in vivo* biocompatibility as well as clinical acceptance of dense HA as artificial teeth roots [112–114]. It was also reported that most of the loaded implants are broken within one year from implantation because of inferior toughness properties. In the absence of the availability of high toughness materials, dense HA ceramics are used only as unloaded tooth roots substitutes. The present study therefore makes a step ahead in developing more mechanically reliable HA composites for dental restorative applications and other biomedical applications requiring high toughness. Importantly, these high toughness materials do not appear to compromise on the biocompatibility aspects. It is noteworthy to mention that HA–10 % Ti composite, which exhibits highest toughness of close to 5 MPa $m^{1/2}$ also supports biomineralization kinetics and similar cell viability and Saos-2 cell proliferation *in vitro*, which is either comparable or better than monolithic HA [115]. In our recent study on cell fate process using flow cytometry, it is demonstrated that HA 10 % Ti also supports the progress of cell division of human fetal osteoblast cells without causing any significant necrosis/apoptosis *in vitro* in a statistically significant manner like monolithic HA [116].

References

1. Begley CT, Doherty MJ, Mollan RAB, Wilson JD. Comparative study of the osteoinductive properties of bioceramic, coral and processed bone graft substitutes. Biomaterials. 1995;16:1181–5.
2. Silva RV, Camilli JA, Bertran CA, Moreira NH. The use of hydroxyapatite and autogenous cancellous bone grafts to repair bone defects in rats. Int J Oral Maxillofac Surg. 2004;10:1–7.
3. Whang PG, Wang JC. Bone graft substitutes for spinal fusion. Spine J. 2003;3:155–65.
4. Khan SN, Fraser JF, Sandhu HS, Cammisa FP, Girardi FP, Lane JM. Use of osteopromotive growth factors, demineralized bone matrix and ceramics to enhance spinal fusion. J Am Acad Orthop Surg. 2005;13:129–37.
5. Summers BN, Eisenstein SM. Donor site pain from the ilium: a complication of lumbar spine fusion. J Bone Joint Sur Br. 1989;71-B:677–680.
6. Giannoudis PV, Dinopoulos H, Tsiridis E. Bone substitutes: an update. Inj Int J Care Inj. 2005;365:S20–7.
7. Grauer JN, Beiber JM, Kwon B, Vaccaro AR. The evolution of allograft bone for spinal applications. Orthopedics. 2005;28:573–7.
8. Vaccaro AR. The role of the osteoconductive scaffold in synthetic bone graft. Orthopedics. 2002;25:571-278.
9. Bonfiglio M, Jeter WS. Immunological responses to bone. Clin Orthop. 1972;18:19–27.
10. GD GDB, Goldberg VM, Zika JM. Immune responses of rats to frozen bone allografts. J Bone Joint Sur, Am. 1983;65-A:239–246.
11. Damien CJ, Parsons JR. Bone graft and bone graft substitutes: a review of current technology and applications. J Appl Biomater. 1991;2:187–208.
12. Pelker RR, Friedlaender GE. Biomechanical aspects of bone autografts and allografts. Orthop Clin North Am. 1987;18:235–9.
13. Goldberg VM, Stevenson S, Shaffer JW. Biology of autografts and allografts. Park Ridge, Illinois: The American Academy of Orthopaedic Surgeons; 1991.
14. Wiles P. The surgery of the osteo-arthritic hip. Br j Sur. 1958;45(193):488–97.
15. Williams DF. On the nature of biomaterials. Biomaterials. 2009;30:5897–909.
16. Basu B, Katti DS, Kumar A. Advanced biomaterials: fundamentals, processing and applications. New Jersey, USA: John Wiley & Sons, Inc.; 2009.
17. Valiathan M, Krishnan V. Biomaterials: an overview. Natl Med J India. 1999;12:270–344.
18. LeGeros RZ, Daculsi G, LeGeros JP. Bioactive bioceramics. Totowa, NJ: Humana Press; 2008.
19. Hench LL, Paschall HA. Direct bonding of bioactive glass ceramics to bone and muscle. J Biomed Mater Res. 1973;4:25–42.
20. Hench LL, Wilson J. Surface-active biomaterials. Science. 1984;226:630–6.
21. Osborn JF, Newesely H. The material science of calcium phosphate ceramics. Biomaterials. 1980;1:108–11.
22. Legeros RZ, Craig RG. Strategies to affect bone remodeling: osteointegration. J Bone Miner Res. 1993;8(S2):S583–96.
23. LeGeros RZ. Properties of osteoconductive biomaterials: calcium phosphate. Clin Orthop Relat Res. 2002;395:81–98.
24. Oonishi H, Hench LL, Wilson J, Sugihara F. Quantitaive comparision of bone growth behaviour in granules of Bioglass^{R}, A-W glass-ceramic, and hydroxyapatite. J Biomed Mater Res. 2000;51:37–46.
25. Urist MR, Silverman BF, Buring K, Dubuc FI, Rosenberg J. The bone induction principle. Clin Orthop. 1967;53:243–83.
26. Geetha M, Singh AK, Asokamani R, Gogia AK. Ti based biomaterials, the ultimate choice for orthopaedic implants-A review. Prog Mater Sci. 2009;54:397–425.
27. Morais JM, Papadimitrakopoulos F, Burgerss DJ. Biomaterials/tissue interactions: possible solutions to overcome foreign body response. Am Assoc Pharaceutical Sci J. 2010;12:188–96.

28. Alliot-Licht B, Gregoire M, Orly I, Menanteau J. Cellular activity of osteoblasts in the presence of hydroxyapatite: an *in vitro* experiment. Biomaterials. 1991;12:752–6.
29. Katti KS. Biomaterials in total joint replacement. Colloids Surf, B. 2004;39:133–42.
30. Biring G, Bankes M, Meswania J, Blunn G. Friction, lubrication, wear and corrosion. Great Britain: Hodder Arnold; 2007. p. 304.
31. Sansone V, Pagani D, Melato M. The effects on bone cells of metal ions released from orthopaedic implants. A review. Clinical cases in mineral and bone. Metabolism. 2013;10:34–40.
32. Kienapfel H, Sprey C, Wilke A, Griss R. Implant fixation by bone ingrowth. J Arthroplasty. 1999;14:355–68.
33. Niinomi M. Recent metallic materials for biomedical applications. Metall Mater Trans A. 2002;33A:477–86.
34. Saenz A, Brostow W, Rivera E, Castaño V. Ceramic biomaterials: an introductory overview. J Mater Educ. 1999;21:297–306.
35. Rivera-Muñoz E, Díaz J, Rodríguez JR, Brostow W, Castaño V. Hydroxyapatite spheres with controlled porosity for eye ball prosthesis: processing and characterization. J Mater Sci Mater Med. 2001;12:305–16.
36. Park J. Bioceramics: properties, characterizations, and applications. New York: Springer; 2008. p. 359.
37. Wolff J. The law of bone remodelling, Translated by P.Maquet and R. Furlong. Heidelberg: Springer; 1986.
38. Prendergast PJ, Huiskes R. The biomechanics of Wolff's law: recent advances. Ir J Med Sci. 1995;164:152–4.
39. Dujovne AR, Bobyn JD, Krygier JJ, Miller JE, Brooks CE. Mechanical compatibility of noncemented hip prostheses with the human femur. J Arthroplast. 1993;8:7–22.
40. Engh CA, Bobyn JD. The influence of stem size and extent of porous coating on femoral bone resorption after primary cementless hip arthroplasty. Clin Orthop Relat Res. 1988;231:7–28.
41. Dowson D. Bio-tribology of natural and replacement synovial joints. New York: Springer; 1992.
42. Murugan R, Ramakrishna S. Nanostructured biomaterials. California: American Scientific Publishers; 2004.
43. Hench LL. Bioceramics. J Am Ceram Soc. 1998;81:1705–28.
44. Audekercke RV, Martens M, Hastings GW. Mechanical properties of cancellous bone. Florida: CRC Press-Boca Raton; 1984.
45. Black J. Orthopedic biomaterials in research and practice. NewYork: Churchill Livingston; 1988.
46. Jee WSS. Histology, cell and tissue biology. New York: Elsevier Science Inc.; 1983.
47. Evans FG, King A. Biomedical studies of the musculoskeletal system. Illinois: Springfield; 1961.
48. Bonfield W, Hastings GW. Elasticity and viscoelasticity of cortical bone. Florida: CRC Press-Boca Raton; 1984.
49. Kumar A, Webster TJ, Biswas K, Basu B. Flow cytometry analysis of human fetal osteoblast fate processes on spark plasma sintered Hydroxyapatite-Titanium biocomposites. J Biomed Mater Res, Part A. 2013;101:2925–38.
50. Kumar A, Dhara S, Biswas K, Basu B. In vitro bioactivity and cytocompatibility properties of spark plasma sintered HA–Ti composites. J Biomed Mate Res Part B-Appl Biomater. 2013;101B:223–36.
51. Chu C, Xue X, Zhu J, Yin Z. In vivo study on biocompatibility and bonding strength of Ti/Ti-20 vol.% HA/Ti-40 vol.% HA functionally graded biomaterial with bone tissue in the rabbit. Mater Sci Eng, A. 2006;429:18–24.
52. Nath S, Tripathi R, Basu B. Understanding phase stability, microstructure development and biocompatibility in calcium phosphate–titania composites, synthesized from hydroxyapatite and titanium powder mix. Mater Sci Eng, C. 2009;29:97–107.

53. Nakahira A, Eguchi K. Evaluation of microstructure and some properties of hydroxyapatite/Ti composites. J Ceram Process Res. 2001;2:108–12.
54. Ning CQ, Zhou Y. In vitro bioactivity of a biocomposite fabricated from HA and Ti powder metallurgy method. Biomaterials. 2002;23:2909–15.
55. Yang Y, Kim K-H, Agrawal CM, Ong JL. Interaction of hydroxyapatite–titanium at elevated temperature in vacuum environment. Biomaterials. 2004;25:2927–32.
56. Dubey AK, A AE, Balani K, Basu B. Multifunctional properties of multi-stage spark plasma sintered HA-BaTiO3 based piezobiocomposites for bone replacement applications. J Am Ceram Soc. 2013.
57. Mallik PK, Basu B. Better early osteogenesis of electroconductive hydroxyapatite–calcium titanate composites in a rabbit animal model. J Biomed Mater Res Part A. 2013:n/a-n/a.
58. Bajpai I, Balani K, Basu B. Spark plasma sintered HA-Fe3O4-based multifunctional magnetic biocomposites. J Am Ceram Soc. 2013;96(7):2100–8.
59. Saha N, Dubey AK, Basu B. Cellular proliferation, cellular viability, and biocompatibility of HA-ZnO composites. J Biomed Mater Res B Appl Biomater. 2012;100B(1):256–64.
60. Mandal S, Sunilkumar B, Kumar A, Basu B. Hot pressed silver doped hydroxyapatite biomaterials with bactericidal properties against magnetotactic bacteria. Mater Technol Adv Perform Mater. 2013.
61. Kong Y-M, Kim S, Kim H-E. Reinforcement of hydroxyapatite bioceramic by addition of ZrO2 Coated with Al2O3. J Am Ceram Soc. 1999;82:2963.
62. Rapacz-Kmita A, Sl′osarczyk A, Paszkiewicz Z. Mechanical properties of HAp–ZrO2 composites. J Eur Ceram Soc. 2006;26:1481–8.
63. Chun Y, Ying-kui G, Mi-lin Z. Thermal decomposition and mechanical properties of hydroxyapatite ceramic. Trans Nonferrous Met Soc China. 2010;20:254–8.
64. Long M, Rack HJ. Titanium alloys in total joint replacement—a materials science perspective: Review. Biomaterials. 1998;19:1621–39.
65. Murugan R, Ramakrishna S. Development of nanocomposites for bone grafting. Compos Sci Technol. 2005;65:2385–406.
66. Kwon S-H, Jun Y-K, Hong S-H, Kim H-E. Synthesis and dissolution behavior of Beta-TCP and HA/Beta-TCP composite powders. J Eur Ceram Soc. 2003;23:1039–45.
67. Kumar A, Biswas K, Basu B. On the toughness enhancement in hydroxyapatite-based composites. Acta Mater. 2013;61(14):5198–215.
68. Reddy KM, Kumar N, Basu B. Innovative multi-stage spark plasma sintering to obtain strong and tough ultrafine-grained ceramics. Scripta Materilia. 2010;62:435–8.
69. Fahami A, Ebrahimi-Kahrizsangi R, Nasiri-Tabrizi B. Mechanochemical synthesis of hydroxyapatite/titanium nanocomposite. Solid State Sci. 2011;13:135–41.
70. Sanosh KP, Chu M-C, Balakrishnan A, Lee Y-J, Kim TN, Cho S-J. Synthesis of nano hydroxyapatite powder that simulate teeth particle morphology and composition. Curr Appl Phys. 2009;9(6):1459–62.
71. Chu C, Lin P, Dong Y, Xue X, Zhu J, Yin Z. Fabrication and characterization of hydroxyapatite reinforced with 20 vol%Ti particles for use as hard tissue replacement. J Mater Sci Mater Med. 2002;13:985–92.
72. Babushkin O, Lindbäck T, Holmgren A, Li J, Hermansson L. Thermal expansion of hot isostatically pressed hydroxyapatite. J Mater Chem. 1994;4:413–5.
73. Lütjering G, Williams JC. Titanium. New York: Springer; 2007. p. 442.
74. Weng J, Liu X, Zhang X, Ji X. Thermal decomposition of hydroxyapatite structure induced by titanium and its oxides. J Mater Sci Lett. 1994;13:159–61.
75. Groot Kd. Bioceramics of calcium phosphate. Boca Raton, Florida: CRC Press; 1983.
76. Ning CQ, Zhou Y. In vitro bioactivity of a biocomposite fabricated from HA and Ti powder by powder metallurgy method. Biomaterials. 2002;23:2909–15.
77. Mondal D, Nguyen L, Oh I-H, Lee B-T. Microstructure and biocompatibility of composite biomaterials fabricated from titanium and tricalcium phosphate by spark plasma sintering. J Biomed Mater Res, Part A. 2013;101A(5):1489–501.

78. Beegan D, Chowdhury S, Laugier MT. Work of indentation methods for determining copper film hardness. Surf Coat Technol. 2005;192:57–63.
79. Kumar A, Dhara S, Biswas K, Basu B. In vitro bioactivity and cytocompatibility properties of spark plasma sintered HA–Ti composites. J Biomed Mater Res: Part B-Appl Biomater. 2012.
80. Zioupos P, Currey JD. Changes in the stiffness, strength, and toughness of human cortical bone with age. Bone. 1998;22(1):57–66.
81. Basu B, Balani K. Advanced structural ceramics. New Jersey, USA: John Wiley & Sons; 2011. p. 474.
82. Lawn B. Fracture of brittle solids. Cambridge, UK: Cambridge University Press; 1998. p. 378.
83. Gang B, Chung-Yuen H. Effects of interface debonding on the toughness of ductile-particle reinforced ceramics. Int J Solids Struct. 1990;26(5–6):631–42.
84. Kaw AK. Machanics of composite materials. Boca Raton, Florida, USA: CRC Press; 2006. p. 466.
85. Weertman J. Dislocation based fracture mechanics. Singapore: World Sci; 1996.
86. Swanson PL. Crack-interface traction: a fracture-resistance mechanism in brittle polycrystals. Westerville, OH: American Ceramic Society; 1988.
87. Faber KT, Evans AG. Crack deflection processes—I. Theory. Acta Metall. 1983;31(4):565–76.
88. Faber KT, Evans AG. Crack deflection processes—II. Experiment. Acta Metall. 1983;31 (4):577–84.
89. Kumar A. Microstructure development, biomineralization and osteoblast cell fate processes on high toughness hydroxyapatite-titanium composites. Kanpur, India: Indian Inst Technol Kanpur; 2013. p. 204.
90. Li Y, Yuan H, von dem Bussche A, Creighton M, Hurt RH, Kane AB, Gao H. Graphene microsheets enter cells through spontaneous membrane penetration at edge asperities and corner sites. Proceedings of the National Academy of Sciences; 2013.
91. Billi F, Campbell P. Nanotoxicology of metal wear particles in total joint arthroplasty: a review of current concepts. J Appl Biomater Biomech. 2010;8(1):1–6.
92. Ashby MF, Blunt FJ, Bannister M. Flow characteristics of highly constrained metal wires. Acta Metall. 1989;37(7):1847–57.
93. Evans AG, McMeeking RM. On the toughening of ceramics by strong reinforcements. Acta Metall. 1986;34:2435.
94. Shou-Zhu Z, Chi-Wei L. Fractal dimension and fracture toughness. J Phys D Appl Phys. 1989;22(6):790.
95. Mandelbrot BB, Passoja DE, Paullay AJ. Fractal character of fracture surfaces of metals. Nature. 1984;308:721–2.
96. Xu L, Xie Z, Gao L, Wang X, Lian F, Liu T, Li W. Synthesis, evaluation and characterization of alumina ceramics with elongated grains. Ceram Int. 2005;31(7):953–8.
97. Charkaluk E, Bigerelle M, Iost A. Fractals and fracture. Eng Fract Mech. 1998;61(1): 119–39.
98. Carpinteri A. Fractal nature of material microstructure and size effects on apparent mechanical properties. Mech Mater. 1994;18(2):89–101.
99. Rishabh A, Joshi MR, Balani K. Fractal model for estimating fracture toughness of carbon nanotube reinforced aluminum oxide. J Appl Phys. 2010;107:7.
100. Zhang X, Gubbels GHM, Terpstra RA, Metselaar R. Toughening of calcium hydroxyapatite with silver particles. J Mater Sci. 1997;32(1):235–43.
101. Tancred DC, McCormack BAO, Carr AJ. A quantitative study of the sintering and mechanical properties of hydroxyapatite/phosphate glass composites. Biomaterials. 1998;19:1735–43.
102. Suchanek W, Yashima M, Kakihana M, Yoshimura M. Hydroxyapatite/hydroxyapatite-whisker composites without sintering additives: mechanical properties and microstructural evolution. J Am Ceram Soc. 1997;80:2805–13.

103. Chaudhry AA, Yan H, Gong K, Inam F, Viola G, Reece MJ, Goodall JBM, ur Rehman I, McNeil-Watson FK, Corbett JCW and others. High-strength nanograined and translucent hydroxyapatite monoliths via continuous hydrothermal synthesis and optimized spark plasma sintering. Acta Biomater. 2011;7(2):791–799.
104. Warren R. Fundamental aspects of the properties of ceramic-matrix composites. London, UK: Blakie and Sons Ltd.; 1992.
105. Becher PF, Lin HT, Hwang SL, Hoffmann MJ, Chen I-W. The Influence of microstructure on the mechanical behavior of silicon nitride ceramics. In: MRS online proceedings library 1992;287:null-null.
106. Wei M, Ruys AJ, Swain MV, Kim SH, Milthorpe BK, Sorrell CC. Interfacial bond strength of electrophoretically deposited hydroxyapatite coatings on metals. J Mater Sci Mater Med. 1999;10(7):401–9.
107. Prokopiev O, Sevostianov I. Dependence of the mechanical properties of sintered hydroxyapatite on the sintering temperature. Mater Sci Eng, A. 2006;431:218–27.
108. Niinomi M. Mechanical properties of biomedical titanium alloys. Mater Sci Eng, A. 1998;243:231–6.
109. Zadra M, Casari F, Girardini L, Molinari A. Microstructure and mechanical properties of cp-titanium produced by spark plasma sintering. Powder Metall. 2008;51:59–65.
110. Ghaith ES, Hayakawa T, Kasuga T, Nogami M. Apatite formation on rutile type TiO2 films formed by laser irradiation. J Mater Sci. 2006;41(8):2521–4.
111. Kumar R, Prakash KH, Cheang P, Khor KA. Microstructure and mechanical properties of spark plasma sintered zirconia-hydroxyapatite nano-composite powders. Acta Mater. 2005;53:2327–35.
112. Ogiso M, Tabata T, Ichijo T, Borgese D. Examination of human bone surrounded by a dense hydroxyapatite dental implant after long-term use. J Long Term Eff Med Implant. 1992;2:235–47.
113. Denissen HW, Kalk W, Veldhuis AAH, van den Hooff A. Eleven-year study of hydroxyapatite implants. J Prosthet Dent. 1989;61(6):706–12.
114. de Putter C, de Groot K, Smitt PAES. Transmucosal implants of dense hydroxylapatite. J Prosthet Dent. 1983;49(1):87–95.
115. Kumar A, Dhara S, Biswas K, Basu B. In vitro bioactivity and cytocompatibility properties of spark plasma sintered HA–Ti composites. J Biomed Mater Res B Appl Biomater. 2013;101B(2):223–36.
116. Kumar A, Webster TJ, Biswas K, Basu B. Flow cytometry analysis of human fetal osteoblast fate processes on spark plasma sintered Hydroxyapatite–Titanium biocomposites. J Biomed Mater Res, Part A. 2013;. doi:10.1002/jbm.a.34603.

Chapter 3
Case Study: Hydroxyapatite Based Microporous/Macroporous Scaffolds

3.1 Relevance of Porous Scaffolds

Over the last few decades, tissue engineering has been considered as a clinically relevant solution for the repair and regeneration of damaged human tissue. In particular, scaffold acts as the matrix which serves as a host for tissue formation in the case of bone tissue engineering. In order to enable tissue formation, scaffolds should have a few basic requirements such as high porosity, sufficiently large pores, specific surface properties that will enable the adhesion differentiation and proliferation of cells. Once the tissue grows into the scaffold, such structure helps to maintain the predetermined tissue structure and biocompatibility. Calcium Phosphate scaffolds are regarded as an interesting material for scaffold application. Calcium Phosphate (CP) based materials are known to support in osteoblast adhesion and proliferation [1, 2]. However, the major disadvantage of Calcium Phosphate based materials is its inability to be used as load-bearing bioceramics, because of brittleness and poor fatigue resistance [3, 4]. It is further pronounced in the case of highly porous bioceramics, where a porosity of greater than 100 μm is considered as the requirement for bone cell colonization. CP-based materials can be prepared from various processing routes [5–8], where biocompatibility and long term stability has been moderately achieved [9]. In general, CP-based bioceramics are characterized by various factors, such as chemical composition (stoichiometry and purity), homogeneity, distribution of phase, grain sizes/shape, crystallinity, size and distribution of porosity and so on. The vast majority of the CP-bioceramics are based on hydroxyapatite (HA), β-Tricalcium Phosphate (β-TCP), α-TCP and/or biphasic Calcium Phosphate (BCP), which is a mixture of β-TCP + HA [10–12] or α-TCP + HA [13, 14]. Various processing routes are attempted to fabricate CP-compounds, which includes uniaxial compaction [15, 16], cold/hot isostatic pressing [17–20], granulation [21], loose packing [22], slip casting [23–25], gel casting [26–28], pressure mold forming [29], injection molding [30], polymer replication [31–33], extrusion [34, 35], slurry dipping and spraying [36]. In

B. Basu and S. Ghosh, *Biomaterials for Musculoskeletal Regeneration*,
Indian Institute of Metals Series, DOI 10.1007/978-981-10-3017-8_3

Fig. 3.1 Fundamental requirements for bone replacement scaffolds

addition, the formation of ceramic sheets by tape casting is also widely employed [37, 38].

As shown in Fig. 3.1, the porous scaffolds for the bone tissue engineering need to meet a range of properties—(a) it must be biocompatible, which enables the cell growth, their attachment to surface and proliferation, (b) the material should induce strong bone bonding, resulting in osteoconduction and osteoinduction, (c) rate of new tissue formation and biodegradability (in case of biodegradable scaffold) should match with each other, (d) the mechanical strength of the scaffolds should be adequate enough to provide mechanical constancy in load bearing sites prior to regeneration of new tissue and (e) porous structure and pore size of more than 100 μm for cell penetration, tissue in growth and vascularization. The interconnected pores allow cells to migrate in multiple directions under real conditions.

Following the above-mentioned criteria, the scaffolds, appropriately designed in terms of structure and properties play an important role [39–41] to direct and provide support to the growth of cells as well as their migration around surrounding tissue [42]. The majority of mammalian cell types are anchorage-dependent, meaning they will not be able to survive in the absence of an adhesion substrate. A number of experimental investigations suggest that the scaffolds with various compositions provide a suitable platform for cell attachment, proliferation, and cell migration [43]. As a consequence, efforts have been placed in the development of porous scaffolds for bone replacement and tissue engineering.

One of the important aspects in the development of bone substitute materials is the processing of scaffolds with an appropriate micro- and macroscopic structural morphology including pore size, pore interconnectivity, biocompatibility, osteoconductivity, mechanical strength, and biodegradability. The idea is that if an implanted porous ceramic is progressively replaced by natural bone, its biomechanical properties should more resemble with of natural bone.

In order to meet the required criteria for fabricating porous scaffold, several processing techniques have been employed. The most common techniques among all are gel casting [44], slip casting [45, 46], fiber compacting [47], freeze casting [8] and gas foaming [48]. Gel casting is a wet ceramic forming technique, which involves the polymerization of a monomer in the presence of a solvent to form a rigid, ceramic-loaded body, which can be machined directly in a complex mold [49]. After gel formation, gel-cast green samples can be easily demolded and then are dried in controlled conditions [50]. Freeze casting is a method in which rapid freezing of a colloidal stable suspension of HA particles in a nonporous mold takes place, followed by sublimation of the frozen solvent under cold temperatures in vacuum.

In the above backdrop, this chapter discusses the fabrication of HA-based scaffolds. Porous HA exhibits strong bonding with the bone and the pores provide a mechanical interlock, leading to a firm fixation of the material. It is well known that bone tissue grows well into the pores, increasing strength of the HA implant. It is realized that size and morphology of pores are crucial factors for an excellent osteointegration [51, 52]. Other important requirements for porous implants are interconnectivity of the pores for the penetration of the osteoblast-like cells inside the pores as well surface roughness for the attachment of cells.

Porous HA has been extensively used as artificial bone substitutes, in particular for repair, regeneration, and reconstruction of lost, damaged or degenerative tissues [53]. Although bone tissue itself shows an excellent ability of bone regeneration, the bone healing process is difficult for critical size bony defect. From design perspective, it is important to match the osteoconductive properties of porous ceramic scaffold in one side with the osteoinductive or osteogenic properties of living bone cells in the other side. This chapter provides an overview of the extensive experimental results to demonstrate the porous scaffold fabrication and their biocompatibility assessments.

3.2 HA-Based Microporous Scaffolds

Hydroxyapatite (HA) was synthesized by widely reported suspension-precipitation route, as described in our earlier work [54–56]. The polymer precursor, i.e. PMMA is used as porogen to synthesize microporous scaffolds. The idea is to mechanically mix HA and PMMA and depending on the size of PMMA, a scaffold of HA can be obtained after sintering with the porogen removal. Therefore, conventional ceramic processing route involving powder mixing followed by high temperature sintering was adopted to obtain HA-based microporous scaffold.

Scanning electronic microscopy revealed microporosity in the as-sintered composite (see Fig. 3.2). It can be seen that pores are present ranging from 1 to 50 μm, where the majority of pores are in the range of 10–20 μm diameter. This kind of structure is considered to be adequate, as the microporosity allows initial cell adhesion.

3.2.1 *Osteoblast-like Cell Adhesion and Spreading*

Cell attachment and proliferation were further investigated using SEM. SaOS2 cells are found to be attached and spread on hybrid composite with morphology, as can be seen in Fig. 3.3a. After 3 days of culturing, the osteoblast-like cells reached confluency on the surface of hybrid composite. Most of the cells were of polygonally shaped and spread with highly extended filipodia and rough dorsal surface characteristic of active cells, as shown in Fig. 3.3a. At higher magnification, an osteoblast cell can be seen intimately associated with an extensive network of fibrillar and globular substances, which were suspected to be the deposition of collagen and other extracellular matrix components synthesized and organized by

Fig. 3.2 Microstructure of microporous HA [119]

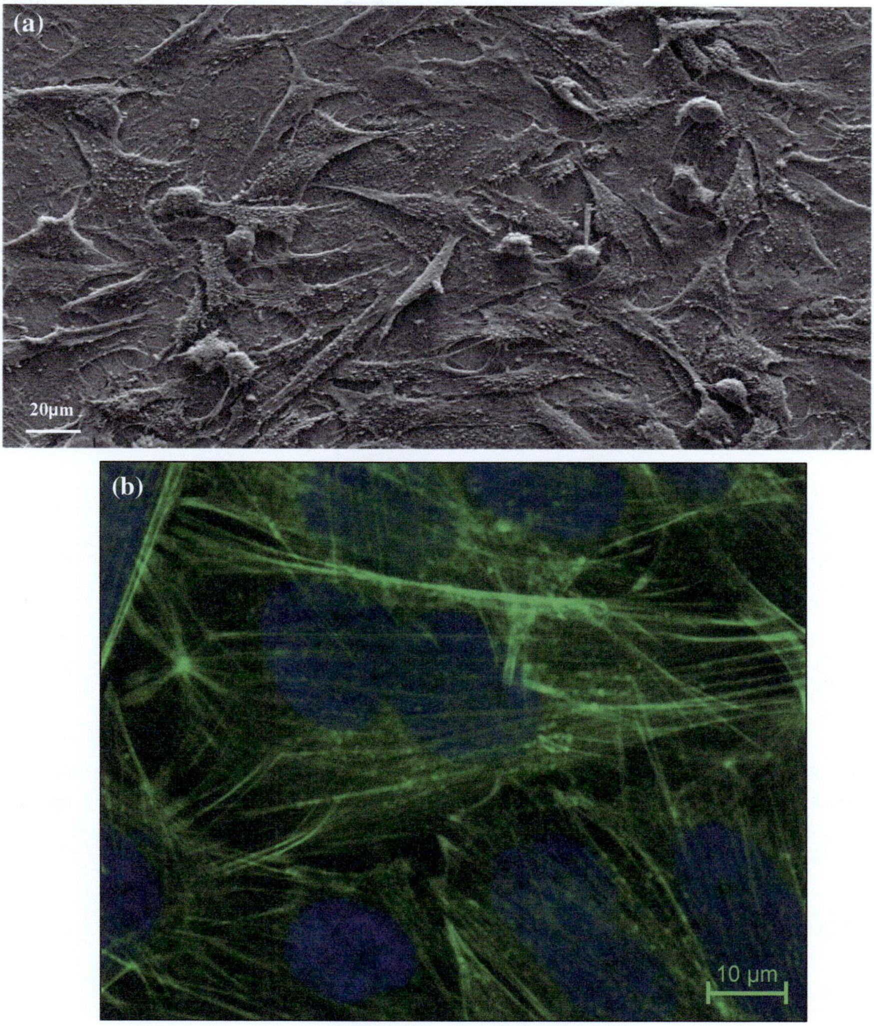

Fig. 3.3 **a** SEM images of SaOS2 human osteoblast-like cells adhered on HA/PMMA hybrid composite, after 72 h of culture. **b** Fluorescence microscopic image revealing well developed actin fibers of SaOS2 human osteoblast like cells [119]

the osteoblast-like cells. Fluorescent imaging of osteoblast cells revealed a well developed cytoskeleton with strong actin stress fibers oriented in the adhered cells (Fig. 3.3b). Such observation is characteristic for highly cytocompatible materials. After 7 days of culture, cells showed a typical morphology presenting a well spread configuration with more lamellopodia connecting to neighboring cells, starting to form a continuous cell layer. After 10 days of incubation, virtually the entire surface was covered by cells, and in some cases, continuous sheet of cell was seen. Similar behaviour was previously observed by Bagamsia and Joos [57].

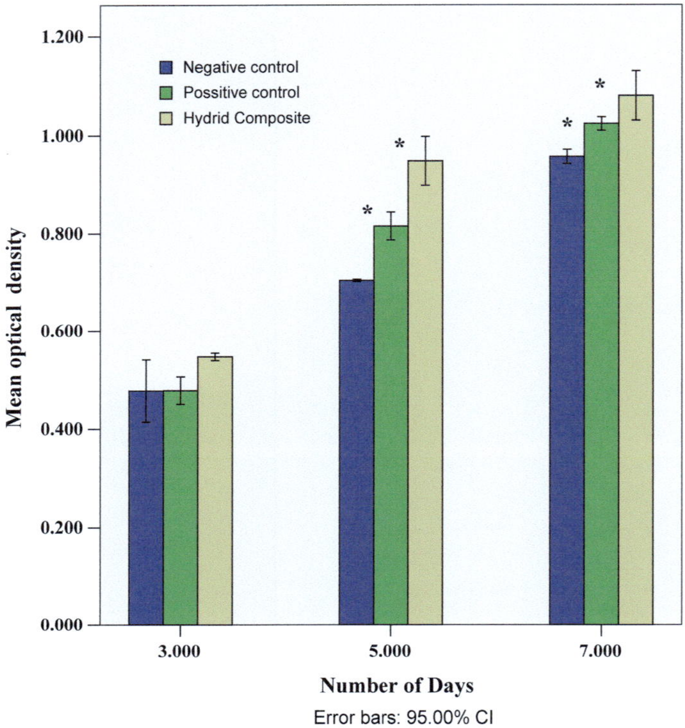

Fig. 3.4 MTT assay results showing the SaOS2 cell on negative control (plastic cover slip), positive control (HA) and microporous HA samples after 3, 5 and 7 days of culture. * represent significant difference at $p < 0.05$ with respect to pure HAp and Error *bars* correspond to ±1.00 SE for number of days of culture [119]

3.2.2 Osteoblast-like Cell Growth and Differentiation

Figure 3.4 shows the MTT assay results obtained after 3, 5 and 7 days of culture, using SaOS2 human osteosarcoma cells. In order to analyze the difference in statistical significance of the biochemical assay results, the commercial SPSS software was used. The statistical analysis was carried out using univariate analysis of variance (ANOVA) method. In particular, the Post Hoc multiple comparisons of the mean of independent groups were made using Tukey test (for homogeneous data) at statistical significant value, $p < 0.05$. The astric mark (*) in Figs. 3.8 and 3.9 represent significant statistical difference at $p < 0.05$ and error bars correspond to ±1.00 SE (standard error). The univariate analysis using Post Hoc test among different investigated materials reveals any statistically significant difference with the control sample. It can be seen that the mean optical density (OD) increases with incubation time period in hybrid composite. Since mean OD directly correlates with the mitochondrially active cells, it is clear that in terms of cell viability among the samples, there was no significant difference on day 3 ($p < 0.05$). In contrast, a statistically significant increase in cell viability was measured for osteoblast-like cells on hybrid composite than on control sample after 5 and 7 days of culture, respectively ($p < 0.05$). These results indicated that hybrid composite can act as good substrate for cell attachment and vitality. The results of MTT assay (Fig. 3.4) demonstrate that the scaffold along with dense HA, stimulated the proliferation of osteoblast cells to a similar extent with similar difference in OD values after 3/5/7 days of incubation. Independent of incubation period, the porous scaffolds exhibited better cell viability in comparison to dense HA. Such observations show efficacy of the porous scaffold. The Post Hoc multiple comparison among the samples revealed that the scaffold showed the significant difference with sintered HA. The viability of the SaOS2 cells was found to be superior in case of scaffold than that of cells cultured on sintered HA. In summary, it can be said that the HA-based microporous scaffold shows good cytocompatibility as well as enhanced cell viability property with respect to incubation period.

Osteoblasts in the culture media containing the HA/PMMA composites expressed good ALP activity throughout the 7 days of the study (Fig. 3.5). From the early incubation period of 3 days, the ALP activity on the hybrid composite was

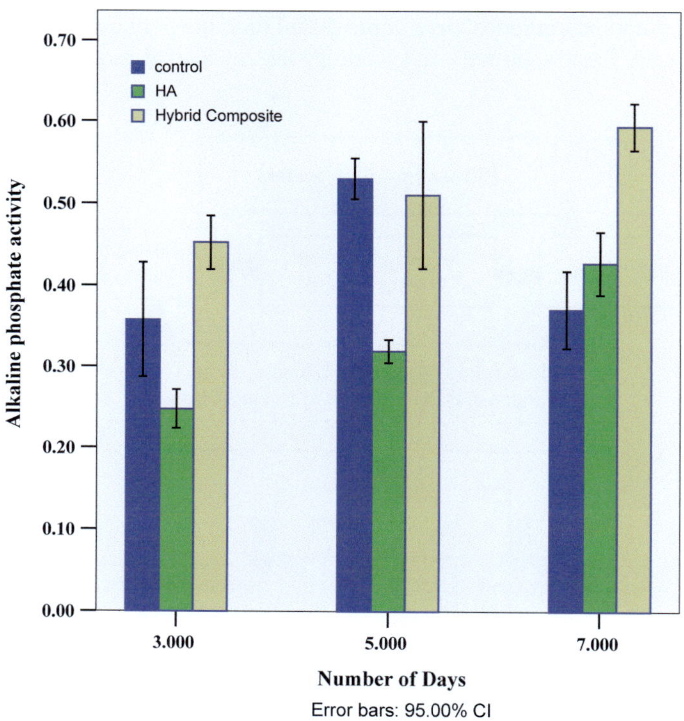

Fig. 3.5 ALP expression of SaOS2 cells on control, HA and Hybrid composite samples after 3, 5 and 7 days of culture [119]

higher and it increases with incubation days i.e. for day 5 and 7. The ALP activity of osteoblast-like cells grown on the porous scaffolds for various culture times was significantly higher than that of the dense HA sample. The results suggested that the porous HA/PMMA hybrid composite can be used as substrate which should facilitate better cell differentiation than sintered HA. It is consistent with the previous reports that the polymer–HA scaffolds are superior to the pure polymer scaffolds for osseous tissue engineering [58], because the presence of HA hydroxyl groups could promote the calcium and phosphate precipitation and improve interactions with osteoblasts [59].

3.3 Polymer-Replication Based Processing of HA-Based Macroporous Scaffold

The scaffolds can be prepared using polymeric sponges infiltrated with ceramic slurry method that integrates the gel-casting and polymer sponge methods. For the results discussed in this section, the scaffolds are prepared using commercially available polyurethane sponges as a template. The methodology, as shown in Fig. 3.6 was adopted for making the scaffold.

Several composition ratios were impregnated onto the polyurethane sponges and were analyzed. Finally, 60 wt% HA loading was considered as the optimum, since

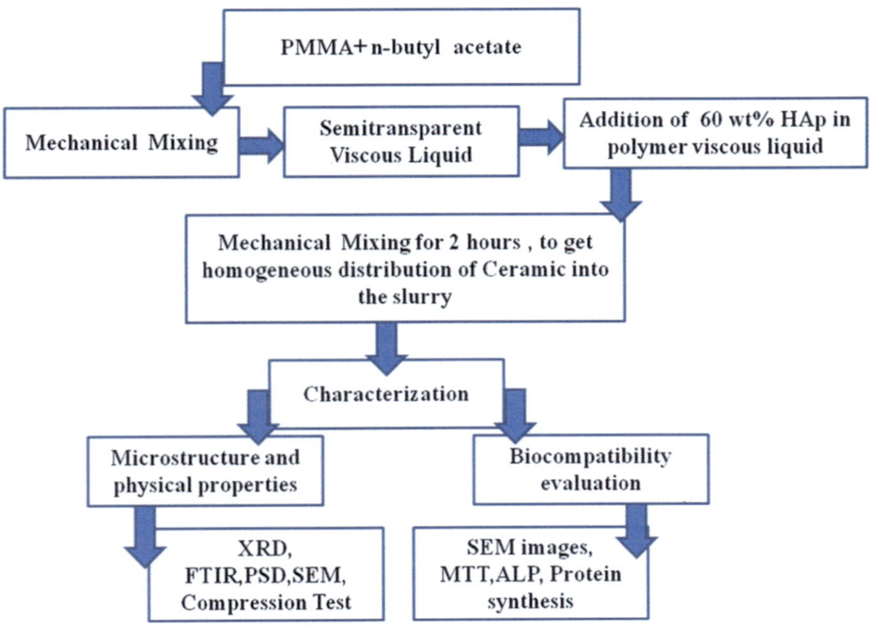

Fig. 3.6 Schematic illustration of preparation and characterization of porous scaffold [118]

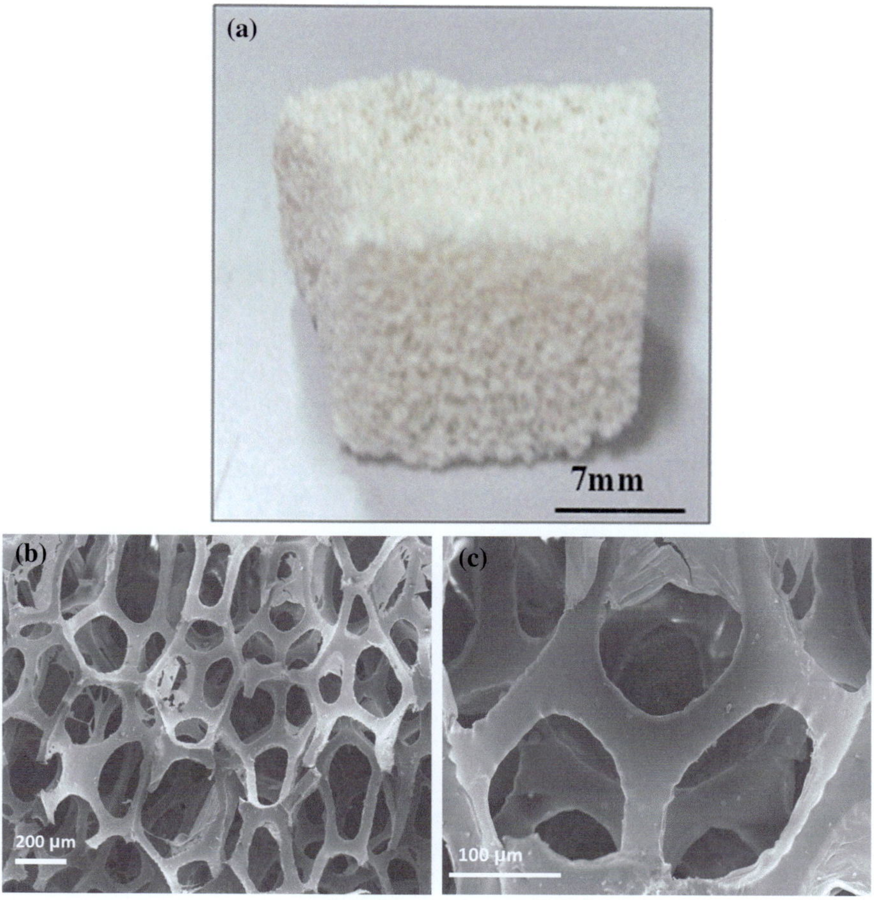

Fig. 3.7 **a** Porous hydroxyapatite body of *rectangula*r shape produced via. Polymeric sponge method using sol-gel derived HA powder and polymer slurry (**b**) and **c** SEM of scaffold [118]

the obtained scaffold has adequate strength to withstand manual handling. The polyurethane sponge was squeezed to remove excess slurry and treated according to the following sintering cycle: heating at 2 °C/min from ambient to 600 °C followed by a 1 h plateau; heating at 3 °C/min from 600 to 1200 °C followed by another plateau of 1 h. For comparison, HA powder were compacted and pressureless sintered at 1200 °C for 1 h to obtain nonporous dense HA (density-98.5 % ρth).

Figure 3.7 shows the micrograph of as synthesized porous HA scaffold. It is seen that there is no sharp faceted geometrical shape. The polymer replication method was useful to produce HA scaffolds of cube like shape or with rectangular cross-section. The largest scaffold was of 25 × 25 × 25 mm dimension. Another important aspect is that this synthesis route to prepare scaffold provides better strength to handle with better reproducibility. The pore size was ranging between

100–300 μm (Fig. 3.7b) and the pore wall thickness was measured as 50 μm (see Fig. 3.7c). Most of the pores are oval shaped and uniformly distributed.

The pore architecture of scaffold directly effects its mechanical strength [60]. Since a higher density usually leads to higher mechanical strength, high porosity provides a favorable biological environment. Therefore, a balance between the porosity and density for a scaffold must be established for the specific application. The measured apparent density and porosity of HA scaffolds prepared with slurries of different HA concentrations using the method shown before (Fig. 3.7) were 0.74 g/cm^3 and about 70 %, respectively. The apparent density of trabecular bone ranges from 0.14 to 1.10 g/cm^3 (average: 0.62 ± 0.11 g/cm^3) [61].

The maximum compressive strength for the prepared scaffold was determined as 1.3 MPa, which resembles with earlier reported data as well as close to strength of natural bone [62]. The compressive strength of porous human bone ranges between 2–12 MPa for cancellous bone and 100–230 MPa for cortical bone [63]. On the other hand, the mechanical strength of porous HA is reported in the range of 1.2–16 MPa [64]. The compressive strength of porous HA after implantation of 3 months is reported as 2–20 MPa [65].

The cytocompatibility of macroporous HA-scaffold was evaluated using SaOS2 osteoblast like cells. SEM results indicated that the scaffold allowed the attachment and spreading of the cells, while keeping a normal cellular morphology. As shown in Fig. 3.8, the cells are well attached and spread on the scaffold, presenting a round shaped configuration with extended filapodia in multiple directions, adhering to the scaffold surface. Figure 3.9a supports better cell viability on macroporous scaffold than on dense HA.

ALP is a representative enzyme of osteoblastic differentiation, and ALP activity is considered as an indicator of osteoblastic differentiation of SaOS2 cells cultured on the scaffolds. As shown in Fig. 3.9b, ALP expression of cells grown on sintered HA and HA scaffold increased continuously over the varying culture duration. After day 3, the ALP activity on the scaffolds was low, and then increased on day 5 and 7. The ALP activity of osteoblast-like cells grown on the porous scaffolds for various culture times was significantly higher than that of dense HA.

In summary, it can be stated that the pore size distribution and surface chemistry provided favorable topographical substratum for cell attachment and ingrowth in the present case. In particular, the scaffold surface favored enhanced surface wetting properties, which could affect cell attachment directly or indirectly through selective adsorption of serum protein required for cell attachment. It is known that the absorbed protein modulated cellular interactions [66].

3.4 HA-Gelatin Based Micro Porous Scaffold

In last one decade, limited efforts are invested to fabricate and characterize gelatin-HA scaffold [67–70]. Such study often reports either the fabrication aspects of these porous scaffolds or the response of osteoblast/human mesenchymal stem

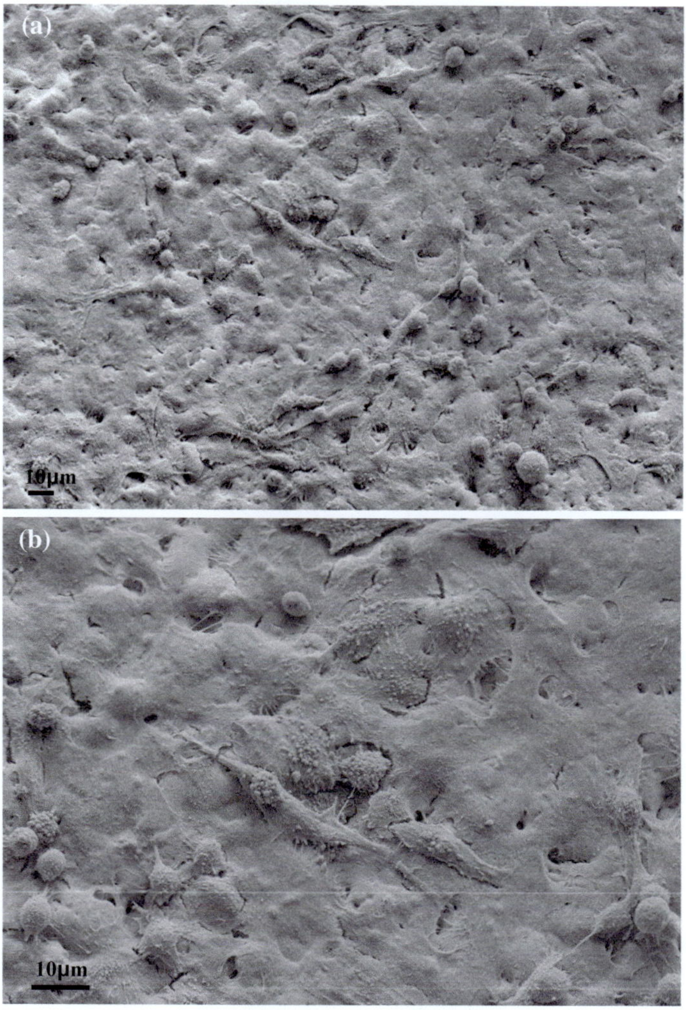

Fig. 3.8 SEM image of osteoblast cells cultured on scaffold **a** after 7 day and **b** after 10 day of culture [118]

cells on the scaffold under static or dynamic culture conditions. For example, Kang et al. [10] reported a freeze drying based approach to develop porous hydrogels through cross-linking of gelatin with glutaraldehyde in aqueous media. They also illustrated how the freezing temperature can influence different pore morphologies of hydrogel (without HA), which is essentially controlled by the size of the ice crystals. Unfortunately, neither any quantified mechanical response nor any cell/protein results are reported in their study. In another study, Landi et al. [11] reported the channel like pore morphologies in gelatin-HA scaffolds (HA/10 % gelatin). Interestingly, such characteristic pore (45–55 vol%) architecture also

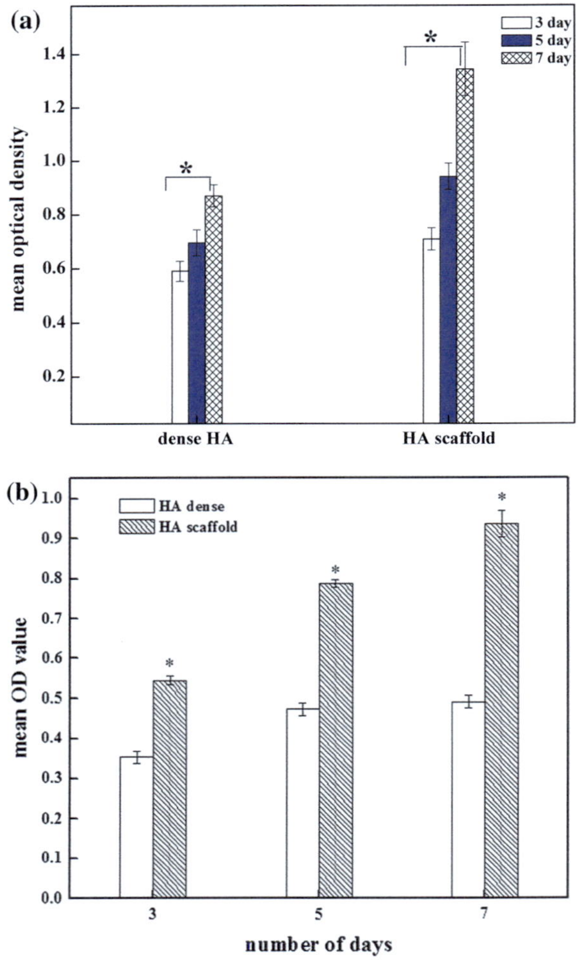

Fig. 3.9 MTT assay (**a**) and **b** results of the SaOS2 cells on dense HA and HA based microporous after 3, 5 and 7 days of culture. * represent significant difference at $p < 0.05$ with respect to compositions and Error *bars* correspond to ±1.00 SE for number of days of culture [118]

resulted in higher compressive strength of more than 15 MPa. In a different work, Kim et al. studied the *in vitro* cytocompatibility of on gelatin-HA scaffolds with osteoblastic cells under dynamic cell culture conditions [12]. The differentiation of the bone cells (MG63), as assessed using ALP (alkaline phosphate) and osteocalcin activity are measured to be higher on the gelatin-HA nanocomposite scaffolds compared to the conventionally processed scaffolds (direct mixing gelatin powder and HA). In a relatively recent study, Zhao et al. reported the cell fate processes of the human mesenchymal stem cells on chitosan-gelatin scaffold [13]. The better stem cell differentiation to bone cells on the scaffolds is attributed to the enhanced calcium adsorption properties of HA on such scaffolds.

This part of the chapter discusses a low temperature based processing approach to make gelatin—HA based porous scaffolds. In particular, a processing strategy to combine freeze drying and cryogenic curing treatment to develop nano hydroxyapatite gelatin based porous scaffold from hydroxyapatite nanorods will be discussed [71]. Freeze casting technique has advantages in terms of retention of original particle size and shape over high temperature process, since the latter favors high aspect ratio conversion to equiaxed morphology and a reduction of their mechanical strength [72, 73]. The compressive strength and pore morphology of such porous scaffolds are close to that of trabecular bone [74].

A common conjecture in biomaterials research is the cellular interaction towards bioimplant surfaces mediated through adsorbed proteins, since proteins are essential biomacromolecules to support cell nutrients, cell growth and attachment [75, 76]. Hence, the assessment of protein interaction with a scaffold is of significance for tissue engineering. Keeping this in view, different *in vitro* analysis such as adsorption/release of BSA protein and biodegradation are carried out in an effort to study the cytocompatibility property and an attempt has been made to correlate between structure and properties of the scaffold. In order to quantitatively and qualitatively characterize the pore architecture, micro computed tomography (micro-CT) technique is used extensively in the present work for protein-loaded and as-fabricated gelatin–HA based porous nanobiocomposite scaffolds.

3.4.1 Freeze Drying—Cryogenic Curing Based Processing Approach

As far as the processing aspect is concerned, a combination of freeze drying and cryogenic curing (liquid N_2) was to develop macroporous scaffolds with interconnected porosity.

In particular, Gelatin-HA scaffolds were prepared from rod morphologies of freeze dried HA powder, as shown in Fig. 3.10. The solid content was optimized to be 40 wt% considering the combination of pore volume, pore diameter and adequate strength. A slurry was made by adding Polyvinyl alcohol (PVA) (10 wt%) and gelatin (15 wt%) with respect to the HA powder used. To start with, solid gelatin (Gelatin granules, mol. wt. ~85000, Fisher Scientific Qualigens) was added to a clear PVA solution, till homogenization of the two additives was achieved. Highly dispersed HA nanorods were added within the mixed solution of gelatin and PVA through continuous magnetic stirring. The solid–liquid mixture turned into highly viscous slurry at 40 °C. The slurry was poured in a cylindrical glass mold and refrigerated at 268 K for 12 h to avoid sudden liquid solidification and evaporation. Semi–dried mass was then freeze dried at 221 K and 77 torr pressure. Demolded freeze dried gelatin-HA cylindrical scaffolds were further cured in liquid nitrogen cryo—chamber for 5 h at atmospheric pressure as discussed in an earlier work [77]. The scaffolds are designated as cryo-treated nano hydroxyapatite–gelatin micro/macroporous scaffold (CHAMPS).

Concerning the mechanical behavior, such scaffolds exhibit moderate strength properties (5 MPa) and progressive yielding behavior with large strain to failure up

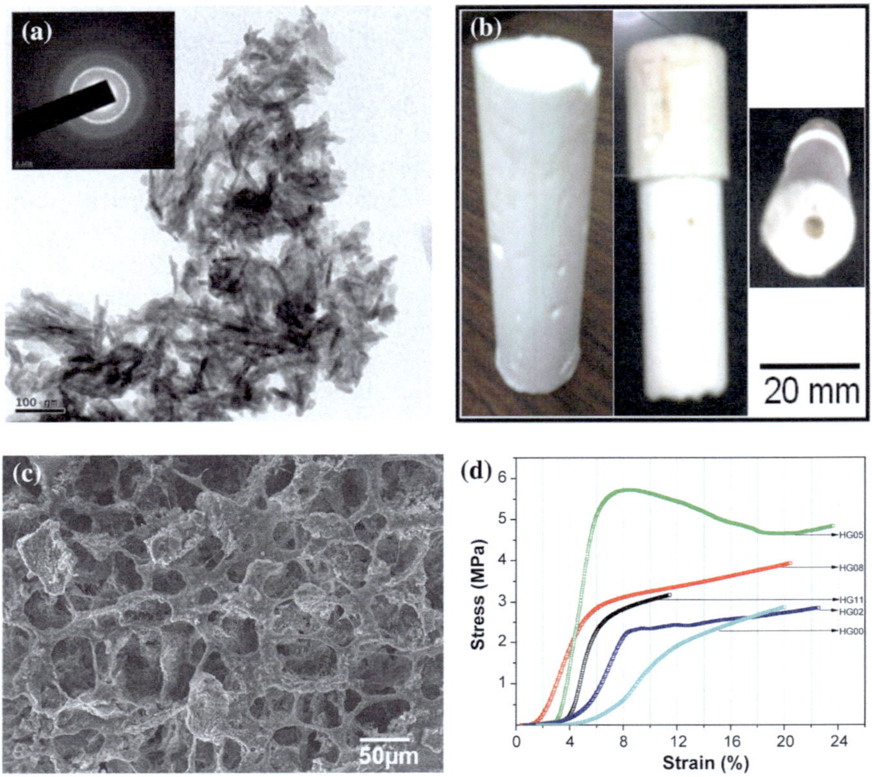

Fig. 3.10 **a** TEM and *inset* SAED pattern for rod shaped hydroxyapatite nanoparticles, **b** Microscopic view of the freeze dried and their cryo treated scaffold, **c** SEM image of the as freeze dried and cryo treated nanoHA—gelatin macroporous scaffold (CHAMPS) and **d** stress-strain response of cryotreated scaffolds under compression [120]

to 24 % (see Fig. 3.10). In particular, the combination of porosity as well as mechanical properties of the synthesized CHAMPS scaffold would enable one to mimic the structure and properties of trabecular bone, respectively. Figure 3.10 also shows large sized scaffold can be fabricated using the above-described processing approach.

3.4.2 *Micro-CT Analysis of Porous Architecture*

Although SEM image, as shown in Fig. 3.11c provide signatures of porous structure in the scaffold, more qualitative and quantitative analysis of 3D porous architecture using micro-CT are now discussed. It is known that micro-CT systems use effectively micro-focal X-ray projections rotated through multiple viewing

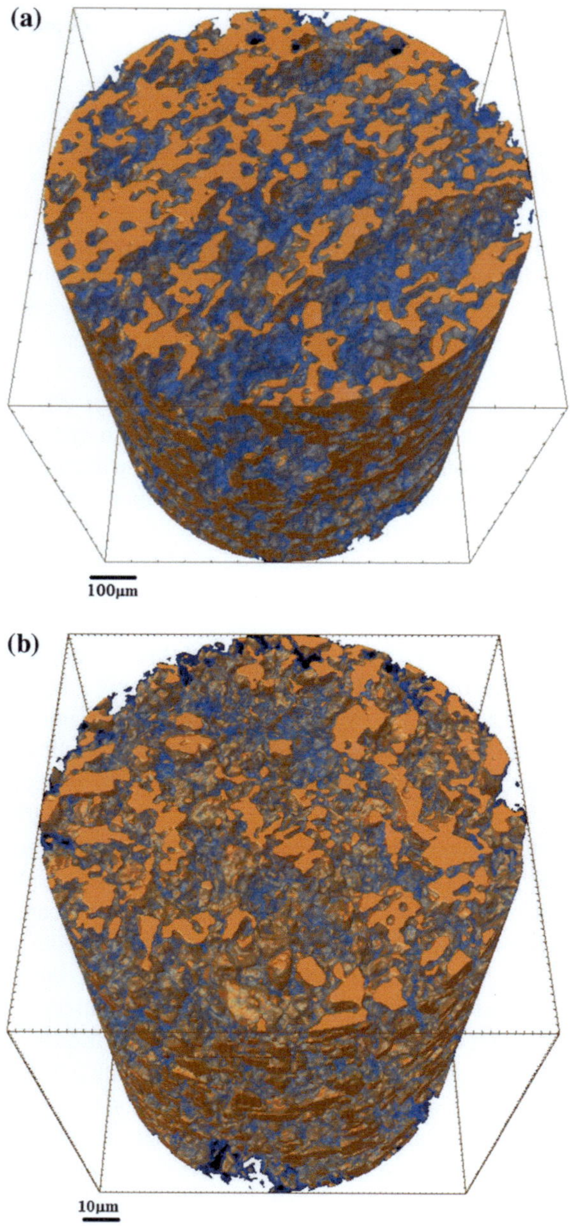

Fig. 3.11 Representative 3D micro–CT scan image of the as-synthesized CHAMPS (**a**) and BSA protein loaded CHAMPS (**b**) [120]

directions in order to obtain 3D reconstructed images, which provide spatial distribution maps of linear attenuation coefficients, determined by the energy of X-ray source and scaffold composition. The as-fabricated nanoHA/gelatine scaffold as well as the protein loaded scaffold were first fixed in 10 % neutral buffered formalin and thoroughly washed with PBS. Subsequently, the test samples (as-synthesized scaffold and protein loaded scaffold) were analyzed using high resolution X-ray micro-computed tomography. The scanning parameters include, X-ray beam energy = 40 kV and current = 250 μA. Under the operating conditions, the pixel size was around 4 μm and typically the scaffold length of 5–6 mm is sectioned into 600–700 parallel slices of equal height during the micro-CT analysis. While analyzing the test scaffolds, the rotation step was 0.25° and 30° rotation was completed. Typical timescale for scanning a test sample was around 4.5 h. The reconstruction of the selected region was made using the commercial NRecon reconstruction program (cone beam convergence/back projection algorithm-based software). Based on the density variation of various regions of the test scaffold, the 3D image of the scaffold was extracted on the basis of the distinction between ROI (region of interest) and VOI (volume of interest). Also, various views (side view, top view, front view) of the 3D images are studied in order to realize the pore interconnectivity or pore distribution along the entire volume of the scaffold and this aspect can only be studied using micro CT technique. Some representative 3D volume rendered images are shown in Fig. 3.11. From the experimentally measured CT database, we have also analyzed multiple 2D slices to quantify the distribution of the pore size and pore volume as well as strut thickness for the entire scaffold volume.

The efficacy of a biomaterial scaffold for biomedical applications depends on the pore size distribution, pore interconnectivity, pore wall thickness etc. In the present case, nano HA/gelatin scaffold is uniquely characterized by highly favorable pore size distribution with the peak size distribution lies at about 50 μm (see Fig. 3.12). Therefore, it can be said that majority of the pores in the scaffold would favor the initial attachment, migration and proliferation of fibroblast and osteoblast like cells. The skewed pore size distribution with sizable fraction of the pores having sizes between 100 and 250 μm would be helpful in the tissue/bone in growth. The fact that moderate compressive strength of 5 MPa can be achieved with such highly porous scaffold indicates good rigidity of the pore wall structures.

The pore wall thickness, like pore size is therefore an important factor. The Gaussian fitting of the measured pore wall thickness data reveals a peak in distribution around 40 μm in the as processed scaffold, which is shifted to 20 μm in the BSA protein loaded scaffold (see Fig. 3.12b). The overall shift of the pore walls thickness distribution towards left essentially reveals favorable protein absorption on the pore wall surfaces in a homogeneous manner. Also the fact that the majority of the pores have pore wall thickness of around 40–50 micron reveals that such pores can support the in growth of biological tissues/neobone.

As far as the complexity of the pore size distribution concerned, the presently investigated scaffold has a highly complex pore structure distribution as can be realized from the fractal dimension (FD) of 2.61 (see Table 3.1). It is known that a less complex porous scaffold is characterized by FD = 2 and for natural cancellous bone, FD = 2.1–2.3 is expected [78]. Another important factor in the context of the pore architecture is degree of anisotropy (DA). Higher value of DA essentially indicates the anisotropic of pore orientation. In case of as processed scaffold,

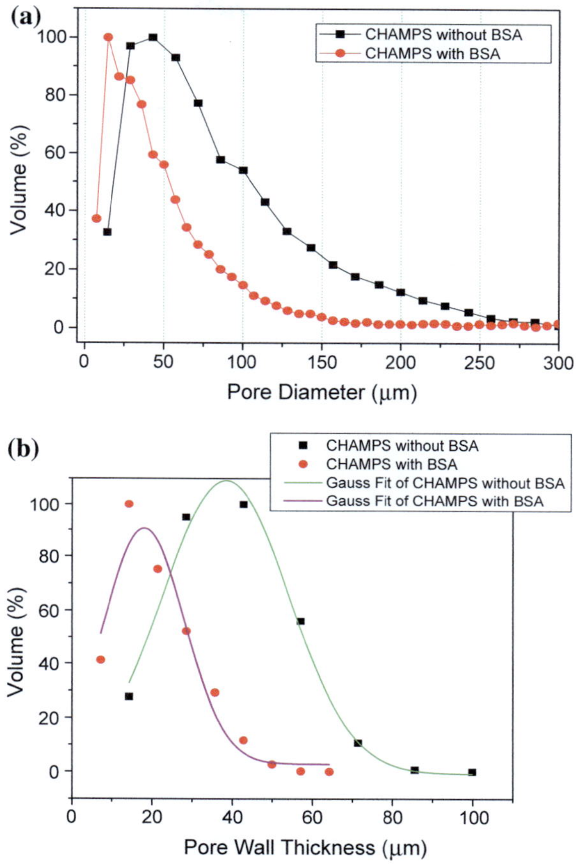

Fig. 3.12 Plot of micro–CT scan analyzed data to quantify the pore size (**a**) and pore wall thickness (**b**) distribution for the as-synthesized CHAMPS scaffold and BSZ protein-loaded CHAMPS scaffold [120]

DA = 5.3 and it is reduced to 2.6 after BSA protein absorption. Considering highly stochastic nature of the tissue/bone in growth, the anisotropic pore orientation would be extremely useful. Also, higher DA values essentially reflect that the CHAMPS scaffold can provide complex physical pathways through the anisotropic pore channels.

As far as the interconnectivity of the pores are concerned, the connectivity density essentially reflects the number of multiple connections within the scaffold volume or number of interconnected solid elements. The large interconnectivity of 3.2×10^3 in case of as processed scaffold as well as 3.3×10^4 for protein loaded scaffold indicates that pores are highly connected in three dimensional and in fact, protein adsorption through complex pore favorably increases the interconnectivity density. It is important to mention here that high interconnectivity density would

Table 3.1 Summary of the physical parameters describing the porous architecture of the CHAMPS scaffold loaded with and without BSA protein, as determined using extensive micro-CT analysis [120]

Physical parameters	CHAMPS without BSA protein adsorption	CHAMPS with BSA protein adsorption after 48 h
Total porosity	63.5 (%)	71.8 (%)
Total volume pore space	1.1×10^{11} (μm^3)	1.9×10^{10} (μm^3)
Fractal dimension	2.61	2.54
Average pore separation distance	86.8 μm	66.4 μm
Degree of anisotropy	5.3	2.6
Connectivity density	3.21×10^{-6} (μm^{-3})	3.30×10^{-5} (μm^{-3})

definitely facilitate the passage of nutrients, blood as well as waste products from the biological cells, which would find easy passage to get into the available pore volume because of highly favorable pore architecture. From the above, it should be clear that highly interconnected porous microstructure with average pore diameter in the range of 40–200 μm facilitated the attachment and delivery of BSA on hydrophobic CHAMPS surfaces and enabled hydrophobic–hydrophobic surface conjugate interaction [79].

3.4.3 BSA Adsorption/Release

The water soluble Bovine Serum Albumin (BSA) was considered as a model protein for adsorption/release study [80]. The crystallized and lyophilized BSA protein, having 69 KDa molecular weight and 4.74 isoelectric points, was dissolved in double distilled deionized water and refrigerated at 4 °C. The identical shaped eight samples of CHAMPS scaffold (0.5 gm, Ø–14 mm, and thickness—2.5 mm, volume—0.38 cm^3) were sliced by steel blade and immersed in 10 ml BSA solution with 5 mg/ml initial concentration. The samples with protein solution were kept in an incubator at 37 °C for different time scale from 6 to 240 h. The supernatant solutions were collected from individual BSA—CHAMPS conjugate specimens through centrifuge at 100 rpm at 4 °C. The amount of BSA protein adsorbed was quantified by taking 0.1 ml of supernatant from each specimen and the supernatant was mixed with 2 ml of bicinchonic acid solution. Similarly, BSA release study was also carried out for the same BSA protein loaded CHAMPS after immersion in 20 ml of phosphate buffer saline (PBS) at pH 7.4 and temperature of 37 °C. UV-vis spectroscopy at 562 nm was used to characterize the absorbance peaks and to estimate of BSA concentration through the use of a predetermined standard concentration—intensity calibration curve. In reference to the

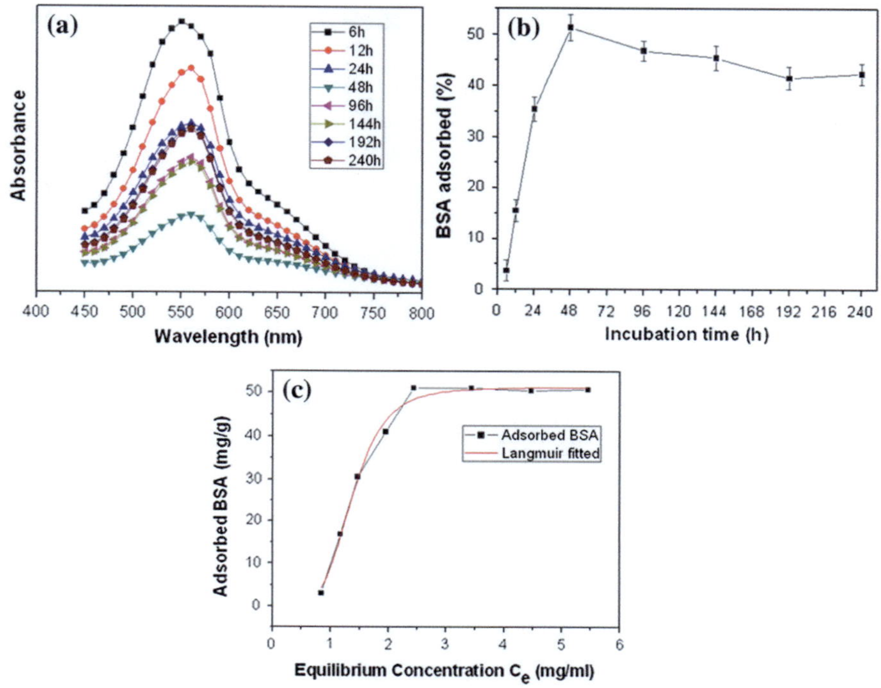

Fig. 3.13 Adsorption study of BSA onto gelatin-HA Scaffold, **a** analysis of maximum absorbance of adsorbed BSA on CHAMPS with time, **b** amount of BSA adsorbed on CHAMPS with time and **c** BSA adsorption isotherm [120]

experimental results shown in Figs. 3.13 and 3.14, the protein adsorption/release of CHAMPS scaffold is analyzed now. A summary of the available literature allows one to make the following observations:

(a) BSA adsorption on HA reaches a steady state at very shorter time scale of around 2 h;
(b) In case of chitosan-gelatin scaffold, the BSA adsorption occurs a overall large time scale and typically reached a steady state over a period of 15 days [13];
(c) The BSA adsorption on collagen derived gelatin HA nanocomposite occurs at relatively much faster rate of within 24 h [12].

In the above perspective, the present study confirms that relatively faster adsorption behavior of BSA proteins in CHAMPS scaffold occurs within 48 h, although the time scale for the attainment for the steady state BSA adsorption is little longer than that reported in another related study [12]. As far as the protein release is concerned, the CHAMPS scaffold clearly outperforms other biodegradable scaffolds as far as the BSA release over 10 days is concerned. For example, Jaklenec et al. measured 40–50 % of BSA release for PLGA micro spheres loaded with 4.3 % BSA *in vitro* over an incubation period of up to 10 days and BSA

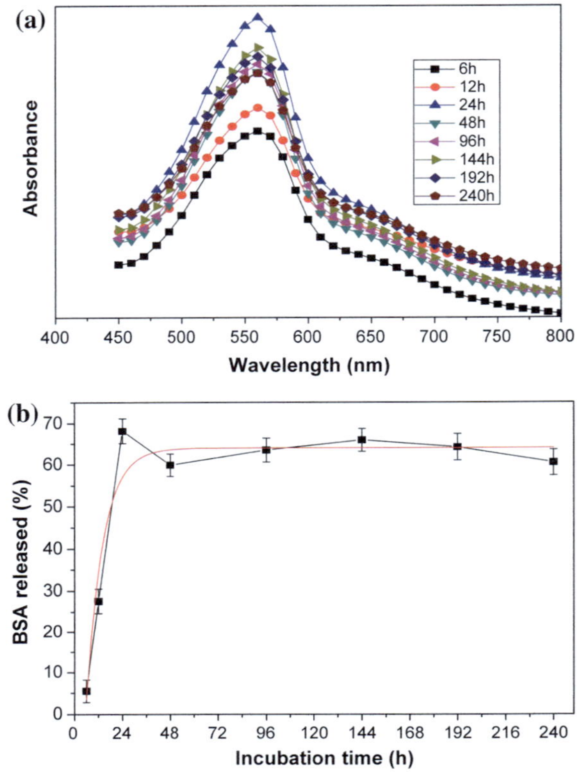

Fig. 3.14 Release study of loaded BSA from CHAMPS, **a** UV analysis of maximum absorbance with time and **b** amount of BSA released from the CHAMPS with time from 6 to 240 h [120]

release does not attain the steady state even up to the period of 20 days. In contrast, the BSA release is around 70 % after 24 h incubation and reaches a steady state behavior of 60 % BSA release over an incubation period of 2–10 days. Such ability of the CHAMPS scaffold to release BSA proteins over a time span of 24 h indicates that these scaffolds can be used for protein delivery applications.

3.5 Challenges in Bone-Mimicking Porous Scaffolds

First, the necessity for developing composite with dispersion of calcium phosphate (CP) in polymeric scaffold can be reiterated. CP bioceramics are brittle. In addition, the mechanical properties decrease significantly with increasing amorphous phase, microporosity and grain size. In addition, high crystallinity, low porosity and small grain size tend to give a higher compressive and tensile strength and a greater fracture toughness. Thus CP has poor mechanical strength and a greater fracture toughness, which forbids its usage in the load-bearing applications [81–83]. Fracture toughness of HA bioceramics does not exceed ~1.2 MPa $m^{1/2}$ [84], where as human natural bone has toughness of 2–12 MPa $m^{1/2}$. With the increasing porosity, the mechanical strength decreases. Bending, compressive and tensile strengths of dense HA bioceramics are in the range of 38–250, 120–900 and 38–300 MPa, respectively, where as those values for the porous HA bioceramics are 2–11, 2–100 and ~3 MPa, respectively [28]. Further, strength was found to increase with increasing Ca/P ratio, reaching a maximum value with stoichiometric ratio and decreases when Ca/P > 1.67 [28]. The strength decreases exponentially with increasing porosity [85, 86]. Also, by changing the pore geometry, it is possible to influence the strength of the bioceramics. It has been noted that the porous HA bioceramics are considerably less fatigue resistant than their dense counterparts. Due to brittleness, CP bioceramics are mostly used in non-load bearing implants. The electrical properties of CP bioceramics have an interesting role in judging their applicability for the biomedical applications. The presence of surface charges on HA bioceramics have significant effect on both *in vitro* and *in vivo* crystallization of apatite phase [87–90]. The negatively charged surface enhances the growth of biomimetic CP and bones rather than the positively charged surfaces [91–94]. The brittleness of CP can be partly circumvented by producing composites with visoelastic matrix, like collagen.

Porosity is another major factor, which provides good mechanical fixation and allows the chemical bonding between the bioceramics and bones [95]. The open porosity is directly related to bone formation and provides the surface and space for cell adhesion and bone ingrowths. Pore interconnection provides the way and migration, as well as for *in vivo* blood vessel formation for bone tissue and remodeling [96–101]. Interconnecting macropores (size > 100 μm) are usually formed due to the porogens in bioceramics. There are several techniques that are used for formation of porosity such as polymer foams by impregnation, dual-phase mixing, particulate leaching, freeze casting, slip casting, stereo lithography. The foaming of gel casting suspensions have been applied to fabricate porous CP bioceramics [102–104]. There are numerous reports about the formation of porous HA bioceramics [105, 106]. The control of pore formation, pore dimensions and internal pore architecture of bioceramics at different length scales is essential in assessing the structure-bioactivity relationship and the rational design of bone forming biomaterials [107–109]. For medical applications, it is important to study the biological properties of fabricated bioceramics and *in vivo* behavior. As the implanted biomaterial will chemically react with their environment, they should not create undesired effects to their adjacent or far tissues. Though there are some reports on the inflammatory reaction by implanting CP bioceramics [110–112], still CP bioceramics with Ca/P ionic ratio within 1.0–1.7 are reported to be non-toxic. Osteoinduction of CP bioceramics is observed in the porous structures or well-defined structures. Scientific studies have showed that an estimation of minimum pore size of ~50 μm for blood vessel formation and a ~200 μm for osteonal in-growth [113]. Both porosity and their architecture are critical in gauging the fluid transport rate through porous bioceramics, which determines the rate and the degree of the bone growth *in vivo* [114].

In the above perspective, an innovative processing scheme will be discussed to develop biocomposites with desired porous network. An important drawback of many of the above-mentioned processes is that the bone-mimicking compositions in terms of the Calcium Phosphate biocomposites can not be effectively processed, as collagen is denatured during high temperature processing. Any process at or near to room temperature is most suitable and this places additional advantage of using freeze isostatic processing (FIP) in fabricating bone-mimicking scaffolds. In this process, a slurry of collagen with HA can be compacted in a test chamber under high hydrostatic pressure of maximum up to 600 MPa at sub-zero temperature with ice crystals acting as porogens. This process requires further investigation.

3.6 Closure

Various processing approaches are discussed in this chapter to fabricate porous scaffolds with desired interpenetrating or interconnected porosity. These include polymer replication, sacrificial method of porogen removal and freeze casting. Other methods, which are not discussed here, include rapid prototyping methods (selective laser sintering, 3D printing etc.), gel casting etc. [115–117]. It has been demonstrated that HA-based macroporous scaffolds are being developed by immersing polymeric spongy foams in polymer-based precursor slurry, followed by high temperature treatment and this route enables one to obtain macroporous scaffolds with interconnected porosity [118]. Such scaffolds have moderate compression strength (1.3 MPa) and also support attachment and proliferation of bone cells. In a different processing approach, microporous HA scaffolds are prepared using PMMA as sacrificial additive [119].

The simple use of PMAA as porogen and adopting conventional ceramic processing route allows one to fabricate HA-based microporous scaffolds with porosity of 1–50 μm, where the majority of pores are in the range of 10–20 μm diameter. In particular, fluorescent imaging of osteoblast cells revealed a well developed cytoskeleton with strong actin stress fibers oriented in the adhered cells.

The discussion in this chapter also illustrates that the porous scaffold with homogeneous pore distribution and pore wall thickness can be fabricated using polymer sponge replication method. While the pore size varies between 100–300 μm, the pore wall thickness was ~50 μm. The apparent density and porosity of HA scaffold were calculated of 0.74 g cm^3 and ~70 %, respectively. The compressive strength of porous scaffold was calculated as 1.3 MPa, which correlates well with the strength of natural cancellous bone.

The results discussed in this chapter reconfirms that porous scaffold could have provided better substrate for cell adhesion and proliferation and differentiation of osteoblast type cells, in comparison to dense sintered HA.

The combination of the freeze drying and cryogenic curing for 5 h in liquid nitrogen allows to develop gelatin HA nanocomposite scaffolds with interconnected porosity and superior protein adsorption/delivery property. On the basis of the micro-computed tomography analysis, the investigated gelatin-nano HA scaffolds are characterized by the skewed pore size with mean at 40 μm and an extended tail of up to 250 μm, as well as pore wall structure thickness varying in the range of 10–100 μm. A systematic shift towards finer pore size/pore wall distribution with BSA adsorption is also recorded. Further, the superior protein adsorption/delivery is attributed to the unique pore interconnectivity and anisotropy in addition to desired pore distribution.

References

1. Davies JE. In vitro modeling of the bone/implant interface. Anat Rec. 1996;245:426–45.
2. Anselme K. Osteoblast adhesion on biomaterials. Biomaterials. 2000;21:667–81.
3. Hench LL. Bioceramics: from concept to clinic. J Am Ceram Soc. 1991;74:1487–510.
4. Suchanek WL, Yoshimura M. Processing and properties of hydroxyapatite-based biomaterials for use as hard tissue replacement implants. J Mater Res. 1998;13:94–117.
5. Rivera EM, Araiza M, Brostow W, Castaño VM, Díaz-Estrada JR, Hernández R, Rodríguez JR. Synthesis of hydroxyapatite from eggshells. Mater Lett. 1999;41:128–34.
6. Lee SJ, Oh SH. Fabrication of Calcium Phosphate bioceramics by using eggshell and phosphoric acid. Mater Lett. 2003;57:4570–4.
7. Balazsi C, Weber F, Kover Z, Horvath E, Nemeth C. Preparation of Calcium-Phosphate bioceramics from natural resources. J Eur Ceram Soc. 2007;27:1601–6.
8. Murugan R, Ramakrishna S. Crystallographic study of hydroxyapatite bioceramics derived from various sources. Cryst Growth Des. 2005;5:111–2.
9. Vallet-Regí M, González-Calbet JM. Calcium Phosphates as substitution of bone tissues. Progr Solid State Chem. 2004;32:1–31.
10. Lecomte A, Gautier H, Bouler JM, Gouyette A, Pegon Y, Daculsi G, Merle C. Biphasic Calcium Phosphate: a comparative study of interconnected porosity in two ceramics. J Biomed Mater Res B Appl Biomater. 2008;84B:1–6.
11. Tancret F, Bouler JM, Chamousset J, Minois LM. Modelling the mechanical properties of microporous and macroporous biphasic Calcium Phosphate bioceramics. J Eur Ceram Soc. 2006;26:3647–56.
12. Bouler JM, Trecant M, Delecrin J, Royer J, Passuti N, Daculsi G. Macroporous biphasic Calcium Phosphate ceramics: Influence of five synthesis parameters on compressive strength. J Biomed Mater Res. 1996;32:603–9.
13. O'Neill WC. The fallacy of the Calcium—phosphorus product. Kidney Int. 2007;72:792–6.
14. da Silva RV, Bertran CA, Kawachi EY, Camilli JA. Repair of cranial bone defects with Calcium Phosphate ceramic implant or autogenous bone graft. J Craniofac Surg. 2007;18:281–6.
15. Rodríguez-Lorenzo LM, Vallet-Regí M, Ferreira JMF. Fabrication of hydroxyapatite bodies by uniaxial pressing from a precipitated powder. Biomaterials. 2001;22:583–8.
16. Nazarpak MH, Solati-Hashjin M, Moztarzadeh F. Preparation of hydroxyapatite ceramics for biomedical applications. J Ceram Proc Res. 2009;10:54–7.
17. Itoh H, Wakisaka Y, Ohnuma Y, Kuboki Y. A new porous hydroxyapatite ceramic prepared by cold isostatic pressing and sintering synthesized flaky powder. Dental Mater. 1994;13:25–35.
18. Gautier H, Merle C, Auget JL, Daculsi G. Isostatic compression, a new process for incorporating vancomycin into biphasic Calcium Phosphate: comparison with a classical method. Biomaterials. 2000;21:243–9.

19. Tadic D, Epple M. Mechanically stable implants of synthetic bone mineral by cold isostatic pressing. Biomaterials. 2003;24:4565–71.
20. Pecqueux F, Tancret F, Payraudeau N, Bouler JM. Influence of microporosity and macroporosity on the mechanical properties of biphasic Calcium Phosphate bioceramics: modelling and experiment. J Eur Ceram Soc. 2010;30:819–29.
21. Viana M, Désiré A, Chevalier E, Champion E, Chotard R, Chulia D. Interest of high shear wet granulation to produce drug loaded porous Calcium Phosphate pellets for bone filling. Key Eng Mater. 2009;396–398:535–8.
22. Reikerås O, Johansson CB, Sundfeldt M. Bone ingrowths to press-fit and loose-fit implants: comparisons between titanium and hydroxyapatite. J Long Term Eff Med Implants. 2006;16:157–64.
23. Rao RR, Kannan TS. Dispersion and slip casting of hydroxyapatite. J Am Ceram Soc. 2001;84:1710–6.
24. Sakka Y, Takahashi K, Matsuda N, Suzuki TS. Effect of milling treatment on texture development of hydroxyapatite ceramics by slip casting in high magnetic field. Mater Trans. 2007;48:2861–6.
25. Zhang Y, Yokogawa Y, Feng X, Tao Y, Li Y. Preparation and properties of bimodal porous apatite ceramics through slip casting using different hydroxyapatite powders. Ceram Int. 2010;36:107–13.
26. Sepulveda P, Ortega FS, Innocentini MDM, Pandolfelli VC. Properties of highly porous hydroxyapatite obtained by the gel casting of foams. J Am Ceram Soc. 2000;83:3021–4.
27. Padilla S, Vallet-Regí M, Ginebra MP, Gil FJ. Processing and mechanical properties of hydroxyapatite pieces obtained by the gel-casting method. J Eur Ceram Soc. 2005;25: 375–83.
28. Sánchez-Salcedo S, Werner J, Vallet-Regí M. Hierarchical pore structure of Calcium Phosphate scaffolds by a combination of gel-casting and multiple tape-casting methods. Acta Biomater. 2008;4:913–22.
29. Fomin AS, Barinov SM, Ievlev VM, Smirnov VV, Mikhailov BP, Belonogov EK, Drozdova NA. Nanocrystalline hydroxyapatite ceramics produced by low-temperature sintering after high-pressure treatment. Dokl Chem. 2008;418:22–5.
30. Kankawa Y, Kaneko Y, Saitou K. Injection molding of highly-purified hydroxylapatite and TCP utilizing solid phase reaction method. J Ceram Soc Jpn. 1991;99:438–42.
31. Kwon SH, Jun YK, Hong SH, Lee IS, Kim HE, Won YY. Calcium Phosphate bioceramics with various porosities and dissolution rates. J Am Ceram Soc. 2002;85:3129–31.
32. Fooki ACBM, Aparecida AH, Fideles TB, Costa RC, Fook MVL. Porous hydroxyapatite scaffolds by polymer sponge method. Key Eng Mater. 2009;396–398:703–6.
33. Sopyan I, Kaur J. Preparation and characterization of porous hydroxyapatite through polymeric sponge method. Ceram Int. 2009;35:3161–8.
34. Velayudhan S, Ramesh P, Sunny MC, Varma HK. Extrusion of hydroxyapatite to clinically significant shapes. Mater Lett. 2000;46:142–6.
35. Yang HY, Thompson I, Yang SF, Chi XP, Evans JRG, Cook RJ. Dissolution characteristics of extrusion freeformed hydroxyapatite—Tricalcium Phosphate scaffolds. J Mater Sci Mater Med. 2008;19:3345–53.
36. Muthutantri AI, Huang J, Edirisinghe MJ, Bretcanu O, Boccaccini AR. Dipping and electrospraying for the preparation of hydroxyapatite foams for bone tissue engineering. Biomed Mater. 2008;3:25009–22.
37. Roncari E, Galassi C, Pinasco P. Tape casting of porous hydroxyapatite ceramics. J Mater Sci Lett. 2000;19:33–5.
38. Tian T, Jiang D, Zhang J, Lin Q. Aqueous tape casting process for hydroxyapatite. J Eur Ceram Soc. 2007;27:2671–7.
39. Dillon GP, Yu X, Sridharan A, Ranieri JP, Bellamkonda RV. The influence of physical structure and charge on neurite extension in a 3D hydrogel scaffold. J Biomater Sci Polym. 1998;9:1049–69.

40. Yu LMY, Leipzig ND, Shoichet MS. Promoting neuron adhesion and growth-Review. Mater Today. 2008;11:36–43.
41. Bhang SH, Lim JS, Choi CY, Kwon YK, Kim BS. The behaviour of neural stem cells on biodegradable synthetic polymers. J Biomater Sci Polym. 2007;18:223–39.
42. Groeneveld EH, Van den Bergh JP, Holzmann P, Ten Bruggenkate CM, Tuinzing DB, Burger EH. Mineralization processes in de-mineralized bone matrix grafts in human maxillary sinus floor elevations. J Biomed Mater Res. 1999;48:393–402.
43. Leony Leon CA. New perspectives in mercury porosimetry. Adv Colloid Interface Sci. 1998;76–77:341–72.
44. Sepulveda P, Ortega F, Innocentini MDM, Pandolfelli VC. Properties of highly porous hydroxyapatite obtained by the gelcasting of foams. J Am Ceram Soc. 2001;83:3021–4.
45. Cyster LA, Grant DM, Howdle SM, Rose F, Irvine DJ, Freeman D, et al. The influence of dispersant concentration on the pore morphology of hydroxyapatite ceramics for bone tissue engineering. Biomaterials. 2005;26:697–702.
46. Fu Q, Rahaman MN, Bal BS, Huang W, Day DE. Preparation and bioactive characteristics of a porous 13–93 glass, and fabrication into the articulating surface of a proximal tibia. J Biomed Mater Res A. 2007;82:222–9.
47. Chang BS, Lee CK, Hong KS, Youn HJ, Ryu HS, Chung SS, et al. Osteoconduction at porous hydroxyapatite with various pore configurations. Biomaterials. 2000;21:1291–8.
48. Harris LD, Kim BS, Mooney DJ. Open pore biodegradable matrices formed with gas foaming. J Biomed Mater Res. 1998;42:396–402.
49. Young AC, Omatete OO, Janney MA, Menchhofer PA. Gelcasting of alumina. J Am Ceram Soc. 1991;74:612–8.
50. Nunn SD, Kirby GH. Green machining of gelcast ceramic materials. Ceram Eng Sci Proc. 1996;17:209–13.
51. Le Huec JC, Schaeverbeke T, Clement D, Faber J, Le Rebeller A. Influence of porosity on the mechanical resistance of hydroxyapatite ceramics under compressive stress. Biomaterials. 1995;16:113–8.
52. Gauthier O, Bouler J, Aguado E, Pilet P. Daculsi G. In vivo bone regeneration with injectable calcium phosphate biomaterial: a three-dimensional micro-computed tomographic, biomechanical and SEM study Biomaterials. 1998;19:133–8.
53. Hench LL. Biomaterials: a forecast for the future. Biomaterials. 1998;19:1419–23.
54. Nath S, Tripathi R, Basu B. Understanding phase stability, microstructure development and biocompatibility in Calcium Phosphate-Titania composites, synthesized from Hydroxyapatite and Titanium Powder mix. Mater Sci Eng C. 2009;29:97–107.
55. Bodhak S, Nath S, Basu B. Friction and wear properties of novel HDPE-HAp-Al_2O_3 composites against alumina counterface. J Biomater Appl. 2009;23:407–33.
56. Nath S, Dey A, Mukhopadhyay AK, Basu B. Nanoindentation response of novel hydroxyapatite-mullite composites. Mater Sci Eng. 2009;513–514:197–201.
57. Bagambisa FB, Joos U. Preliminary studies on the phenomenological behavior of osteoblasts cultured on hydroxyapatite ceramics. Biomaterials. 1990;11:50–6.
58. Ma PX, Zhang R, Xiao G, Franceschi R. Engineering New bone tissue *in vitro* on highly porous poly(alpha-hydroxyl acids)/Hydroxyapatite composite scaffolds. J Biomed Mater Res. 2001;54(2):284–93.
59. Ramires PA, Romito A, Cosentino F, Milella E. The influence of titania/hydroxyapatite composite coatings on *in vitro* osteoblasts behaviour. Biomaterials. 2001;22(12):1467–74.
60. Sepulveda P, Ortega FS, Innocentini MDM, Pandolfelli VC. Properties of highly porous hydroxyapatite obtained by the gelcasting of foams. J Am Ceram Soc. 2000;83:3021–4.
61. Evans LA, Macey DJ, Webb J. Calcium biomineralization in the regular teeth of the chiton, acanthopleura hirtosa. Calcif Tissue Int. 1992;51:78–82.
62. Zhang Y, Zhang M. Three dimensional macroporous calcium phosphate bioceramics with nested chitosan sponges for load bearing implants. J Biomed Mater Res. 2002;61:1–8.
63. Hench LL, Wilson J, editors. In: An introduction to bioceramics. Singapore: World Scientific. 1993; p. 12–26.

64. Yoshikawa H, Myoui A. Bone tissue engineering with porous hydroxyapatite ceramics. J Artif Organs. 2005;8:131–6.
65. Sopyan I, Sulaiman NS, Gustiono D, Herdianto N. Porous hydroxyapatite-gelatin composites with functions of bone substitutes and drug releasing agents: a preliminary study, BioMEMS and Nanotechnology II (SPIE, The International Society for Optical Engineering), Vol. 6036, C1-C11. 2006.
66. Zeng H, Chittur KK, Lacefield WR. Analysis of bovine serum albumin adsorption on calcium phosphate and titanium surfaces. Biomaterials. 1999;20:377–84.
67. Kang H, Tabata Y, Ikada Y. Biomaterials. 1999;20:1339.
68. Landi E, Valentini F, Tampieri A. Acta Biomater. 2008;4:1620.
69. Kim H, Kim HE, Salih V. Biomaterials. 2005;26:5221.
70. Zhao F, Grayson WL, Ma T, Bunnell B, Luc WW. Biomaterials. 1859;2006:27.
71. Swain SK, Dorozhkin SV, Sarkar D. Mater. Sci. Engg. C. 2012;32:1237.
72. Swain SK, Sarkar D. Ceram Int. 2011;37:2927.
73. Verron E, Bouler JM, Guicheux J. Acta Biomater. 2012;8:3541.
74. Van der Meulen MC, Jepsen KJ, Mikic B. Bone. 2001;29:101.
75. Cerroni L, Filocamo R, Fabbri M, Piconi C, Caropreso S, Condo SG. Biomol Eng. 2002;19:119.
76. Saltzman MW, Baldwin SP. Adv Drug Deliv Rev. 1998;33:71.
77. Swain SK, Sarkar D. Mater Lett. 2013;92:252.
78. Uchiyama T, Tanizawa T, Muramatsu H, Endo N, Takahashi HE. Bone. 1999;25:487.
79. Takaoka K, Koezuka M, Nakahara H. J Orthop Res. 1991;9:902.
80. Arends J, Schuthof J, Christoffersen J. Caries Res. 1986;20:337.
81. Hench LL. Bioceramics: from concept to clinic. J Am Ceram Soc. 1991;74:1487–510.
82. Cao W, Hench LL. Bioactive materials. Ceram Int. 1996;22:493–507.
83. Hench LL. Bioceramics. J Am Ceram Soc. 1998;81:1705–28.
84. Ramesh S, Tan CY, Sopyan I, Hamdi M, Teng WD. Consolidation of nanocrystalline hydroxyapatite powder. Sci Technol Adv Mater. 2007;8:124–30.
85. Tancret F, Bouler JM, Chamousset J, Minois LM. Modelling the mechanical properties of microporous and macroporous biphasic Calcium Phosphate bioceramics. J Eur Ceram Soc. 2006;26:3647–56.
86. Bouler JM, Trecant M, Delecrin J, Royer J, Passuti N, Daculsi G. Macroporous biphasic Calcium Phosphate ceramics: Influence of five synthesis parameters on compressive strength. J Biomed Mater Res. 1996;32:603–9.
87. Itoh S, Nakamura S, Kobayashi T, Shinomiya K, Yamashita K, Itoh S. Effect of electrical polarization of hydroxyapatite ceramics on new bone formation. Calcif Tissue Int. 2006;78:133–42.
88. Iwasaki T, Tanaka Y, Nakamura M, Nagai A, Hashimoto K, Toda Y, Katayama K, Yamashita K. Rate of bonelike apatite formation accelerated on polarized porous hydroxyapatite. J Am Ceram Soc. 2008;91:3943–9.
89. Itoh S, Nakamura S, Kobayashi T, Shinomiya K, Yamashita K. Enhanced bone ingrowth into hydroxyapatite with interconnected pores by electrical polarization. Biomaterials. 2006;27:5572–9.
90. Kobayashi T, Itoh S, Nakamura S, Nakamura M, Shinomiya K, Yamashita K. Enhanced bone bonding of hydroxyapatite-coated titanium implants by electrical polarization. J Biomed Mater Res A. 2007;82A:145–51.
91. Bodhak S, Bose S, Bandyopadhyay A. Role of surface charge and wettability on early stage mineralization and bone cell-materials interactions of polarized hydroxyapatite. Acta Biomater. 2009;5:2178–88.
92. Nakamura S, Kobayashi T, Yamashita K. Highly orientated calcification in newly formed bones on negatively charged hydroxyapatite electrets. Key Eng Mater. 2005;284–286: 897–900.

93. Kato R, Nakamura S, Katayama K, Yamashita K. Electrical polarization of plasma-spray-hydroxyapatite coatings for improvement of osteoconduction of implants. J Biomed Mater Res A. 2005;74A:652–8.
94. Nakamura S, Kobayashi T, Nakamura M, Itoh S, Yamashita K. Electrostatic surface charge acceleration of bone ingrowth of porous hydroxyapatite/β-Tricalcium Phosphate ceramics. J Biomed Mater Res A. 2010;92A:267–75.
95. Tancred DC, McCormack BA, Carr AJ. A synthetic bone implant macroscopically identical to cancellous bone. Biomaterials. 1998;19:2303–11.
96. Bucholz RW, Carlton A, Holmes R. Interporous hydroxyapatite as a bone graft substitute in tibial plateau fractures. Clin Orthop. 1989;240:53–62.
97. Cavagna R, Daculsi G, Bouler JM. Macroporous Calcium Phosphate ceramic: a prospective study of 106 cases in lumbar spinal fusion. J. Long Term Eff. Med. Implants. 1999;9:403–412. J Funct Biomater. 2010;1:87.
98. Lu JX, Flautre B, Anselme K, Hardouin P, Gallur A, Descamps M, Thierry B. Role of interconnections in porous bioceramics on bone recolonization *in vitro* and *in vivo*. J Mater Sci Mater Med. 1999;10:111–20.
99. Ayers RA, Simske SJ, Nunes CR, Wolford LM. Long-term bone ingrowth and residual microhardness of porous block hydroxyapatite implants in humans. J Oral Maxillof Surg. 1998;56(Suppl. 5):1297–302.
100. Jones AC, Arns CH, Sheppard AP, Hutmacher DW, Milthorpe BK, Knackstedt MA. Assessment of bone ingrowth into porous biomaterials using MICRO-CT. Biomaterials. 2007;28:2491–504.
101. Karageorgiou V, Kaplan D. Porosity of 3D biomaterial scaffolds and osteogenesis. Biomaterials. 2005;26:5474–91.
102. Gauthier O, Bouler JM, Weiss P, Bosco J, Daculsi G, Aguado E. Kinetic study of bone ingrowth and ceramic resorption associated with the implantation of different injectable Calcium-Phosphate bone substitutes. J Biomed Mater Res. 1999;47:28–35.
103. Hing KA, Best SM, Bonfield W. Characterization of porous hydroxyapatite. J Mater Sci Mater Med. 1999;10:135–45.
104. Carotenuto G, Spagnuolo G, Ambrosio L, Nicolais L. Macroporous hydroxyapatite as alloplastic material for dental applications. J Mater Sci Mater Med. 1999;10:671–6.
105. Charriere E, Lemaitre J, Zysset P. Hydroxyapatite cement scaffolds with controlled macroporosity: fabrication protocol and mechanical properties. Biomaterials. 2003;24: 809–17.
106. Zhou L, Wang D, Huang W, Yao A, Kamitakahara M, Ioku K. Preparation and characterization of periodic porous frame of hydroxyapatite. J Ceram Soc Jpn. 2009;117:521–4.
107. Munch E, Franco J, Deville S, Hunger P, Saiz E, Tomsia AP. Porous ceramic scaffolds with complex architectures. JOM. 2008;60:54–9.
108. Yan X, Yu C, Zhou X, Tang J, Zhao D. Highly ordered mesoporous bioactive glasses with superior *in vitro* bone-forming bioactivities. Angew Chem Int Ed Engl. 2004;43:5980–4.
109. Izquierdo-Barba I, Ruiz-González L, Doadrio JC, González-Calbet JM, Vallet-Regí M. Tissue regeneration: A new property of mesoporous materials. Solid State Sci. 2005;7: 983–9.
110. Nagase M, Baker DG, Schumacher HR. Prolonged inflammatory reactions induced by artificial ceramics in the rat pouch model. J Rheumatol. 1988;15:1334–8.
111. Rooney T, Berman S, Indersano AT. Evaluation of porous block hydroxylapatite for augmentation of alveolar ridges. J Oral Maxillof Surg. 1988;46:15–8.
112. Prudhommeaux F, Schiltz C, Lioté F, Hina A, Champy R, Bucki B, Ortiz-Bravo E, Meunier A, Rey C, Bardin T. Variation in the inflammatory properties of basic Calcium Phosphate crystals according to crystal type. Arthritis Rheum. 1996;39:1319–26.
113. LeGeros RZ. Properties of osteoconductive biomaterials: Calcium Phosphates. Clin Orthop Rel Res. 2002;395:81–98.

114. Gauthier O, Bouler JM, Aguado E, Pilet P, Daculsi G. Macroporous biphasic Calcium Phosphate ceramics: influence of macropore diameter and macroporosity percentage on bone ingrowth. Biomaterials. 1998;19:133–9.
115. Leong KF, Chua CKC. Biomaterials. 2003;24:2363.
116. Lee M, Dunn JCY, Wu BM. Biomaterials. 2005;26:4281.
117. Ramay HR, Zhang M. Biomaterials. 2003;24:3293.
118. Tripathi G, Basu B. Ceram Int. 2012;38:341.
119. Tripathi G, Basu B. Int J Adv Eng Sci Appl Math. 2012;2:161.
120. Basu B, Swain SK, Sarkar D. Cryogenically cured Hydroxyapatite-gelatin nanobiocomposites with interconnected porosity as a suitable scaffold for adsorption/ delivery of BSA protein of upto 60%. RSC Adv. 2013;3:14622–14633.

Chapter 4
Case Study: Osseointegration of Strontium Containing Glass Ceramic

4.1 Introduction

In the last few decades, glass ceramics has been widely considered for a variety of biomedical applications such as restorative material in dentistry [1], orthopaedic implants for replacement of damaged/fractured bone [2] and as bone cements (in orthopaedic surgery) [3]. Such an array of beneficial effects are envisaged due to the ability of some of the investigated glass ceramic composition in eliciting reasonable biocompatibility and osteoconductivity, excellent adhesion behaviour with enamel and dentin, commendable bone-tissue integration with strong interfacial bonding and non-immunogenic over a long period of implantation [4]. Some inherent limitation of other materials, e.g. polymer-ceramic composites (low elastic modulus/wear resistance) and calcium phosphate ceramics (limited contact with host tissues) has been a reason for the increased interest in developing glass ceramics based material as an alternative choice for osteoconductive bone implants.

Even though, glass ceramics are considered to be bioactive in nature, they exhibit limited bonding to the surrounding osseous tissue, which greatly affects the bone formation [5]. In continuous surge towards developing glass-ceramics with better osseointegration properties, Sr-containing glass ceramics are recently being investigated, due to its ability to promote bone healing [6]. Strontium, being one of the trace element involved in bone homeostasis, has been proven to enhance the replication of preosteoblastic cells and to stimulate bone formation by inhibiting the osteoclast activity, a mechanism initiated through the calcium-sensing receptors, located in the membrane of osteoblasts and osteoclasts [7–9]. Overall, the structure, physical and *in vitro* biocompatibility properties of Sr-doped/substituted glass ceramics is widely reported. For example, the influence of Sr-substitution on the structure and properties of HA or various bioactive glass or glass ceramics has largely been investigated by Hill and co-workers [10–12] and lately in the research group of Ferreira [13] and Sampath Kumar [14]. In particular, Hill and co-workers used Raman spectroscopy and XRD-based Rietveld analysis to characterize wet-chemically synthesized

B. Basu and S. Ghosh, *Biomaterials for Musculoskeletal Regeneration*,
Indian Institute of Metals Series, DOI 10.1007/978-981-10-3017-8_4

Sr-substituted hydroxyapatites, $(Sr_x\ Ca_{1\text{-}x})_5(PO_4)_3(OH)$, where x is varied systematically between values of 0 and 1 [12]. The mechanical or biocompatibility property is not reported in this study. In a recent study, Sampathkumar and co-workers reported the *in vitro* cytocompatibility and antimicrobial properties of Sr-substituted and Ca deficient hydroxyapatite (HA) nanoparticles [14]. Importantly, non-cytotoxic behaviour towards human periodontal ligament fibroblast cells as well as bactericidal property against E. coli and S. aureus were established in their work. The significant reduction of the number in bacterial colony count of S. aureus in contact with $Ca_{8.69}Sr_{0.96}\ (PO_4)_6(OH)_2$ nanoparticles was measured. In a different work, Goel et al. [13] reported that the Sr-substituted glass ceramic with (36.07-x) CaO–xSrO–19.24MgO–5.61P_2O_5–38.49SiO_2–0.59 CaF_2 [X = 0–10] composition can result in a maximum of a sevenfold decrease in biodegradation in Tris-HCl medium as well as can exhibit flexural strength of around 100 MPa or more. In a different study, Hill et al. [15] utilised magic angle spinning nuclear magnetic resonance (MAS-NMR) to probe into the changes in local ultrastructural environment upon the substitution of Ca by Sr in 4.5SiO_2–3Al_2O_3–1.5P_2O_5–3CaO–2CaF_2 glass. The critical analysis of Si, [16] Al, [17] P and [15] F MAS-NMR spectra reveals the formation of strontium fluoroapatite in the fully Sr-substituted glass-ceramic, while a mixed Ca–Sr apatite formation is favored with intermediate Sr-containing glasses. It is to be further noted that the castable fluoroapatite glass ceramics in the SiO_2–Al_2O_3–P_2O_5–CaO–CaF_2 multicomponent system can have a good combination of mechanical properties (fracture toughness ~ 2.7 MPa $m^{1/2}$ and strength > 260 MPa) [15, 18, 19]. The importance of the heat treatment cycle on 1.5SiO_2–Al_2O_3–0.5P_2O_5–CaO–0.67CaF_2 glass was studied in reference to the amorphous phase separation and the evolution of $[Ca_5(PO_4)_3F]$—fluoroapatite phase during crystallization using the real-time small-angle neutron scattering (SANS) and neutron diffraction techniques. Taken together, using various state-of-the-art characterization tools, the structure of the glass-ceramics of relevance to biomedical applications is relatively well established and the physical properties along with some *in vitro* properties are also investigated widely in literature. However, the *in vivo* biocompatibility, in particular long term implant stability and osseointegration study is clearly lacking.

Motivated from the recent use of strontium ranelate (SrR) under the trade name, Protelos for the treatment of osteoporosis, a number of research groups have investigated the biocompatibility of Sr-containing glass-ceramics and bioactive glasses [20–23]. Strontium incorporated into bone cements has been found to improve bone formation as well as reduce bone resorption *in vivo*. Therefore, there has been a considerable research interest to develop Sr-containing bone substitute materials. Although, the *in vivo* biocompatibility of Sr-containing glass ceramics has not been comprehensively studied, several studies reported the *in vitro* biocompatibility of multi-component glass-ceramics. For example, SiO_2–P_2O_5–Na_2O–CaO based glass ceramics with 0–100 % Sr-substitution for Ca, supported increased cell growth of osteoblasts (human osteosarcoma cell line, Saos-2) as well as significantly higher alkaline phosphatase expression (differentiation marker) with the highest being measured for both 50 and 100 % Sr-substitution [24]. Most importantly, a typical resorption activity of osteoclasts was also observed in these glass ceramics. Based on the histomorphometrical analysis to compare the efficacy of Sr-doped hydroxyapatite (HA) on fast resorption of allograft and bone growth as bone graft extender at Ti alloy/host bone interface, Vestermark et al. reported that Sr–HA can induce 1.2-fold increase in bone growth along with 1.4-fold increase in neobone formation in comparison to Sr-free HA based bone graft extender after 4 weeks of osseointegration in a dog animal model [22]. It was suggested that $Ca_{9.51}Sr_{0.49}(PO_4)_6(OH)_2$ enhanced the bone defect healing with increased neobone formation as well as delaying the allograft resorption after 4 weeks.

In this chapter, the *in vivo* biocompatibility property of the 100 % Sr substituted LG26 glass ceramic composition systematically investigated in rabbit animal model over 26 weeks of implantation is summarised. A comparison is made with that of HA-bioglass based bioactive ceramic, which is known to be highly bioactive. In view of the fact that 100 % Sr-substituted glass ceramics of SiO_2–Al_2O_3–P_2O_5–SrO–SrF_2 composition can exhibit Sr-fluorapatite crystals and such glass ceramics are being proven to be *in vitro* cytocompatible in earlier research [1–6, 25–28], this specific glass ceramic composition is being investigated for long term *in vivo* biocompatibility in the present work. A detailed analysis using micro-computed tomography, fluorescence microscopy, confocal microscopy of the *in vivo* tested implant/tissue samples has been made to obtain the bone regeneration process both quantitatively and qualitatively.

4.2 Materials and Methods

4.2.1 Materials

The glasses with the composition $4.5SiO_2$–$3Al_2O_3$–$1.5P_2O_5$–$3SrO$–$2SrF_2$ were prepared by a melting-quench route, as described earlier [16]. Appropriate amounts of analytical grade oxide powders supplied by Sigma-Aldrich of silica (SiO_2), alumina (Al_2O_3), phosphorus pentoxide (P_2O_5), strontium carbonate ($SrCO_3$) and strontium fluoride (SrF_2), all supplied by Sigma Aldrich were ball milled together and melted in a platinum crucible at 1475 °C for 2 h. The melt was then quenched in deionised water in order to obtain amorphous frit glass. The frit was then ball milled in a planetary ball mill using agate jar-agate ball and acetone as a grinding medium for 12 h at a speed of 250 rpm. The ball milled powders were dried overnight at 100 °C and the dried powders were sieved in order to obtain a particle size of less than 45 μm. The pressureless sintering of the powder compact was carried out at 1200 °C for 2 h in air in a conventional sintering furnace. All the samples were found to be around 95 % dense. After sintering, the glass was crystallised to a strontium fluorapatite phase and a Sr-celsian phase (feldspar) as identified by XRD analysis (not shown).

The above processed glass-ceramics were used as test implants and designated as LG26Sr throughout the chapter. Commercially available *Biograft® HABG (*IFGL Bioceramics, India) was used as the control implant. This new generation bone graft substitute material, processed by a sol–gel based technique, contains a patented composition of synthetic hydroxyapatite (bone mineral) and calcium phosphate silicate. This is a unique bioactive composite, which is one of the osteoinductive and resorbable bioceramics, currently being manufactured by IFGL Bioceramics, India. Recent studies reported a faster healing of bone defects using HABG [29, 30]. Biograft® (designated throughout this document as HABG), is specially designed for repairing intra-bony defects including bone cavities and defects arising due to cyst, tumour or trauma. The surface of samples, Sr-containing glass ceramics and HABG appeared to be irregularly porous.

4.2.2 in vivo Implantation

As illustrated in Fig 4.1, the cylindrical shaped (6 mm X 2 mm) test and control implants of $4.5SiO_2$–$3Al_2O_3$–$1.5P_2O_5$–3SrO–$2SrF_2$ and *Biograft® HABG* respectively, were implanted in rabbits to study bone healing. The implants were subjected to ultrasound cleaning in distilled water for 10 min, dried overnight at 70 °C in a hot air oven (Memmert GmbH) and sterilized by autoclaving at a pressure of 15 psi at 121 °C for 15 min. All the animal experiments in the study were carried out with prior approval of the Institute Animal Ethics Committee (IAEC).

Both test and control materials were implanted in the femur of 10 New Zealand White Rabbits (Sctb: Nncl NZW) of body weight more than 2 kg and of either sexes as per ISO 10993-6. The experimental animals were anaesthetized with xylazine (5 mg/kg body weight), ketamine (35 mg/kg body weight), and midazolam (0.3 mg/kg body weight) and maintained by continuous propofol infusion (0.4 mg/kg body weight/min). The animals were controlled on lateral recumbency and a linear incision was made on the cranio-lateral aspect of the thigh under standard aseptic precautions. The shaft of the femur was exposed after a blunt dissection through the junctions of the tensor Fascia Lata and Vastus Lateralis muscle. Three cortical defects of approximately 2.5–3 mm diameter were created approximately 10 mm apart in the mid shaft of each femur with a 1.5 mm drill using surgical micromotor (SUNI TM, Expert system SATELEC, France) with continuous saline irrigation. Each experimental animal received 3 test implants (Sr-containing glass ceramic designated as LG26Sr) in the left femur and 3 control implants (designated as HABG) in the right femur. A total of 30 numbers of LG26Sr and 30 numbers of HABG implants were implanted. Each implant was pre-soaked with saline and press fitted into the defect (Fig. 4.1). Implant position was confirmed by radiography. The surgical wound was closed in layers. Both ceftriaxone (20 mg/kg body weight) and meloxicam (0.25 mg/kg body weight) were administered intramuscularly for 5 days post-implantation.

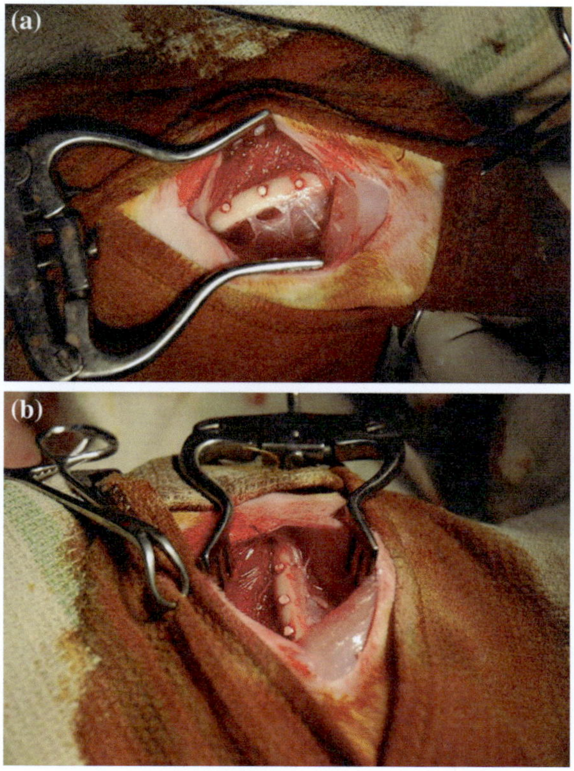

Fig. 4.1 Gross images of a rabbit femur immediately after the implantation of **a** HA-bioglass composite (control) and **b** strontium containing glass ceramic (test) [58, 69]

At 4 and 12 weeks post implantation, 5 animals were euthanized by an overdose of thiopentone sodium at each time period (total of 15 LG26Sr and 15 HABG implants at 4 and 12 weeks). Also, the implantation sites were grossly examined for healing of bone defects. The femurs with the implant materials were removed and fixed in 10 % neutral buffered formalin.

4.2.3 Explantation and Histopathological Analysis

At different timepoint of rabbit sacrifice (4/12/26 weeks), all the experimental animals were euthanized by an overdose of thiopentone sodium. Also, the implantation sites were grossly examined for healing of bone defects. The femurs with the implant materials were removed and fixed in 10 % neutral buffered formalin. In order to prepare thin sections for the histological analysis, the cross-sectional blocks of bone with the test/control implant were cut using a low speed saw (ISOMET 2000, Beuhler). Subsequently, the blocks were further fixed in 10 % neutral buffered formalin and they were further dehydrated in ascending grades of alcohol. Following this, the blocks were cleared in alcoholic acetone (1:1 v/v) mixture and immersed in two changes of 100 % alcohol. Prior to embedding the blocks in Polymethyl methacrylate resin, the bone blocks were washed twice using methyl-methacrylate. A number of serial sections (100–150 μm thickness) of implant with adjacent bone were machined from resin blocks using a high speed precision diamond saw (ISOMET 5000, Beuhler). The sections were further ground and polished using a variable speed grinder-polisher (ECOMET 3000, Beuhler). Subsequently, all the thin sections were stained with Stevenel's blue, followed by counter staining with Van Gieson's Picro-Fuchsin. Several images from bone/material interface were captured from the stained thin sections using a trinoccular light microscope (NIKON Eclipse 600, Japan)., attached with a digital camera (DXM1200, Nikon).

4.2.4 Bone Labeling

The polychromatic fluorescence labeling of mineralizing tissues during the process of osteogenesis *in vivo* was performed using sequential administration of three different fluorescent dyes, namely Xylenol orange, Alizarin red and Calcein blue in a measured manner (see Table 4.1). It is known that during mineralization process, these fluorochromes, at the timepoint of administration are permanently deposited in tissues by the process of chelation [31]. According to dosage and timeline mentioned in Table 4.1, one animal received intramuscular injections of the fluorescent dyes in test and control group, respectively. For the purpose of identifying the amount and location in order to analyze the time sequence of the regeneration process within 26 weeks of follow-up period, the thin histology sections were mounted on glass slides for fluorescence microscopic analysis.

The flurochrome labeled new bone deposition was identified by observing selected unstained sections under a fluorescence microscope (Nikon E 600, Tokyo, Japan) and the fluorescence images were captured using a DXM 1200 camera attached to the microscope. The G-2A filter with specific filter set up for red

Table 4.1 Flurochrome dosage schedule used in vital staining during the implantation in the present work

Experimental animal groups	Time point and fluorochrome dosage		
	2 weeks	4 weeks	8 weeks
4 weeks group	Xylenol orange (25 mg/kg)	–	–
12 weeks group	Xylenol orange (25 mg/kg)	Calcein blue (7 mg/kg)	Alizarin red (10 mg/kg)
	2 weeks	12 weeks	20 weeks
26 weeks group	Xylenol orange (25 mg/kg)	Alizarin red (10 mg/kg)	Calcein blue (7 mg/kg)

(excitation: 530–560 nm, emission: 580 nm) and orange (excitation: 546 nm, emission: 580 nm) was used in order to excite the used fluorochromes at their optimal wavelength.

4.2.5 Micro-CT Analysis

The retrieved control and test implant samples along with host bone tissue after 26 weeks of implantation were first fixed in 10 % neutral buffered formalin and thoroughly washed with PBS. Subsequently, the implants with host bone tissue were placed at the center of the measurement chamber of high-resolution X-ray micro-computed tomography (μCT 40, Scanco Medical, Switzerland) and scanned. The scanning parameters used were as follows: slice thickness/slice increment—16 μm; X-ray beam energy—70 kV and X-ray intensity—114 μA. In particular, the specific area of the scanned region containing one implant was selected and the reconstruction of the selected region was made using the cone beam convergence/back projection algorithm-based software. Based on the density variation of the host bone and the implant, the 3D image of the implant was extracted on the basis of the distinction between ROI (region of interest) and VOI (volume of interest). Overall scan time was 3,00,000 microseconds and the entire scan distance varied typically in the range of 12,000–16,000 μm. From the experimentally measured scanned region of the implant and neighboring host bone, a cylindrical region of interest (ROI) was selected for analysis and this corresponds to the cortical bone during implantation. In order to evaluate bone regeneration within the defect, ROI is further sectioned transversely as top, medium and bottom sectioned and radially as outer, inner shells and core. Typical dimension of the ROI around each of the investigated implant for the BV/TV calculation and other histomorphometrical analysis is, 16.384 mm (X-dimension) × 16.384 mm (Y-dimension) × 4.528 mm (Z-dimension). ROI can be easily distinguished on the basis of density variation between the implant and the host bone. Following the convergence/back projection algorithm, thresholds were suitably applied to images of each sample in order to segment newly formed bone from residual implant. After thresholding, the bone volume (BV) was determined by counting the total number of bone voxels and multiplying by their known volume, while total volume (TV) by counting bone and non-bone voxels. Apart from bone volume fraction (BV/TV), the bone mineral density (BMD) was also measured. Bone mineral density, or BMD is defined as the amount of bone within a mixed bone-soft tissue region within a medullary volume of interest containing trabecular bone. The in-built algorithm protocol of the application software supplied with the μCT 40 machine was used for the thresholding and subsequently obtaining BMD values across the implant/neobone using the established formula, which incorporates the attenuation coefficient. Briefly, two rods of CaHA with varying density of 0.25 and 0.75 gm/cc were used for calibration purpose and the diameter of the rods was 2 mm to match the implant diameter used in the present study.

4.2.6 Statistical Analysis

The experimental results were analyzed with commercially available software, SPSS 19 (IBM, USA) to assess the statistically significant difference between the properties of different samples at $p < 0.05$ and $n = 5$. Dunnett's t test (2 sides) was used to find the significant difference between the samples, and the results were compared with the early stage of osseointegration, i.e. 4 weeks.

4.2.7 Bone Morphometric Analysis

In the present work, the analysis of the micro-CT results has been made in reference to the following aspects: (a) morphology of neobone and implant material in three dimensions after careful reconstruction of a number of slices and extracting the host bone from the reconstructed image, (b) grey scale image revealing how the implant was placed into the cylindrical defect at the intramedullary region, (c) bone mineralization of implant as measured by the bone mineral density on 2D micro-CT images and (d) 3D colour images to evaluate the thickness variation of the tissues on and around the implant surface. As far as the first aspect is concerned, the 3D morphology of the implant after 4 and 12 weeks of implantation revealed a rough surface morphology with a high ratio of bone volume (BV) to total volume (TV). It was clear from extensive micro-CT analysis that both the HABG and LG26Sr implants were in physical contact with the neighboring host bone. Surface morphology of both implants at both time periods revealed a rough surface. As shown in Fig. 4.6, the BV/TV ratio significantly increased from 0.62 to 0.81 for LG26Sr implant with increasing implantation time from 4 to 12 weeks. On the other hand, in the case of HABG control implant, the BV/TV ratio increased only from 0.77 to 0.80 with increasing implantation time from 4 to 12 weeks, respectively. Such observations clearly reveal excellent bone regeneration at 4 weeks for the HABG control implant, while bone regeneration was comparable after 12 weeks for both LG26Sr and HABG (Fig. 4.6c, d). Overall, the CT images suggested that the LG26Sr implant showed a good integration with the bone within the cylindrical defects, irrespective of the implantation time (see also Fig 4.7).

The quantification of the bone mineralization density (BMD) of the newly formed bone and the BMD distribution graph across the 2D slices showed extremely high values of around 2800 mg HA/ccm, irrespective of the implantation time of up to 12 weeks. However, the baseline BMD value remained almost similar for 4 and 12 weeks of implantation in the case of the control implant (not shown). Additionally, in the case of the control implant, a perfect continuity of the newly formed bone with the surrounding host bone was observed at all time points. In contrast, a sharp increase in BMD over a few micrometer of width as one crosses from the host tissue to the test implant (LG26Sr) was recorded at all the time points (not shown). This increase as well as much higher BMD in the case of the test implants must be due to a higher mineral density of the test implants (prior to implantation), used in the present study. Nevertheless, the density (BMD) of the newly formed bone on both test and control implants was uniform throughout the tissue construct.

In order to illustrate more details, 3D colour tomography images of the test implants after 4 and 12 weeks are shown in Fig. 4.8. After 4 weeks of post-implantation, the test implant was found to be covered with tissues of around 1 mm of thickness with focal thickness of 1.5 mm or above (Fig. 4.8a), indicating good integration in the osseous system. Similarly, the entire LG26Sr implant in the intermedullary cavity of the cylindrical defect was covered uniformly by relatively thicker tissue of 2 mm or above after 12 weeks post-implantation (Fig. 4.8b). As far as the first aspect is concerned, the 3D morphology of the implant after 26 weeks of implantation is characterized by a rough surface morphology with a high ratio of bone volume (BV) to total volume (TV). It was clear from extensive micro-CT analysis that both the HABG and LG26Sr implants were in physical contact with the neighboring host bone (see Fig. 4.8). As shown in Fig. 4.8, both the control and test implants were in physical contact with the neighboring host bone. Tissue growth around the implant was suggested by some finer features present in the CT-images (Fig. 4.8). The color scheme attached to 3D color tomographs also indicate tissues of various thickness of up to around 2 mm were attached to both the test and control implants. Overall, the micro-CT analysis suggested that the LG26Sr implant showed a good integration with the host bone within the cylindrical defects after 26 weeks of implantation.

In order to illustrate the robustness of micro CT technique over conventional histological analysis, some representative 2D histology slices are provided in Fig. 4.9 and the regions of new bone formation are sufficiently delineated with dotted lines on Figs. 4.9a and 4.9c. When one compares histology images with micro computed tomograms, one can clearly realise the ability of the latter to reliably quantify bone regeneration. Further, Figs. 4.8 and 4.10 reveal the grey scale and colour contrast images of the implant surface in 3D space after different time of implantation. On the basis of extensive micro CT analysis BV/TV ratio measured with test implant (LG26Sr) and control implant (HABG) are plotted in Fig. 4.11. In order to investigate the statistical significance in terms of the bone regeneration among the samples, Dunnett's t test was used and Fig. 4.11 reveals the statistically significant increase in BV/TV ratio after 8 and 12 weeks of implantation w.r.t early stage osseointegration (4 weeks) in case of LG26Sr test implant at the 95 % confidence level. However, the significant increase has been recorded only at 12 weeks post-implantation for HABG control implant, when compared to 4 week post-implantation at $p < 0.05$ and no significant difference between the results obtained after 4 weeks or 12 weeks. Overall, the systematic increase in bone volume fraction with implantation time was more pronounced in case of LG26Sr test implant. Importantly, the bone regeneration for both the implant groups in terms of BV/TV ratio was comparable and the difference was not statistically significant at $p < 0.05$. Summarizing, the BV/TV ratio in the case of LG26Sr increased significantly with implantation time, implying greater bone regeneration capability with time. The much more new bone deposition at the LG26Sr interface in comparison to that deposited to the HABG interface as early as 2 weeks post implantation was also evident by flurochrome labeled images (Figs. 4.4 and 4.5). Furthermore, the radio opacity of LG26Sr implants was an added advantage for clinical identification.

4.3 *in vivo* Osseointegration

After 26 weeks of implantation, all the experimental rabbits recovered well from the previously mentioned surgical procedure, as revealed from physical gross examination. Also, all the animals gained weight and overall, the wound healing was found to be uneventful. Both the test and control implants were intact and the periosteal surface was completely covered with new bone, making identification of implants difficult by gross examination in both groups (see Figs. 4.2a, b). A closer look at the X-ray radiograph in Fig. 4.2b clearly revealed the outline of three test implants, indicating stable placement at the cylindrical femoral defects. However,

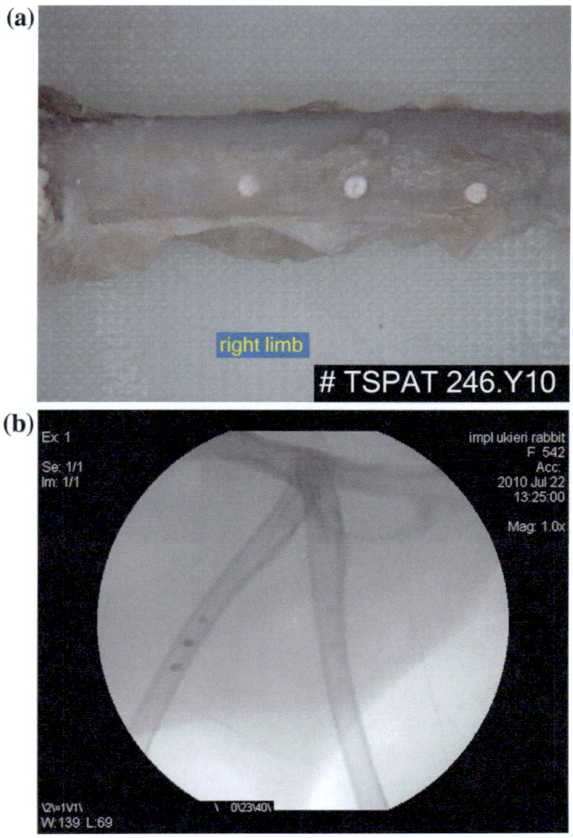

Fig. 4.2 **a** Gross images of rabbit femur with 100 % Sr-substituted glass ceramics after 26 weeks post-implantation **b** X-ray radiograph showing radio-opaque test implants (LG26Sr glass ceramic samples) on the *left* limb and less radio-opaque control (HABG) on *right* limb of an experimental rabbit [69]

such visual difference between control implant and neighboring bone is not clear in Fig. 4.2. It is possible that newly formed bone is completely integrated with host bone and also underwent mineralization over 26 weeks of implantation. The mineralized new bone has similar density to host bone in the defect region, leading to the difficulty in distinguishing the control implant from host bone.

4.3.1 Histological Observations

All implants were found to be press fit in the cavities in cortical bone. In all the histology sections, cortical and trabecular bone were oriented in a vertical direction from the host bone to the implant surface. All implants were rectangular and irregularly porous and present in the cortical bone and extended into the marrow space. The HABG implants appeared yellowish brown (see Fig. 4.3a) and LG26Sr implants were greyish white under bright field microscopy. Both in case of HABG

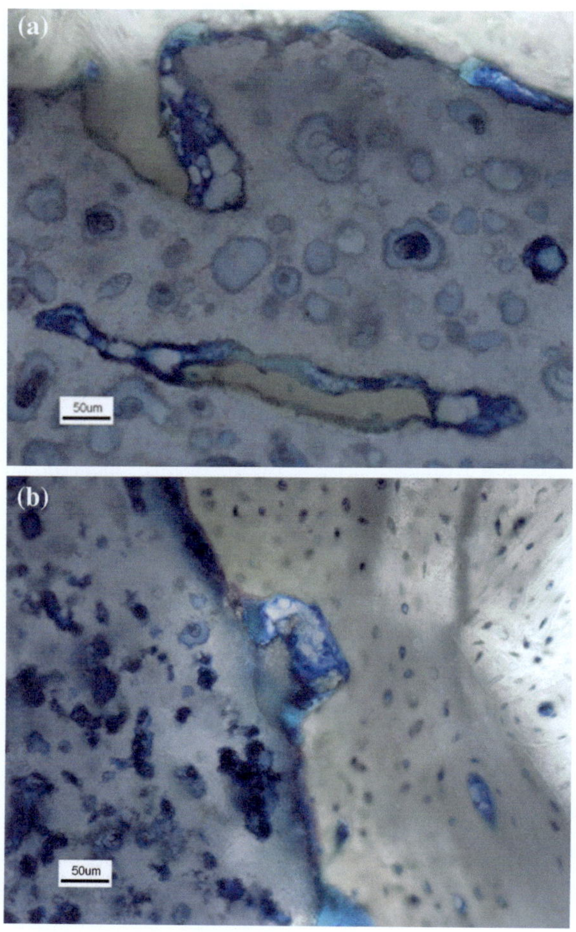

Fig. 4.3 Histological images of host bone/implant interface in the case of test implant (LG26Sr) after implantation of **a** 4 weeks and **b** 12 weeks [58]

and LG26Sr implant site, foci of bone remodeling with rosettes of osteoblasts and osteoclasts were noted at the interface. Foci of ossified bone matrix, osteoid and cells were observed in the pores and grooves of both the implant groups (Fig. 4.3). Mature new woven and lamellar bone with Haversian systems were observed both at the periosteal and endosteal aspects of the cavity with new bone deposition with osteoid and osteoblasts extending onto the implant in marrow space in both the implant groups. Thin layers of woven bone were also seen on the implant surface in marrow space in both the implant groups. There was no evidence of inflammation or bone necrosis at the implant site in both the implant groups. Multiple areas of indistinguishable apposition were noticed between the bone and the implant. It may be of interest to mention here that in a recent study carried out by Tripathi et al., osseointegration of polymer-ceramic hybrid biocomposites with similar histological features was observed in a rabbit model at the bone-implant interface [17].

For the Sr-substituted glass ceramic, the foci of ossified bone matrix, osteoid and cells were observed (Fig. 4.3). Apart from other similar observation to that of control implant, some additional histological observations are as follows. The osteoblasts were observed at the bone implant interface. Bone ingrowth was evidenced by foci of ossified bone matrix, osteoid and cells inside the implant in both groups. Importantly, very close apposition of the implant with host bone was clearly observed.

4.3.2 Bone Labeling

The identification of the time point of bone formation was made using polychromatic fluorochrome labeling. It can be reiterated here that the polychromatic fluorescent stains using calcium-binding flourchromes are deposited at various active sites of biomineralization *in vivo* with different fluorochrome colors providing sequential information regarding accretion and progression of bone formation [32, 33]. In Figs. 4.4 and 4.5, all the intravitally administrated fluorochrome labels were clearly visible with alizarin red providing red signatures. The accumulation sequence of these fluorochrome labels indicated that initial bone formation started at the periphery of the bone defects and the active ingress of new bone. A clear enhancement of bone formation on the surface of the LG26Sr ceramics was clearly observed and this is comparable with the control implant (Fig. 4.4). The first signature of ossification was observed on the implant surface after 4 weeks of implantation, indicating that the bone formation started as early as 2 weeks post-implantation (see Table 4.1). In Fig. 4.5, the intravitally administrated bone seeking fluorochrome labels are clearly visible with alizarin complexion providing red fluorescence. The accumulation sequence of these fluorochrome labels indicate that initial bone formation started at the periphery of the bone defects and the active ingress of new bone is observed in Fig. 4.5. The abundant bone formation with progressive spreading of mineralized tissues was observed in case of both LG26Sr (test) implants. Importantly, any sign of adverse tissue reaction or any osteolytic process could be recorded in any of the thin sections. A clear observation of the

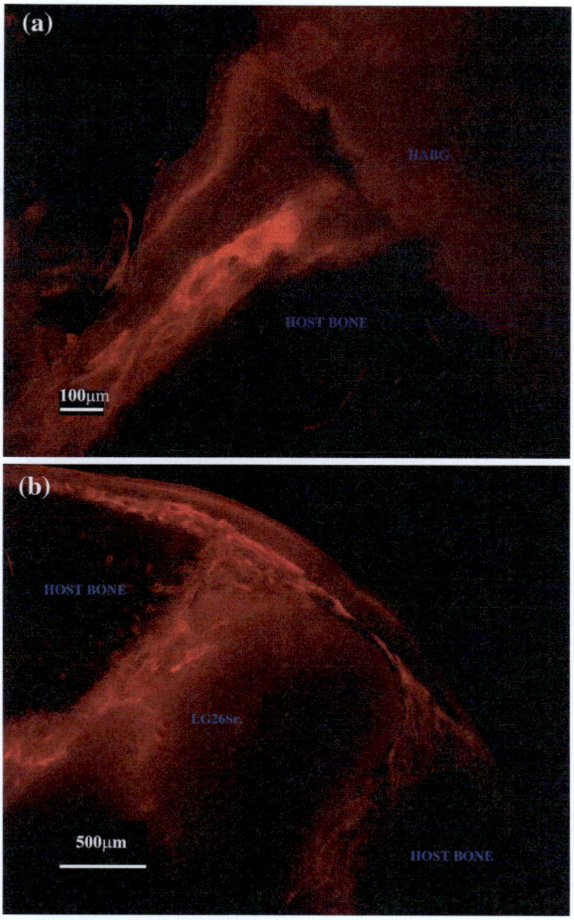

Fig. 4.4 Fluorescence images of host bone-implant interface, at 12 weeks post-implantation to illustrate the deposition of new bone at the interface with HABG (**a**) and LG26Sr glass-ceramic (**b**), as evident from the flurorescent labeling/staining with xylenol orange [58]

fluorescence images in Fig. 4.5 reveals that 12-week label in case of 26-weeks samples could be observed more extensively throughout the samples, demonstrating bone apposition throughout the implant. Since we do not find any significant difference in the fluorochrome label patterns between the HABG and LG26Sr, it may be worthwhile to state that the investigated implant composition is as bioactive as the control implant in rabbit model.

It was clear that thin region of neobone spanning over 100 μm or larger width exists in case of both control or test implant. The irregular boundary marking the interface of neobone with host bone was another feature to be noted. More importantly, the neobone formation around the test implant (LG26Sr) was observed, as could be realized by the distribution of red fluorescent regions. In summary, the histological, fluorescence and confocal microscopy of thin bone/implant sections together qualitatively establish the excellent bone regeneration ability of LG26Sr implant over 26 weeks of implantation, which was as comparable as that of a known bioactive implant (HABG).

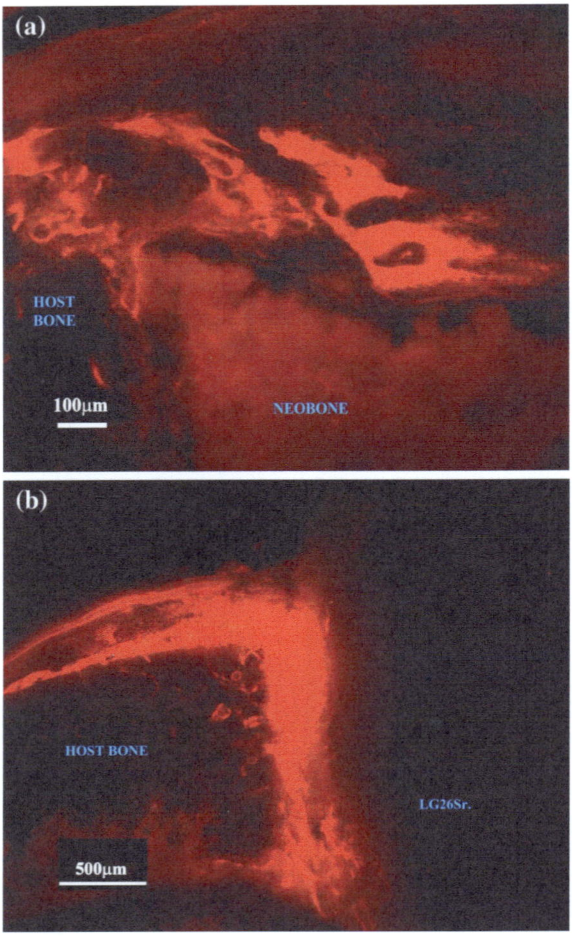

Fig. 4.5 Fluorescence images of host bone-implant interface after 26 weeks of implantation, stained with xylenol: implant site on control sample (**a**) and neobone formation around the test implant (**b**). The *red* stained region indicates the biomineralised bone [69]

The studies related to the *in vivo* biocompatibility of calcium phosphate-based biomaterials [34–37] are much more in comparison to glass-ceramics [38–40]. Gauthier et al. [39] studied the efficacy of biphasic calcium phosphates (BCP) with two different pore sizes of 80–200 μm or 200–500 μm using implantation experiments for 6 weeks in critical sized bone defects in a rabbit animal model. An extensive use of micro-CT analysis revealed a high interconnectivity of the newbone structure with more quantified bone volume being measured with BCP of 80–200 μm size (BV/TV = 38.5 ± 5.5) in comparison to BCP of 200–500 μm (BV/TV = 24.8). Vogel et al. reported the bone regeneration ability of bioglass particles of three different compositions (45S5:$45SiO_2$–$24.5Na_2O$–24.5CaO–$6P_2O_5$;52S:$52SiO_2$–$21Na_2O$–21CaO–$6P_2O_5$ and 55S;$55SiO_2$–$19.5Na_2O$–19.5CaO–$6P_2O_5$, all compositions in wt%), when implanted in the distal femoral epiphysis in rabbits for 7, 28 and 84 days [41]. A major observation of this study was the *in vivo* degeneration of the

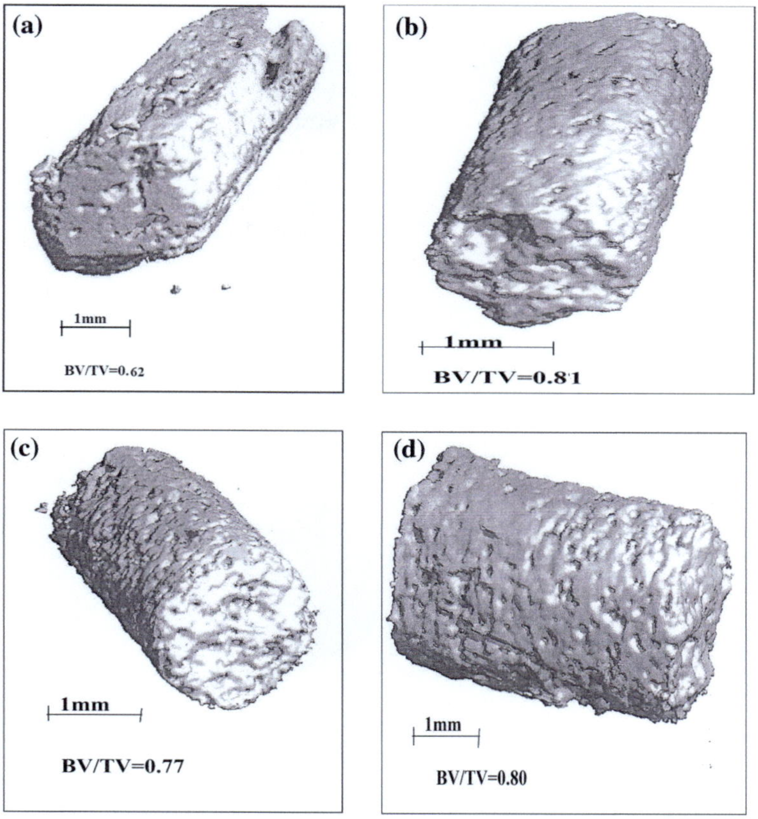

Fig. 4.6 Microcomputed tomographs showing three-dimensional morphology of newly formed bone over residual LG26Sr based test implant after implantation of **a** 4 weeks and **c** 12 weeks as well as HA-bioglass based control implant after implantation of **b** 4 weeks and **d** 12 weeks [58]

investigated bioglass particles either to Sr-rich remnants or to CaP-rich shells as well as the finding of higher number of multinuclear giant cells (MNGC) on the degraded bioglass remnants. In another study, the same research group reported an increased number of MNGC with increasing implantation time from 28 days to 84 days with the highest number of MNGCs being measured in the case of the widely researched 45S5 bioglass [40]. In a relatively recent study, Gorustovich et al. reported the effectiveness of Sr-containing 45S5 bioglass for bone tissue repair in a rat model [38]. The addition of 6 wt% SrO to substitute for CaO in 45S5 bioglass did not cause any statistically significant difference in bone bonding behavior in terms of affinity index, being measured as the neobone length in contact with the implant surface (expressed as % of total length of implant) after 30 days of implantation in rat tibia. The overall affinity index ($\sim$89 %) of 45S5-6 % SrO bioglass is quite notable after implantation in rat tibia for 4 weeks. The histomorphometric analysis revealed an increase in the affinity index from 73.6 $\pm$ 3.5 % after 3 months to 85.2 $\pm$ 2.7 % after 6 months of

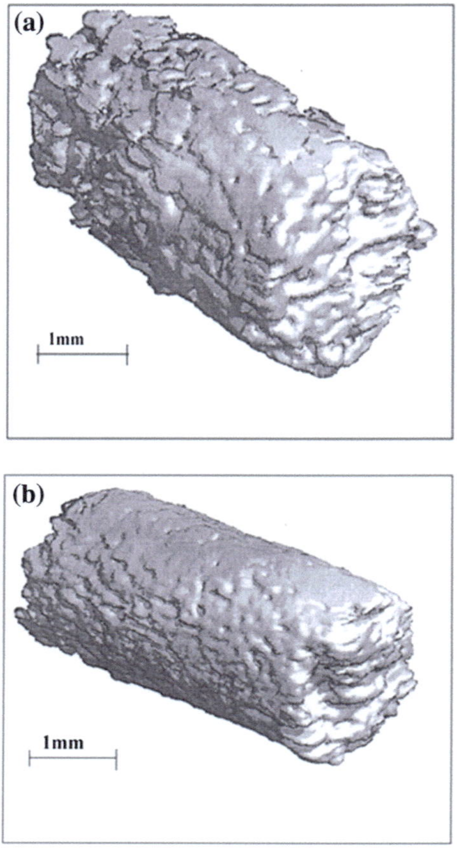

Fig. 4.7 Representative Microcomputed tomographs showing three-dimensional morphological features of newly formed bone over residual HA-bioglass based control implant (**a**) and LG26Sr based test implant (**b**) in the cylindrical defects, post-implantation for 26 weeks in rabbit model[69]

implantation. Importantly, 45S5 bioglass had much larger affinity index in the rat model than that measured (30–60 %) with an identical glass composition in a rabbit model after similar implantation period, as reported by Vogel et al. [40]. In another study, Li et al. reported the progression of bone bonding of Sr–HA based bioactive bone cement into the cancellous bone of the iliac crest of rabbits [42].

4.4 Discussion

The long term stability of an implant in an appropriate animal model as well as evaluation of bone regeneration over longer implantation time (>6 months) is an important issue in the field of bone-tissue engineering applications [43, 44]. The longevity of orthopaedic prostheses depends upon the implant fixation and interfacial stability. Aseptic loosening and failure of an implant are often reported due to

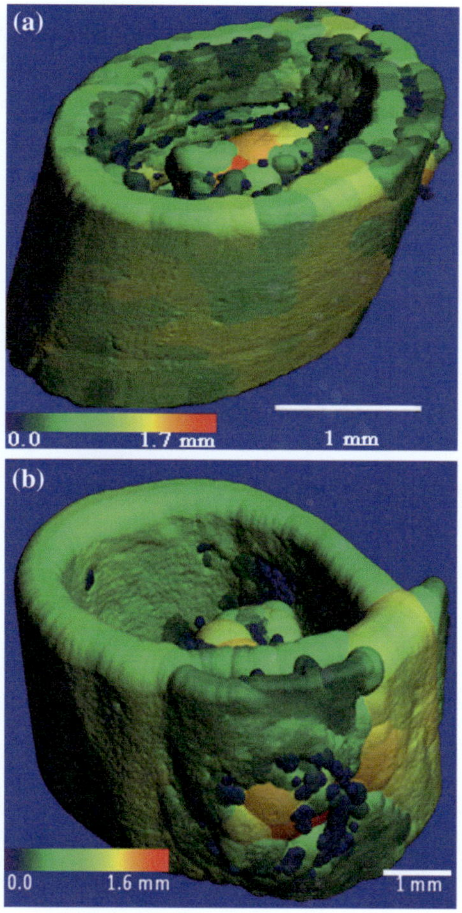

Fig. 4.8 3D micro CT tomographs in color contrast of the neobone around LG26Sr based test implant at two different time points: **a** 4 weeks and **b** 12 weeks. The color scheme shows the variable thickness of around 1 mm higher in the neobone area [58]

instability and incomplete anchorage at the bone/implant interface [45]. There are different approaches to improve the stability and anchorage and these approaches include bioactive fixation using compositionally tailored biomaterials, cemented fixation, mechanical bone implant interlocking, etc. The results summarized in this chapter illustrate the bioactive fixation of a Sr-containing glass ceramic, based on the glass composition $4.5SiO_2$–$3Al_2O_3$–$1.5P_2O_5$–3SrO–$2SrF_2$. From the implant perspective, a non-toxic surface is essential for the migration, attachment and proliferation of the osteoblasts *in vivo* [46, 47] and this has been assessed in the present work primarily using histology images and micro-CT analysis. Both the implants, i.e. control (HABG) and test group (LG26Sr glass ceramic) are irregularly porous. New bone deposition with osteoid also extended from the periosteal and endosteal aspect of cortical bone onto the implant in bone marrow space in both groups (see Fig. 4.9).

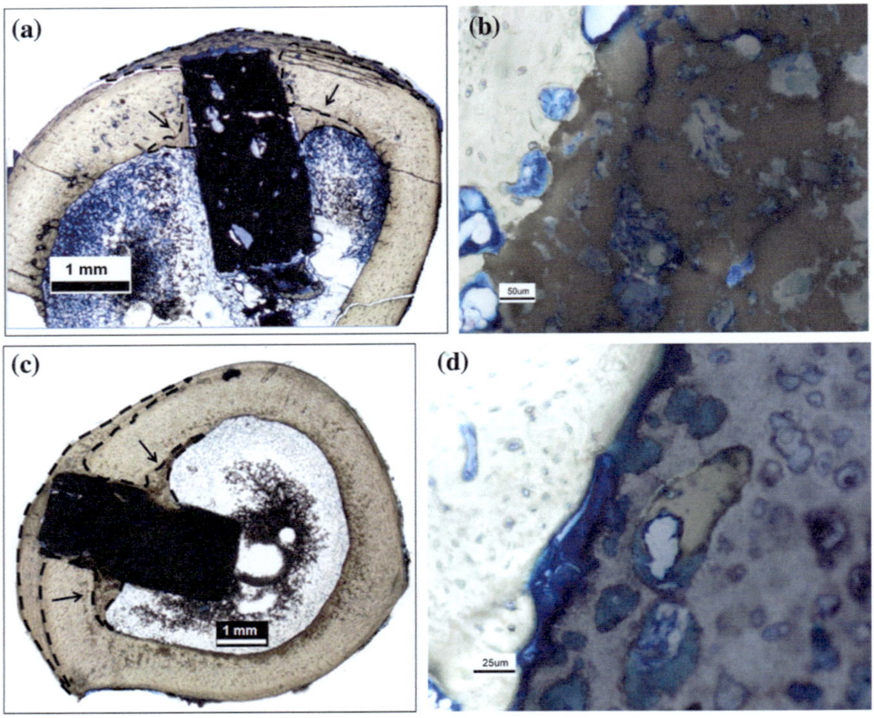

Fig. 4.9 Representative photomicrographs of thin histology sections of host bone-implant interface after 26 weeks of implantation: implant site on control implant, HABG (**a**, **b**) and test implant, LG26Sr (**c**, **d**). The region marked with *dotted lines* and *bold arrows* in (**a** and **c**) indicates significant neobone formation along either side of the implant with large amount of 'cap-like' neobone on the periosteal surface [69]

As far as the *in vivo* biocompatibility is concerned, only 45S5 bioglass received some attention. For example, Blencke et al. [43] reported the long-term stability of the ceravital glass ceramics (SiO_2–CaO–P_2O_5–Na_2O–MgO–K_2O) in rat model for implantation period of up to 30 months. The absence of osteoclastic activity or giant cells was noted. In a different study, the histomorphometrical analysis established good bone formation around NiTi shape memory alloys *in vivo* during 26 weeks of periosteal implantation in rat model [44]. Höland et al. [48] reported the interfacial reactions between host bone and SiO_2–Al_2O_3–MgO–CaO–K_2O/Na_2O–F–P_2O_5 glass ceramics in guinea pig model. Based on electron microscopy coupled with XRD analysis after implantation of up to 16 weeks, it was concluded that the interfacial reaction involves an initial slight solubility of glass ceramic, followed by solid state epitaxial layer-like growth of apatite on host bone. Also, the typical width of interfacial reaction zone is around 5–10 μm. In a later study, Kitsugi et al. [49] extensively used SEM-EPMA analysis of the interfacial region between various types of apatite containing glass ceramics and rabbit tibia at various post-implantation period of up to 12 months (48 weeks). Interestingly, the

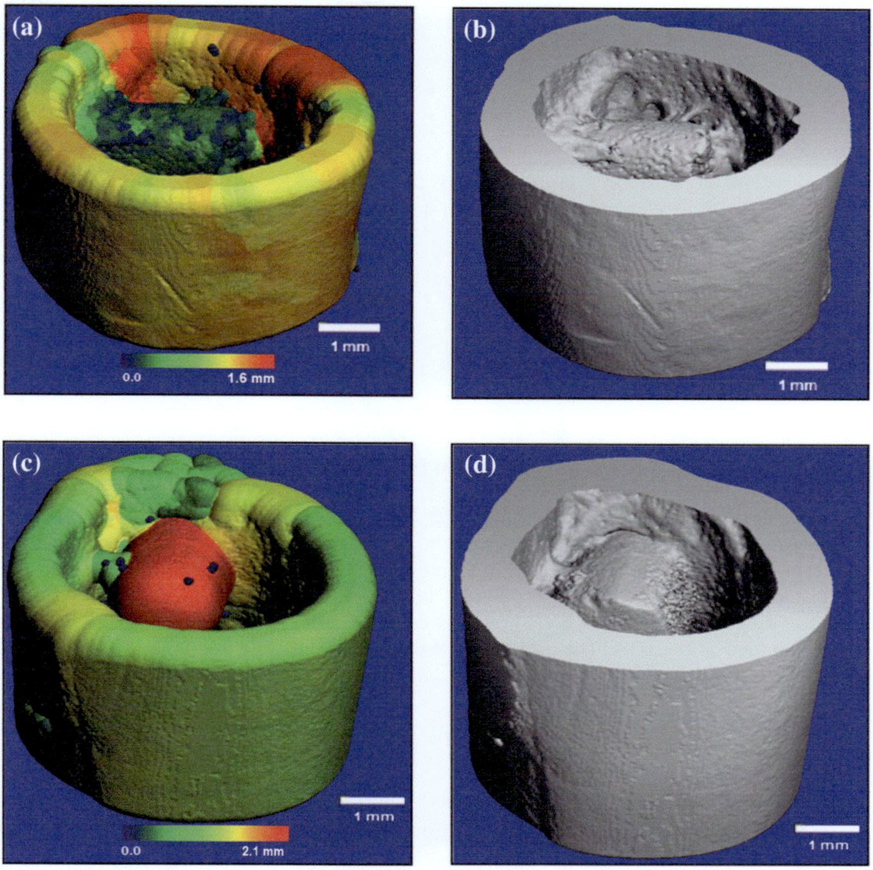

Fig. 4.10 Representative 3D micro-CT tomographs in color contrast (**a**) and grey scale (**b**) showing implant as well as regenerated tissue around HABG control implant; 3D micro-CT tomographs in color contrast (**c**) and *grey scale* (**d**) of implant as well as neotissue around LG26Sr test implant, all after 26 weeks of implantation in rabbit model [69]

width of reactive zone does not increase beyond 6 months and a maximum interfacial zone width of 70–80 μm was measured with apatite-wollastonote based glass ceramics. Apart from the above mentioned studies on bulk glass ceramics, the glass ceramic coated metallic implants are also investigated for *in vivo* biocompatibility property. In one such study, Wheeler et al. [50] reported greater bone growth around 45S5 glass coated Ti than HA-coated Ti over implantation of up to 16 weeks in unloaded distal femurs of juvenile rabbits. A better osteointegration is also associated with superior histomorphometric property (~3.2 μm/day mineral apposition rate for porous 45S5 coating) and biomechanical shear strength properties (4.8 N/mm^2 for 45S5 coatings). However, long term stability of 45S5 bioglass coatings on metallic implant was not investigated. Overall, long term implantation for 26 weeks in rabbit model is not commonly carried out and that too

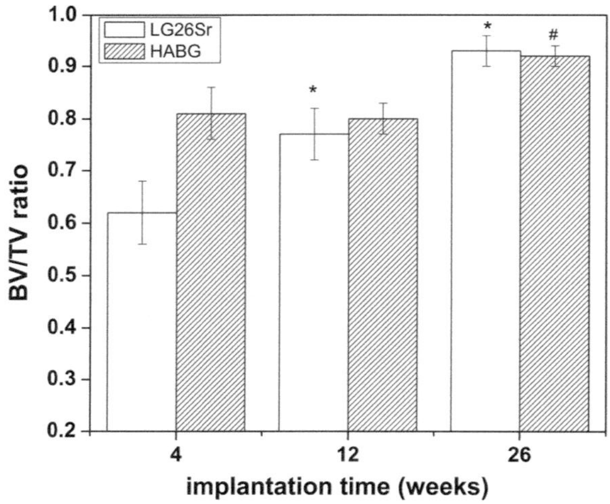

Fig. 4.11 Plot of the ratio of bone volume (BV) to total volume (TV) for the test implant (LG26Sr) and control implant (HABG), as measured using micro-CT analysis after implantation in rabbit femurs for different timeperiods. (*): Significant difference (statistical) in terms of BV/TV ratio in case of LG26Sr implant with respect to 4 weeks implantation data at $p < 0.05$. (#): Significant difference in terms of BV/TV ratio in case of HABG implant with respect to 4 weeks implantation data at $p < 0.05$. The data are represented as mean $\pm$ SE [69]

rarely reported for a glass ceramics. This places major relevance for the present study.

Among various glass ceramics, some material compositions are found to be either unsuitable in producing an optimum balancing effect between biological response, physico-mechanical behaviour *in vivo* or a lack of machinability/castability to desired shapes [51]. The cytotoxicity and mechanical integrity of the implants predominantly depend on the stability of its surface, while the release of the different ions and molecules also inhibits the appropriate host response and bone regeneration [52]. It is reported that slow leaching of the limited amount of fluoride stimulates the cell growth and even the radiological examination supports that present glass ceramics enhances osteoblastic activities during *in vivo* bone growth [53]. One of the reasons for this, probably the generation of fluroapatite phases, which exhibits significantly good biocompatibility behaviour due to the formation of apatite layer on material surface and this, is essential for bone tissue-material bonding. However, an excessive amount of fluoride release may exhibit cytotoxic response. Previously, there have been several studies on different glass ceramic systems [54, 55] to investigate the biomechanical response, but no *in vivo* study related to calcium alumino silicate glass ceramics and more specifically for LG26 composition series ($4.5SiO_2$–$3Al_2O_3$–$1.5P_2O_5$–(3-x)CaO–(2-x)CaF_2) has ever been reported. The present study therefore adds the long term stability and osseointegration property of 100 % Sr-substittued glass ceramics. It may be further added that the control implant (HABG) used in the present work is a unique bioactive composite, characterized by osteoinductive properties. Some recent studies reported a faster healing of bone defects using HABG and based on such studies, the efficacy of such implant materials in repairing intra-bony defects including bone cavities and defects arising due to cyst, tumour or trauma is established [56, 57]. From the above discussion, it is therefore apparent that the LG26Sr test implant is as bioactive as the control implant. It can be further mentioned that the early stage osseointegration of the presently investigated LG26Sr implant over the short term implantation of up to 12 weeks in the rabbit model is reported [58]. This study together with the present work therefore establishes the *in vivo* biocompatibility of the glass ceramics with the nominal composition of $4.5SiO_2–3Al_2O_3–1.5P_2O_5–3SrO–2SrF_2$ over a broad spectrum of implantation period.

In attempts to characterise bone regeneration property, researchers have used various tools, apart from histopathological analysis. It is worth mentioning that multiple characterisation techniques, including histology, micro-CT, fluorescence microscope are effectively utilised to evaluate the influence of Sr- doping/substitution on bone regeneration capability in a qualitative/quantitative manner in the present work. It may be worthwhile to mention that the purpose of incorporating Sr in the glass ceramics was to confer radiopacity. Apart from this aspect, Sr^{2+} ions, when substituted for Ca^{2+} does not disrupt the glass structure and understood to have a negligible influence on fluoride release [59]. Moreover, strontium predominantly contributes to the bactericidal activity of the implant [60, 61] with sustained release of Sr^{2+} ions *in vivo*. In a unique study to characterize the fine scale compositional changes at the cortical or cancellous bone with Sr–HA cement, Ni et al. utilized the time-of-flight secondary ion mass spectrometry (ToF-SIMS) [62]. The important results of this study are worthwhile to mention. An extremely thick interface of 70 μm at cancellous bone/Sr–HA cement is characterized by high ionic intensity of Ca, P, Sr, Na and O, whereas a much thinner interface of 1 μm (low ionic intensity) was observed in the interface with cortical bone. The morphological and compositional differences indicated the precipitation coupling mechanisms at the host bone/implant interface. It can be speculated that similar phenomenon may take place at the host bone/test implant interface in the present case.

The influence of 100 % Sr substitution on good *in vivo* biocompatibility can be additionally explained on the basis of the following consideration. Systemic administration of strontium ranelate (SR) has been found to improve fixation of hydroxyapatite (HA) —coated titanium screws in ovariectomized (OVX) rats [63]. The present incorporation of Strontium in glass ceramic assures a regular delivery of strontium to host and new bone, thus facilitating good bone deposition. This would definitely improve fixation of joint prosthesis and concurrently reduce activation of osteoclasts and prevent long term host bone resorption. Further *in vitro* studies with osteoclasts and long term *in vivo* studies in animal models of osteoporosis are required to find out the beneficial effect of strontium in older patients with joint replacement. At the closure, the relevance of the results discussed in the present chapter can be reiterated. The aseptic loosening of joint prosthesis over long term implantation has resulted in the release of ultrafine particles, which initiates a series of events beginning with the influx/activation of a large number of inflammatory cells, release of cytokines and stimulation of host bone remodeling cells, the osteoclasts, which together leads to bone resorption. Osteoporosis, a commonly reported disease among post-menopausal women and the ageing population, is characterized by an imbalance in the functionality of osteoblasts (bone forming) and osteoclast (bone resorption) cells, leading to the disruption of bone remodelling [23]. This consequently makes the bone porous and prone to fracture. The prevention of such bone resorption is a primary requirement for maintenance of the implant in the elderly patients, who otherwise have to undergo revision surgery. In future, studies should be conducted with Sr-bioglass using osteoporotic animal model.

It is worthwhile to mention here that the commercially available SrR (Strontium ranelate) drug combines two atoms of stable strontium with the organic moiety, ranelic acid. The biophysical role of SrR appears to be through the strontium cations themselves, which have been shown to work both *in vivo* and *in vitro* by stimulating osteoblasts to make new bone and preventing osteoclasts from resorbing bone, leading to the restoration of the normal bone remodelling balance [64]. It has also been proposed that Sr, like Ca, can stimulate the Ca–sensing receptors at the cell membrane of osteoblast cells. These, in turn, influence RANKL production and reduce the osteoblast proliferation or even can cause apoptosis. The enhanced proliferation and differentiation of osteoblasts in the presence of Sr is believed to increase the neobone formation and this is the result of the stimulation of homeostatic local hypercalceuria around the implant [65–68]. Further study using osteoclasts *in vitro* and large animal study with the presently investigated test implants is warranted to realize the clinical efficacy.

4.5 Closure

This chapter presents a case study to demonstrate the *in vivo* biocompatibility 100 % Sr-substituted glass ceramic based implants with nominal composition of $4.5SiO_2–3Al_2O_3–1.5P_2O_5–3SrO–2SrF_2$, based on the histopathological, polychrome sequence labeling, micro-CT analysis. Overall, the experimental results summarized in this chapter establishes the following aspects: (a) non-biodegradable and osteoconductive nature, (b) long term implant stability and osseointegration, (c) dynamic bone healing and progressive bony reconstruction/bone formation both around periosteum and endosteum regions of bone defect (d) direct biological contact between host bone and implant material. Overall, this chapter discusses the long term *in vivo* biocompatibility of 100 % Sr-substituted glass ceramics, emphasizing the fact that the abundant presence of strontium does not adversely affect the bone regeneration and remodeling processes [58, 69].

References

1. Friedma S, Löst C. Malaekeh Zarrabian and Martin Trope; evaluation of success and failure after endodontic therapy using a glass ionomer cement sealer. J Endod. 1995;21:384–90.
2. Brook IM, Craig GT, Hatton PV, Jonck LM. Bone cell interactions with a granular glass-ionomer bone substitute material: *in vivo* and *in vitro* culture models. Biomaterials. 1992;13:721–5.
3. Jonck LM, Grobbelaar CJ, Strating H. Biological evaluation of glass-ionomer cement (Ketac-0) as an interface material in total joint replacement. A screening test. Clin Mater. 1989;4:201–224.
4. Brook IM, Hatton PV. Glass-ionomers: bioactive implant materials. Biomaterials. 1998;19:565–71.
5. Ducheyne P, Qiu Q. Bioactive ceramics: the effect of surface reactivity on bone formation and bone cell function. Biomaterials. 1999;20:2287–303.
6. Zhao Y, Guo D, Hou S, Zhong H, Yan J, Zhang C, Zhou Y. Porous allograft bone scaffolds: doping with strontium. PLoS ONE. 2013;8:e69339.
7. Isaac J, Nohra J, Lao J, Jallot E, Nedelec JM, Berdal A, Sautier JM. Effects of strontium-doped bioactive glass on the differentiation of cultured osteogenic cells. Eur Cell Mater. 2011;21:130–43.
8. Dahl SG, Allain P, Marie PJ, Mauras Y, Boivin G, Ammann P, Tsouderos Y, Delmas PD, Christiansen C. Incorporation and distribution of strontium in bone. Bone. 2001;28(4):446–53.
9. Andersen OZ, Offermanns V, Sillassen M, Almtoft KP, Andersen IH, Sorensen S, Jeppesen CS, Kraft DC, Bottiger J, Rasse M et al. Accelerated bone ingrowth by local delivery of strontium from surface functionalized titanium implants. Biomaterials 2013;34:5883–5890.
10. Hill R, Calver A, Stamboulis A, Bubb N. Real-time nucleation and crystallization studies of a fluorapatite glass- ceramics using small-angle neutron scattering and neutron diffraction. J Am Ceram Soc. 2007;90:763–8.
11. Hill RG, Stamboulis A, Law RV, Clifford A, Towler MR, Crowley C. The influence of strontium substitution in fluorapatite glasses and glass-ceramics. J Non-Cryst Sol. 2004;336:223–9.
12. Odonnell MD, Fredholm Y, Rouffignace A, Hill R. Structural analysis of a series of strontium- substituted apatites. Acta Biomater. 2008;4:1455–64.
13. Goel A, Rajagopal R, Ferreira JMF. Influence *of* strontium *on* structure, sintering *and* biodegradation behaviour *of* $CaO\text{-}MgO\text{-}SrO\text{-}SiO_2\text{-}P_2O_5\text{-}CaF_2$ glasses. ActaBiomaterialia. 2011;7:4071–80.
14. Devi ND, Balu R, Sampath Kumar TS. Strontium-substituted calcium deficient hydroxyapatite nanoparticles: synthesis, characterization, and antibacterial properties. J Am CerSoc. 2012;95:2700–8.
15. Hill R, Stamboulis A, Law RV, Matusuya SA. MAS-NMR study of the crystallisation process of apatite–mullite glass-ceramics. J Phys Chem. 2004;45:127–133.
16. Hill RG, Stamboulis A, Law RV, Clifford A, Towler MR, Crowley C. The influence of strontium substitution in fluorapatite glasses and glass-ceramics. J Non-Cryst Solids. 2004;336(3):223–9.
17. Tripathi G, Gough JE, Dinda A, Basu B. In vitro cytotoxicity and in vivo osseointegration property of compression moulded $HDPE\text{-}HA\text{-}Al_2O_3$ hybrid biocomposites. J Biomed Mater Res A. 2013;101(6):1539–49.
18. Rafferty Aran, Clifford Anton, Hill Robert, Wood David. BisserkaSamuneva, Margarita Dimitrova-Lukacs. Influence of fluorine content in apatite mullite glass-ceramics. J Am Ceram Soc. 2000;83:2833–8.

19. Boyd D, Towler MR, Law RV, Hill RG. An investigation into the structure and reactivity of calcium-zinc-silicate ionomer glasses using MAS-NMR spectroscopy. J Mater Sci: Mater Med. 2006;17:397–402.
20. Ravi ND, Balu R, Sampath Kumar TS. Strontium-substituted calcium deficient hydroxyapatite nanoparticles: synthesis, characterization, and antibacterial properties. J Am Ceram Soc. 2012;95(9):2700–8.
21. Goel A, Rajagopal RR, Ferreira JM. Influence of strontium on structure, sintering and biodegradation behaviour of CaO–MgO–SrO–SiO_2–P_2O_5–CaF_2 glasses. Acta Biomater. 2011;7(11):4071–80.
22. Vestermark MT, Hauge EM, Soballe K, Bechtold JE, Jakobsen T, Baas J. Strontium doping of bone graft extender. Acta Orthop. 2011;82(5):614–21.
23. Hamdy NA. Strontium ranelate improves bone microarchitecture in osteoporosis. Rheumatology 2009;48:iv9–13.
24. Gentleman E, Fredholm YC, Jell G, Lotfibakhshaiesh N, O'Donnell MD, Hill RG, et al. The effects of strontium-substituted bioactive glasses on osteoblasts and osteoclasts *in vitro*. Biomaterials. 2010;31(14):3949–56.
25. Kenny SM, Buggy M. Bone cements and fillers: a review. J Mater Sci Mater Med. 2003;14:923–38.
26. Davidson CL. Advances in glass-ionomer cements. J Appl Oral Sci. 2006;14:3–9.
27. Brentegani LG, Bombonato KF, LamanoCarvalho TL. Histological evaluation of the biocompatibility of glass-ionomer cement in rat alveolus. Biomaterials. 1997;18:137–40.
28. Nair MB, Varma H, Shenoy SJ, John A. Treatment of goat femur segmental defects with silica-coated hydroxyapatite–one-year follow-up. Tissue Eng Part A. 2010;16:385–91.
29. Acharya NK, Mahajan CV, Kumar RJ, Varma HK, Menon VK. Can iliac crest reconstruction reduce donor site morbidity?: a study using degradable hydroxyapatite-bioactive glass ceramic composite. J Spinal Disord Tech. 2010;23(4):266–71.
30. Nair MB, Suresh Babu S, Varma HK, John A. A triphasic ceramic-coated porous hydroxyapatite for tissue engineering application. Acta Biomater. 2008;4(1):173–81.
31. Pautke C, Vogt S, Tischer T, Wexel G, Deppe H, Milz S, Schieker M, Kolk A. Polychrome labeling of bone with seven different fluorochromes: enhancing fluorochrome discrimination by spectral image analysis. Bone. 2005;37:441–5.
32. Pautke C, Vogt S, Tischer T, Deppe H, Milz S, Schieker M, Kolk A. Polychrome labeling of bone with seven different fluorochromes: enhancing fluorochrome discrimination by spectral image analysis. Bone. 2005;37:441–5.
33. Oest ME, Dupont KM, Kong HJ, Mooney DJ, Guldberg RE. Quantitative assessment of scaffold and growth factor-mediated repair of critically sized bone defects. J Orthop Res. 2007;25(7):941–50.
34. Simank HG, Stuber M, Frahm R, Helbig L, van Lenthe H, Muller R. The influence of surface coatings of dicalcium phosphate (DCPD) and growth and differentiation factor-5 (GDF-5) on the stability of titanium implants in vivo. Biomaterials. 2006;27(21):3988–94.
35. Rai B, Oest ME, Dupont KM, Ho KH, Teoh SH, Guldberg RE. Combination of platelet-rich plasma with polycaprolactone-tricalcium phosphate scaffolds for segmental bone defect repair. J Biomed Mater Res A. 2007;81(4):888–99.
36. Shao XX, Hutmacher DW, Ho ST, Goh JC, Lee EH. Evaluation of a hybrid scaffold/cell construct in repair of high-load-bearing osteochondral defects in rabbits. Biomaterials. 2006;27(7):1071–80.
37. Nath S, Basu B, Mohanty M, Mohanan PV. In vivo response of novel calcium phosphate-mullite composites: results up to 12 weeks of implantation. J Biomed Mater Res B Appl Biomater. 2009;90(2):547–57.
38. Gorustovich AA, Steimetz T, Cabrini RL, Porto Lopez JM. Osteoconductivity of strontium-doped bioactive glass particles: a histomorphometric study in rats. J Biomed Mater Res A. 2010;92(1):232–7.

39. Gauthier O, Muller R, von Stechow D, Lamy B, Weiss P, Bouler JM, et al. In vivo bone regeneration with injectable calcium phosphate biomaterial: a three-dimensional micro-computed tomographic, biomechanical and SEM study. Biomaterials. 2005;26 (27):5444–53.
40. Vogel M, Voigt C, Knabe C, Radlanski RJ, Gross UM, Muller-Mai CM. Development of multinuclear giant cells during the degradation of bioglass particles in rabbits. J Biomed Mater Res A. 2004;70(3):370–9.
41. Vogel M, Voigt C, Gross UM, Muller-Mai CM. In vivo comparison of bioactive glass particles in rabbits. Biomaterials. 2001;22(4):357–62.
42. Li YW, Leong JC, Lu WW, Luk KD, Cheung KM, Chiu KY, et al. A novel injectable bioactive bone cement for spinal surgery: a developmental and preclinical study. J Biomed Mater Res. 2000;52(1):164–70.
43. Blencke BA, Bromer H, Deutscher KK. Compatibility and long-term stability of glass-ceramic implants. J Biomed Mater Res. 1978;12:307–16.
44. Ryhanen J, Kallioinen M, Tuukkanen J, Lehenkari P, Junila J, Niemela E, Sanvik P, Serlo W. Bone modeling and cell-material interface responses induced by nickel-titanium shape memory alloy after periosteal implantation. Biomaterials. 1999;20:1309–17.
45. Sundfeldt M, Carlsson LV, Johansson CB, Thomsen P, Gretzer C. Aseptic loosening, not only a question of wear: a review of different theories. Acta Orthop. 2006;77(2):177–97.
46. Basle MF, Rebel A, Grizon F, Daculsi G, Passuti N, Filmon R. Cellular response to calcium phosphate ceramics implanted in rabbit bone. J Mater Sci Mater Med. 1993;4:273–80.
47. Boss JH. Osseointegration. J Long Term Eff Med Impl. 1999;18:1–10.
48. Holand W, Vogel W, Naumann K, Gummel J. Interface reactions between machinable bioactive glass-ceramics and bone. J Biomed Mater Res. 1985;19:303–12.
49. Kitsugi T, Nakamura T, Yamamura T, Kokubu T, Shibuya T, Takagi M. SEM-EPMA observation of three types of apatite-containing glass-ceramics implanted in bone. The variance of a Ca-P-rich layer. J Biomed Mater Res. 1987;21:1255–71.
50. Wheeler DL, Montfort MJ, McLoughlin SW. Differential healing response of bone adjacent to porous implants coated with hydroxyapatite and 45S5 bioactive glass. J Biomed Mater Res. 2001;55:603–12.
51. Kokubo T, Ito S, Sakka S, Yamamuro T. Formation of a high strength bioactive glass-ceramic in the system $MgO–CaO–SiO_2–P_2O_5$. J Mater Sci. 1986;21:536–40.
52. Freeman CO, Brook IM. Bone response to a titanium aluminium nitride coating on metallic implants. J Mater Sci Mater Med. 2006;17:465–70.
53. Sasanaluckit P, Albustany KR, Doherty PJ, Williams DF. Biocompatibility of glass ionomer cements. Biomaterials 1993;4:906–916.
54. Kokubo T, Ito S, Sakka S, Yamamuro T. Formation of a high strength bioactive glass-ceramic in the system $MgO–CaO–SiO_2–P_2O_5$. J Mater Sci. 1986;21:536–40.
55. da Rocha Barros VM, Salata LA, Sverzut CE, Xavier SP, Van Noort R, Johnson A, Hatton PV. *In vivo* bone tissue response to a canasiteglass-ceramic. Biomaterials. 2002;23:2895–900.
56. Acharya NK, Mahajan CV, Kumar RJ, Varma HK, Menon VK. Can Iliac Crest reconstruction reduce donor site morbidity?: a study using degradable hydroxyapatite-bioactive glass ceramic composite. J Spinal Disord Tech. 2010;23:266–71.
57. Nair MB, Suresh Babu S, Varma HK, John A. A triphasic ceramic-coated porous hydroxyapatite for tissue engineering application. Acta Biomater. 2008;4:173–81.
58. Sabareeswaran A, Basu B, Shenoy SJ, Jaffer Z, Saha N, Stamboulis A. Early osseointegration of a strontium containing glass ceramic in a rabbit model. Biomaterials. 2013;34:9278–86.
59. Vogel M, Voigt C, Knabe C, Radlanski RJ, Gross UM, Müller-Mai CM. Development of multinuclear giant cells during the degradation of Bioglass particles in rabbits. J Biomed Mater Res A. 2004;70(3):370–9.
60. Vestermark MT. Strontium in the bone-implant interface. Dan Med Bull. 2011;58:B4286.
61. Guida A, Towler MR, Wall JG, Hill RG, Eramo S. Preliminary work on the antibacterial effect of strontium in glass ionomer cements. J Mater Sci Lett. 2003;22:1401–1403.

62. Ni GX, Lu WW, Xu B, Chiu KY, Yang C, Li ZY, et al. Interfacial behaviour of strontium-containing hydroxyapatite cement with cancellous and cortical bone. Biomaterials. 2006;27:5127–33.
63. Bain SD, Jerome C, Shen V, Dupin-Roger I, Ammann P. Strontium ranelate improves bone strength in ovariectomized rat by positively influencing bone resistance determinants. Osteoporos Int. 2009;20(8):1417–28.
64. Gentleman E, Fredholm YC, Jell G, Lotfibakhshaiesh N, O'Donnell MD, Hill RG, et al. The effects of strontium-substituted bioactive glasses on osteoblasts and osteoclasts *in vitro*. Biomaterials. 2010;31:3949–56.
65. Brennan TC, Rybchyn MS, Green W, Atwa S, Conigrave AD, Mason RS. Osteoblasts play key roles in the mechanisms of action of strontium ranelate. Br J Pharmacol. 2009;157:1291–300.
66. Buehler J, Chappuis P, Saffar JL, Tsouderos Y, Vignery A. Strontium ranelate inhibits bone resorption while maintaining bone formation in alveolar bone in monkeys (Macaca fascicularis). Bone. 2001;29:176–9.
67. Canalis E, Hott M, Deloffre P, Tsouderos Y, Marie PJ. The divalent strontium salt S12911 enhances bone cell replication and bone formation *in vitro*. Bone. 1996;18:517–23.
68. Fromigue O, Hay E, Barbara A, Petrel C, Traiffort E, Ruat M, et al. Calcium sensing receptor-dependent and receptor-independent activation of osteoblast replication and survival by strontium ranelate. J Cell Mol Med. 2009;13:2189–99.
69. Basu B, Sabareeswaran A, Shenoy SJ. Biocompatibility property of 100% strontium substituted $4.2SiO_2$-$3Al_2O_3$-$1.5P_2O_5$-3CaO-$2CaF_2$ glass ceramics over 26 weeks of implantation in rabbit model: Histology and micro-Computed Tomography analysis. J Biomed Mater Res B. 2015;103:1168–1179.

Chapter 5
Microstructure and Composition Dependent Physical and Cytocompatibility Property of Glass-Ceramics for Dental Restoration

5.1 Background—Materials for Dental Restorations

In last few decades, glass ceramics (GC) are being developed for repair and replacement of damaged tissue [1, 2]. In the development of GC, the base glass is heat treated in a desired temperature—time window and crystals of various morphology and amount can form. Because of the presence of crystalline phase, glass ceramics have more favorable mechanical and biological properties than pure glass. Numerous clinical trials have shown intergrowth between glass-ceramics and human bone [3]. The important clinical applications of glass ceramic are related to the repair of skeletal system, which is composed of bone, joints and teeth. In order to achieve better combination of physical and biological properties, a number of researchers attempted to tailor composition and properties in various GC systems, which include SiO_2–CaO–Na_2O–P_2O_5, SiO_2–CaO–MgO–P_2O_5–F, SiO_2–Al_2O_3–MgO–CaO–Na_2O–K_2O–P_2O_5–F. An important group of these materials i.e. mica containing glass ceramics received wider application due to their high machinability, which is attributed to the unique microstructure of an interlocking array of plate-like mica crystals, dispersed throughout a glassy matrix [4, 5].

Hench's study first demonstrates that bioactive glasses (45S5 series) and bioactive glass-ceramics belong to two different classes of bioactive implants, i.e. Class A and Class B, respectively, due to large differences in the rate of bone bonding [6]. Machinable bioactive glass-ceramics containing microcrystalline phases of mica and apatite can be considered as class B bioactive implants [7]. Also, machinable glass ceramics are potential materials for dental restorations due to possibility of achieving a combination of high machinability, matching mechanical properties with natural teeth, bioactivity and finally, good aesthetics [8–11]. In the development of bioactive materials, Bioglass [12, 13], Ceravita [14], Apatite–Wollastonite (AW) glass-ceramic [15, 16] and dense hydroxyapatite ceramic [17, 18] have been well characterized for clinical applications. These materials either have a close chemical resemblance to bone mineral, or they provide an intimate contact with the

B. Basu and S. Ghosh, *Biomaterials for Musculoskeletal Regeneration*,
Indian Institute of Metals Series, DOI 10.1007/978-981-10-3017-8_5

surrounding tissue. However, owing to their brittle nature, their applications are limited to cases, where the mechanical stresses, particularly under low tensile stress. Therefore, a better combination of high strength, hardness, appropriate E-modulus and good *in vitro* biocompatibility property is an essential requirement for biomaterial; not only because they ensure reliability and long lifetime, but also because they are the indispensable factors that govern the performance and osteoconduction [19] between the implant and living tissue, when used under load-bearing conditions *in vivo*. Among the bioactive materials mentioned above, A-W glass-ceramic has been reported to have the best mechanical properties and also shows excellent bioactivity in prosthetic applications [20–22]. Quinn et al. [23] reported the influence of microstructure and chemistry on the fracture toughness of various ceramics. However, all the materials mentioned above are not machinable and they do not have a good combination of mechanical properties. To this end, it is easier to process mica-based glass-ceramics with good mechanical properties into surgical parts with various complex shapes by using normal clinical machining methods [24].

Like other glass-ceramics, the properties of machinable glass-ceramics depend on the type, sizes and volume fraction of the primary crystalline phases [7]. Also, the microstructures as well as properties are sensitive towards base glass composition (e.g. fluorine content). The presence of fluorine in the base glass enhances phase separation, and phase separation usually precedes the crystallization. In the glass system of the MAS type (MAS: $MgO \cdot Al_2O_3 \cdot SiO_2$), fluorine has the role of a nucleating agent [25]. Radonjić and his co-workers [25] demonstrated that when fluorine content in glass is high, nucleation starts at relatively lower temperature. Cheng et al. [26] also reported that fluorine source has an effect on the crystallization behavior of the glass ceramics (GC) system. They established that with the NaF as the source of fluorine, only surface crystallization occurs. In contrast, with MgF_2 as the source of fluorine, uniform bulk crystals can form.

Apart from crystallization/microstructure development aspect, the *in vitro* dissolution/leaching of glass ceramics also depends on glass composition or crystal morphology. In a study on the dry unlubricated sliding behavior, Park et al. [27] demonstrated the dominating influence of the amount and orientation of wollastonite phase on the hardness and wear of the Apatite–Wollastanite (A–W) GC. The wear resistance of the material decreased as the wollastonite amount is decreased. In different work, Xiao et al. observed that the frictional properties of CaO–MgO–Al_2O_3–SiO_2 (CMAS) glass-ceramics were related to the contact pressure and sliding speed, while plastic deformation and recrystallization were identified as dominant wear mechanisms during sliding tests in dry conditions [28]. The tribological study with Dicor glass-ceramic against alumina revealed higher COF of 0.8 in dry conditions [29]. Based on the Hertzian contact studies, it was argued that the wear of mica-containing Macor GC is controlled by the short-crack toughness as well as by the size and volume fraction of mica plates [29].

GC is also a good candidate material for the dental implant, and the primary requirement for successful use of GC for dental applications is the anchoring and attachment with bone tissue. Therefore, it is necessary to use different kinds of connective tissue cell line like, L929 fibroblast and osteoblast like cell Saos-2 and such cells must be interfaced with the implant material. Harrison et al. studied the differentiation of embryonic stem cell (ES) on hydroxyfluorapatite [11]. It was also reported that fluroapatite mullite coating on biomaterial surface has capacity of osteoblast differentiation [30]. *in vitro* study of cell viability/proliferation on the surface of a biomaterial is an indicator of biocompatibility and can be evaluated by MTT assay and cell adhesion test [11, 31].

Liu and co-workers studied behavior of mica/apatite glass-ceramics in simulated body fluid (SBF) solution and found good bioactivity [32]. *in vitro* bioactivity of wollastonite glass-ceramic materials were carried out by Alemany et al. and the formation of Hydroxyapatite (HA) layer on the material surface was observed after immersion in SBF [33]. The HA forming ability of $CaMgSi_2O_6$–Ca_5 $(PO4)_3F$–$CaAl_2SiO_6$ glass ceramics has also been reported by Salama and co-workers [34]. Verne et al. found that antimicrobial effect of Ag based bioglass is due to leaching of Ag from the glass matrix, change in pH and ionic strength. Brook et al. studied the *in vitro* effect of leached ions on viability of cells and conclude that fluoride ions exhibited mild toxicity effect on the osteoblast and fibroblast cells [35]. Moreover, leached ions significantly affect the microbial growth and adhesion on sample surface. Soljanto et al. concluded that bacterial growth inhibition depends on the concentration and size of the material, such as fluorapatite and glass beads etc., in the culture medium [36].

The present chapter is a summary of our ongoing research in the area of glass ceramics. In our recent work, the development of some unusual spherulitic-dendritic crystals as well as their *in vitro* dissolution properties are reported. One of the sensitive element in K_2O–B_2O_3–Al_2O_3–SiO_2–MgO–F glass compositions is fluorine and therefore, this present study investigates the aspects of microstructure development as well as properties when fluorine content in base glass is varied over a narrow range. This chapter will demonstrate the following aspects: (a) how the characteristically different crystal morphologies can be developed with variation in F^- content in the narrow window of 1–4 %, (b) how one can develop GC with superior combination of hardness, strength and E-modulus by designing both base glass combination and heat treatment condition and (c) *in vitro* dissolution and Ca–P rich layer formation and *in vitro* cytocompatibility.

Three compositions with varying fluorine and B_2O_3 content in the K_2O–MgO–Al_2O_3–B_2O_3–SiO_2–F were investigated in author's research group and discussed in this chapter. (Table 5.1).

Table 5.1 Compositions of base glass (in wt%) [9]

Starting materials	Precursor constituent	M1	M2	M3
Qurtz powder	SiO_2	47.98	48.94	42.57
White tabular alumina	Al_2O_3	17.62	16.29	17.81
MgO powder	MgO	19.36	17.45	18.80
K_2CO_3	K_2O	8.25	7.15	7.81
Boric acid (H_3BO_3)	B_2O_3	5.17	5.25	10.02
NH_4F/MgF_2	F^-	1.08	3.85	2.53

5.2 Microstructure-Mechanical Property Correlation

Glass batches with desired compositions were mixed using agate and mortar and thereafter, melted in a platinum crucible at 1550 °C for 2 h using electrical furnace. The glass melts were cast into a cast iron mould to form plates, and then annealed for 2 h in the temperature range of 600–650 °C and such annealing conditions were selected in order to avoid nucleation at this stage. Two-stage heat-treatment has been done to crystallize the glass and to convert it into glass ceramics. Nucleation was done in the temperature range of 750–850 °C. The crystallization experiments were conducted in the temperature range of 1000–1120 °C, employing varying time of holding from 4–24 h. XRD analysis of the temperature variation batches (heat treated for 12 h in temperature range of 1000–1120 °C) for the M1 sample revealed predominant presence of fluorophlogopite in all the samples. Variation in peak intensity of various crystalline phases was observed with different temperatures during crystallization heat-treatment of sample M1 (Fig. 5.1). Similar results were also obtained for M2 and M3 glass ceramics, heat treated at 1040 °C (not shown). A closer observation and comparison of XRD spectra in Fig. 5.1 suggest that the large diffuse and broad peak of base glass, as noticed in the 2θ range of 15–35°, has been decomposed to a number of sharp peaks with varying intensity. Such peaks were identified as fluorophlogopite peaks. In addition to such observation, X-ray peak at lower $2\theta = 9°$ is recorded in M1 GC compositions. Besides this, many peaks, identified for Fluorophlogopite (FPP) phases have also been detected in the higher 2θ angle range (See Fig. 5.1).

The detailed SEM analysis was carried out to study crystal shapes, size and size distribution in various heat-treated glass-ceramic samples of varying composition. In the following, some characteristic crystal morphology for some selected samples has been discussed to illustrate the influence of base glass composition and/or heat treatment conditions on crystallization behavior. In Fig. 5.2, SEM images of M1 glass-ceramic, crystallized at 1040 °C for 12 h are provided. The presence of large rod like crystals, aligned in specific directions can be clearly observed in Fig. 5.2a, b. In Fig. 5.2a, the presence of residual glass can also be observed. In addition, the boundary between two microstructural regions having rod like crystals, aligned in two different orientations has also been identified by dotted line in Fig. 5.2a. From Fig. 5.2b, it is clear that the characteristic crystals have 3–5 μm in width and also a large aspect ratio. EDS analysis of the residual glass in Fig. 5.2a indicates strong O-peak, as well as peaks of Si and Mg of equal intensity. In contrast, the EDS analysis of the crystal show Mg-peaks of highest intensity (Fig. 5.2c), followed by Si and Al (both having equal intensity). Clearly, the relative intensity of O^- peak from crystals is much less than that of residual glass. From such observations, it should be clear that the crystals are relatively Mg rich. SEM observation of Fig. 5.2b reveals that crystals are unidirectionally oriented and glassy droplets are dispersed on crystals. EDS analysis of the glassy phase demonstrates characteristic stronger peaks of oxygen, with Si and Mg peaks of comparable intensity. In Fig. 5.2d, SEM images of the sample reveal the presence of crystal domain boundary, which are formed by

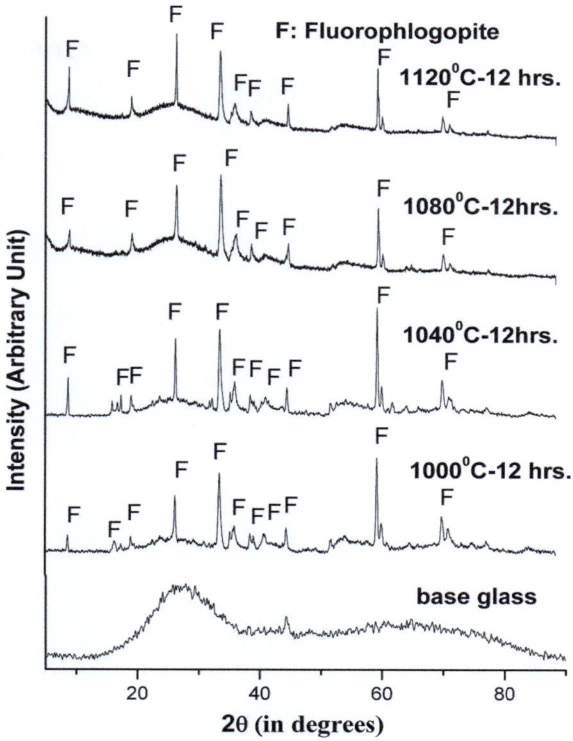

Fig. 5.1 XRD plot of the heat-treated M1 glass ceramic samples, crystallized at varying temperatures with a constant time of 12 h [9]

the intersection of crystals growing from different directions. These crystal domain boundaries are distinctly visible in Fig. 5.2d and the boundary between different crystals is marked with dotted line.

As far as the crystallization is concerned, highly oriented unidirectional crystals were obtained for the M1 samples, under all heat-treatment conditions. The analysis of the chemical composition of the crystallized M1 samples reveals lower content of fluorine (1.08 %), as compared to other glass samples of M2 and M3. Fluorine acts as the nucleating agent in the investigated system. As the composition of the glass samples of M1, M2 and M3 is suitable for FFP crystal formation, it is expected that the phase-separated glass has grown to form fluorophlogopite crystals. With reference to the sheet like structure of FPP, the composition of the system was selected in such a manner to create deficiency of cross-linking species potassium, for facilitating unidirectional growth. With less amount of nucleating species F^- and therefore, having very less number of nuclei sites in the phase-separated glass, the unidirectional growth mechanism of crystals is enhanced. In glass ceramics based on M1 composition, the unidirectional growth of crystals can be related to the inherent crystal structure and base glass compositions.

In the following, the kinetics of crystallization behavior is analyzed. The crystal volume fraction were calculated from the SEM images of all the samples using image analysis and adopting basic stereological concepts (point counting method on

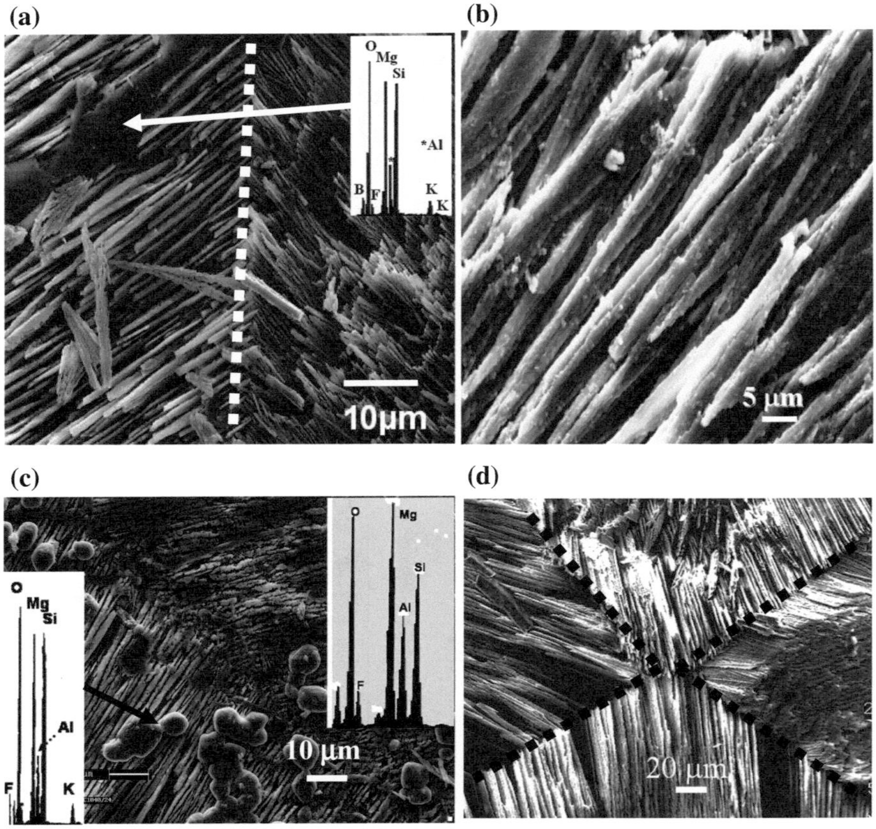

Fig. 5.2 SEM micrographs of M1 glass ceramics crystallized at 1040 °C for 12 h. **a** Oriented mica crystals meeting at a line to form a crystal domain boundary. **b** Orientation of Individual mica rods of diameter about 2–3 μm. **c** Oriented crystals of different morphology meeting at a domain boundary. **d** Highly oriented crystals are meeting at a point forming envelope like crystal morphology. The crystal domain boundaries are identified by the *dotted lines* in (**a**) and (**d**) [9]

a grid paper superimposed on a number of SEM images), the data reveal that M1 sample heat treated at 1040 °C for 12 h has the maximum average crystal volume of 85 % and M2 sample heat treated at 1040 °C for 12 h has the average crystal content of 71 % and M3 sample heat treated at 1040 °C for 12 h has average 75 % crystals. Most important findings from the present observation are that with an increase in fluorine content in the glass composition, the amount of crystal content decreases. The variation of crystal content with duration of crystallization time for M1, M2 and M2 sample has been provided in Table 5.2.

Among many theoretical models, the most widely used John-Mehl-Avarami (JMA) equation [37, 38] has been used to determine the activation energy for isothermal crystallization of mica glass ceramics:

Table 5.2 Variation in crystalline content (measured using conventional stereology technique) obtained with different crystallization heat treatment experiments [9]

Sample	Heat treatment condition	Crystal volume fraction (%)
M1	1040 °C, 12 h	85
	1040 °C, 6 h	75
M2	1040 °C, 12 h	71
	1040 °C, 6 h	58
M3	1040 °C, 12 h	75
	1040 °C, 6 h	55

$$-\ln(1-x) = (Kt)^n \quad (5.1)$$

where x is the crystallized fraction, t is the crystallization time, n is Avarami index; and K is a constant, which can be expressed by Arhenius equation,

$$K = \upsilon \exp(-E_c/RT) \quad (5.2)$$

where T is the absolute temperature of crystallization, E_c is the activation energy of the crystallization and υ is the frequency factor, assumed to be as $2.88 \times 10^{11}\ s^{-1}$ (Table 5.3).

Based on the experimental results provided in Table 5.2, the Avarami index (n) was found to be 1.39 for M1 samples, 0.51 for M2 samples and 0.80 for M3 samples. It can be mentioned here that much higher value of n = 3.4 is reported for nucleation and crystallization of tetrasilisic fluormica crystals in Dicor glass ceramics. Additionally, the activation energies were found to be 272, 403 and 401 kJ/mol for M1, M2 and M3 samples, respectively. Such values are higher compared to the activation energies of Dicor glass ceramics (203 kJ/mol). Among the three samples, for M1 sample, Avarami index was maximum and also the activation energy for crystallization was lower than that of other two samples. This indicates that the rate of growth is higher than the other two samples. The activation energies for crystallization are comparable for M2 and M3 samples, despite differences in base glass composition.

The hardness of the heat-treated glass ceramic samples is measured by acquiring micro indentation at an indent load of 40 gm and the average diagonal length of the hardness impression is calculated. For selected glass-ceramic samples i.e. M1, M2 and

Table 5.3 Composition of Artificial Saliva. AR means analytical reagent grade [9]

Material	Amount (gm) per 1 L of distilled water	Comment
NaCl	0.4	99.9 % pure, AR grade
KCl	0.4	99.5 % pure, AR grade
$CaCl_2 \cdot 2H_2O$	0.795	99.5 % pure, AR grade
$NaH_2PO_4 \cdot 2H_2O$	0.78	99 % pure, AR grade
$Na_2S \cdot 9H_2O$	0.005	99.9 % pure, AR grade
Urea	1	99.5 % pure, AR grade

M3 composition, heat treated at 1040 °C for 12 h, the micro hardness measurements were carried out and the hardness as well as crystal volume fractions are plotted together in Fig. 5.3a. The SEM image of indent impression was taken on M1 sample with 40 gm load has been provided in Fig. 5.3b. SEM image reveals clear indent impression on the sample surface and the propagation of radial cracks from indent corners. The maximum hardness value of 8.2 GPa is found for the M1 sample. Whereas, maximum hardness values of the other two samples M2 and M3 were found to be 6.5 and 4.9 GPa, respectively. These hardness values are well above the previously reported hardness of Macor, which is not more than 2–3 GPa [19]. As expected, highest hardness is measured for GC with maximum crystal volume fraction, which is in agreement with the findings of the previous researches [33]. Other

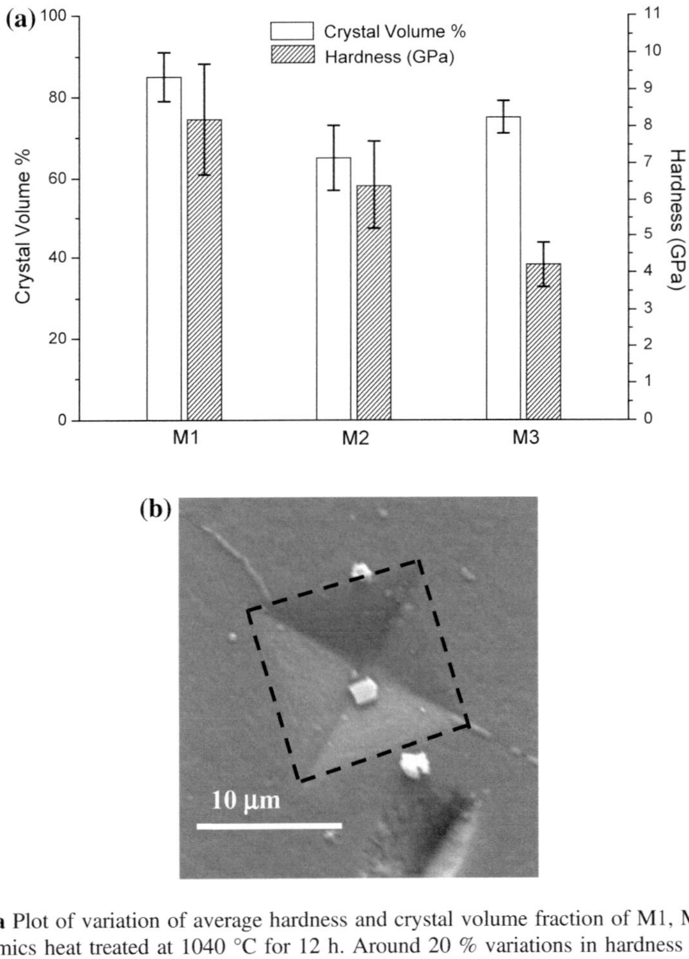

Fig. 5.3 **a** Plot of variation of average hardness and crystal volume fraction of M1, M2 and M3 glass-ceramics heat treated at 1040 °C for 12 h. Around 20 % variations in hardness value were measured and **b** SEM image of indent impression on M1 sample with 40 gm load. The *dotted lines* in (**b**) indicate the indent edges [9]

than the amount of crystal present, the aspect ratio and interlocking between the crystals also play an important role in determining the hardness. Crystals with larger aspect ratio i.e. longer crystals are able to form an effective interlocking [33]. More amount of interlocking in the glass ceramic material actually helps in stopping the cracks more effectively, thereby enhancing the plastic deformation of the material. In the present case, M1 sample, due to the presence of highly oriented crystals with higher aspect ratio compared to the M2 and M3 samples, exhibit high hardness. For M2 and M3 samples, hardness values do not commensurate well with the crystal volume fraction. This is because of the fact that the crystal morphologies in these glass ceramics are different from M1 glass ceramic.

The basic mechanical properties i.e. flexural strength (3-point) and E-modulus of selected samples (M1 and M2, both heat treated at 1040 °C for 12 h) were measured. The experimental results reveal that a combination of relatively higher strength (94 MPa) and lower E-modulus (57 MPa) could be obtained in M1 samples. In contrast, a little lower strength of 80 GPa with better E-modulus property (69 GPa) was measured with M2 sample (see Table 5.4).

As mentioned in Sect. 3.2, the evaluation of basic mechanical property reveals better hardness and strength property of M1 GC compared to that of M2 GC (see also Table 5.2). From the microstructure-mechanical property correlation point of view, it appears that higher volume fraction of crystalline phase and highly oriented 'envelop' like crystals are beneficial to obtain such better mechanical properties. However, E-modulus of M1 GC is lower than that of M2 GC. In many of the published research articles on GC, such a comprehensive measurement of all basic mechanical properties is not presented. Nevertheless, some published data on

Table 5.4 A comparison of mechanical property with earlier developed glass-ceramics/bioglass [9]

Material	Hardness (GPa)	Strength	E-modulus (GPa)
Mica-apatite glass-ceramics	2.8–4.8	–	–
Macor GC	~3	–	–
SiO_2–Al_2O_3–MgF_2 (MAS)	4–7	–	–
MgO–Al_2O_3–SiO_2–TiO_2	9.5	–	–
Dicor	3.5	127 (biaxial)	66.9 GPa
MGC glass ceramics	–	–	49.5–70.5
Dental enamel	3.4–4.6		66.2–91.1
45S5 bioglass	–	40–60 (bending)	30–50
A/W glass-ceramic	–	215 (bending)	35
KGy213 ceravital	–		77–88
M1	8.2 ± 1.5	94.9 ± 14.0 (3 point flexural)	57.6 ± 2.8
M2	6.4 ± 1.2	80.6 ± 7.7 (3 point flexural)	69.7 ± 2.9

hardness or E-modulus have been collected and a comparison with mechanical properties of various other glass-ceramics is made in Table 5.4. The properties of Dicor glass ceramics (commercially produced dental implant material) and natural enamel has also been compared with the glass ceramics developed in the present study. From the data provided in the Table 5.4, it is quite clear that strength and E-modulus properties obtained in our system is comparable with both the Dicor and dental enamel. However, hardness property measured in our system is much higher i.e. almost double than that of the Dicor/enamel. It is also clear that the mechanical property (flexural strength, E-modulus) measured in the present system can be compared with other glass ceramics/Bioglass (from Table 5.4). But little higher hardness of 9.5 GPa has been reported only for $MgO–Al_2O_3–SiO_2–TiO_2$ (MAST) system. The major crystalline phases in those glass-ceramics include Mg–Alumino silicate, Mg–Alumino titanate etc., and in the presence of such phases in the MAST system, machinability is expected to be very poor. Other mechanical property data is not reported for MAST system. However, comparative hardness in the range of 4–7 GPa were measured by Goswami et al. [39] in magnesium aluminium silicate glass-ceramics system. From the above discussion, it should be clear that the heat treatment of $K_2O–B_2O_3–Al_2O_3–SiO_2–MgO–F$ glass ceramics composition at 1040 °C for 12 h, in the present study, enables us to obtain glass-ceramics with an exceptionally higher hardness of around 8.2 GPa and better combination of strength and E-modulus.

Summarizing, it has been demonstrated here that how higher crystal content and envelop-like crystal morphology can attribute to the high hardness and mechanical properties in the present machinable glass ceramics system. Our observations also reconfirm earlier literature results that the mechanical properties of glass-ceramics are not only dependent on the crystalline phase and their volume fraction, but also highly related to the spatial arrangement of the crystals.

5.3 Biomineralisation in Artificial Saliva

The *in vitro* dissolution of the heat-treated samples was assessed in artificial saliva. M2 samples (heat treated at 1040 °C for 12 h) were selected for *in vitro* dissolution test for various reasons (see Table 5.1 for composition). Firstly, this glass ceramic sample contains least crystal volume fraction of about 71 % and lesser amount of crystal content in a glass ceramic has been reported to give better dissolution property. Secondly, its hardness value was moderate (6.4 GPa) compared to human dental enamel (4 GPa) and therefore is more suited for dental restoration purpose. SEM-EDS analysis of the surface of the sample immersed in AS for different time periods of 2, 3, 6 and 8 weeks were carried out in order to assess the extent of Ca–P layer formation on the surface (Fig. 5.4). The observation of the SEM images in Fig. 5.4a reveals the formation of thin surface layers after testing for 3 weeks. With increase in further leaching time for 6 weeks, spherical particles of diameter 2–3 μm were observed to disperse densely and uniformly on the surface (Fig. 5.4b). EDS

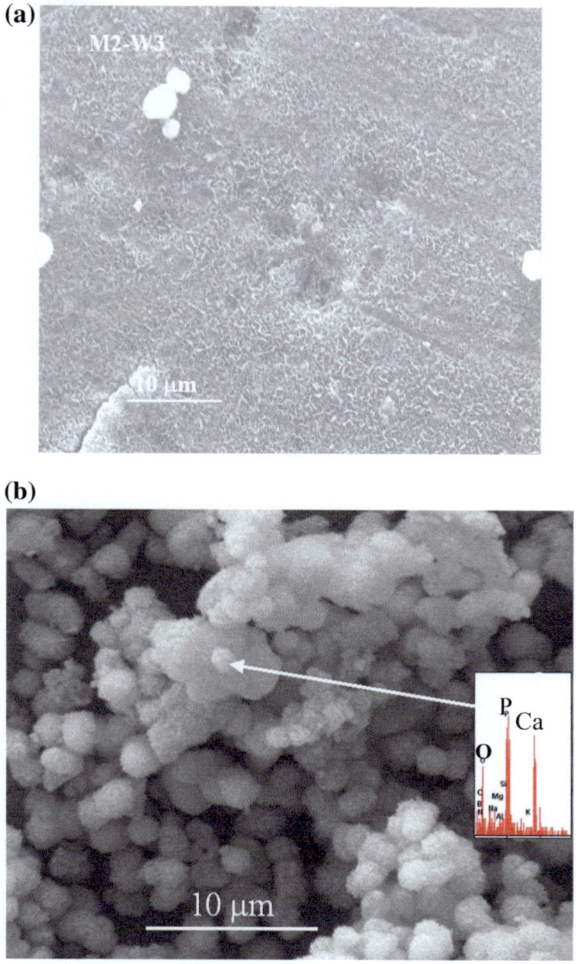

Fig. 5.4 SEM observation of glass-ceramic surface (M2 sample, heat treated at 1040 °C for 12 h) after immersing in AS solution for **a** 3 weeks and **b** 6 weeks. EDS analysis of the brighter particles in (**b**) has been provided as *inset* [9]

analysis of spherical particles indicates the strong peaks of Ca, P and O, thereby confirming the formation of Ca–P compounds. Further, the analysis of EDS data indicates that the composition of those spherical brighter contrasting particles, which cover the majority of surface area, can be characterized by Ca–P ratio of 1.57 (average). Ca–P compounds are highly bioactive and therefore, the formation of these layers on the glass ceramics sample is a good indication of the bioactive nature of the developed glass ceramic.

5.4 Wear Resistance in Artificial Saliva

The tribological experiments were performed using a computer-controlled fretting machine, which produces linear relative oscillating motion with ball-on-flat configuration. The fretting tests were carried out in ambient conditions in air without using any lubricant (dry conditions), and using AS medium with selected parameters of 1 N load, 8 Hz oscillation frequency and 100 μm (gross slip regime) linear stroke length. The composition of AS is given in Table 5.3. The experiments were done for different test durations of 5000, 10,000, 50,000 and 100,000 cycles.

The wear rate data for the investigated GC materials is presented in Fig. 5.5. Irrespective of environment, the wear rate systematically decreases with the fretting test duration. It appears that the wear rate for the investigated glass ceramics in the selected tribological environment varies in the range of 10^{-4}–10^{-5} mm^3/Nm. Considering the application of extremely low load (1 N) and also low sliding velocity (much less than 0.1 m/s), the wear rate of the investigated GC against steel is one order of magnitude higher than that of various structural ceramic materials (typical wear rate $\sim 10^{-6}$ mm^3/Nm), investigated earlier in our group [40]. Such difference is obvious to expect because of much lower hardness of the presently investigated GC materials. However, when compared with other GC materials, the wear rate of the investigated GC is at least one order of magnitude lower. Further, it is evident from Fig. 5.5 that wear rate is less in AS medium than in dry conditions at any given test duration. The wear rate decreases systematically from 3.5×10^{-4} to 1.2×10^{-5} mm^3/Nm during fretting of GC in dry or AS environment. After fretting for 100,000 cycles, a maximum wear rate of 9×10^{-5} mm^3/Nm is measured for GC in dry conditions, while a minimum wear rate of 1×10^{-5} mm^3/Nm for GC in AS medium. From the above observation of lower magnitude of wear rate and decreasing trend in wear rate with test duration, longer durability or better wear

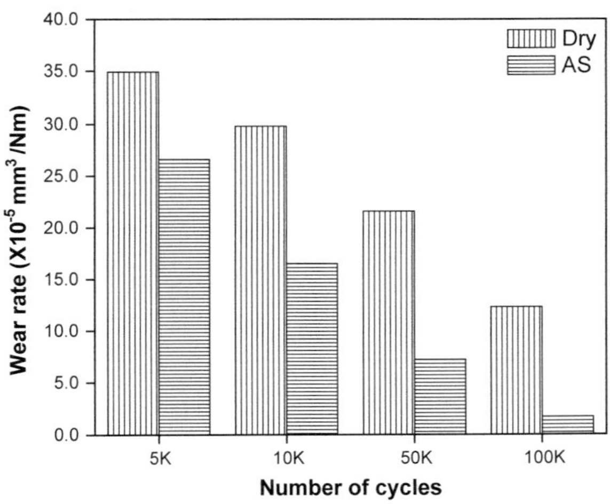

Fig. 5.5 Wear rate of glass ceramics after fretting in dry and AS medium for different time duration. Around 10–15 % deviation around the reported data were measured in our experimental results [11]

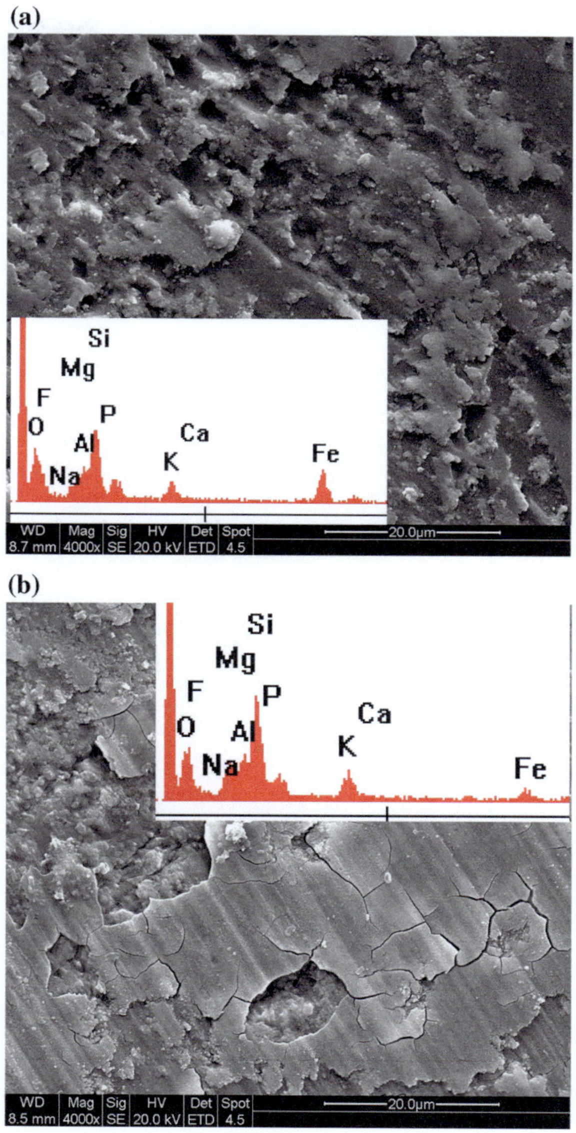

Fig. 5.6 SEM image illustrating the overall fretting damage experienced by glass-ceramic plate, after tested against steel ball at 1 N load for 100,000 cycles, 8 Hz, 100 μm stroke length in **a** dry and **b** AS medium. The *doubly pointed arrows* indicate fretting directions [11]

resistance during long-term application of the presently investigated GC could be expected.

In order to understand the nature of dominant mechanisms during the wear process, detailed SEM-EDS analysis of worn surfaces of investigated GC samples was conducted after fretting against steel in different environment. Some representative SEM images of the worn surfaces at different stages (number of cycles) of fretting are presented in Fig. 5.6.

The severity of the pull-outs is further increased with increase in number of cycles (100 k) in dry conditions (Fig. 5.6). Further, Fig. 5.6b reveals the formation of thin layer fragments and their smearing on the worn surface after 100 k cycles. The characteristics of the worn surface after 100 k cycles in AS medium include the formation of potential tribolayer protecting the base surface, the dispersion of fine debris around the periphery of the wear scar and the brittle nature of the tribolayer with number of cracks. In addition, the tribochemical layer is observed to have fine grooves in the fretting direction, indicating mild abrasion due to the sliding of hard debris particles (see Fig. 5.6d).

Although the friction data and details of wear mechanisms at different timepoints of the interrupted fretting tests are not discussed in this chapter, a comparison of such properties with that of other glass-ceramics/bioglass is provided in Table 5.5. Looking at this table, it is apparent that the frictional properties of M2 grade glass-ceramics is comparable with that of the natural tooth, when both can be slided against steel counterbody. As far as wear resistance is concerned, the M2 grade glass-ceramics has a lower wear rate than that of Dicor, although a difference in counterbody needs to be considered (see Table 5.5).

Table 5.5 Summary of the tribology test results obtained with some of earlier developed glass ceramics as well as human teeth and comparison with the presently investigated GC. The variation in friction/wear rate depends on the variation in operating conditions [11]

Tribocouple	Operating conditions	COF	Wear rate (mm^3/N/m)	Wear mechanisms
Human teeth versus steel	20 N, dry/AS 0.002 m/s	0.8–1.2 (dry) 1.0 (AS)	–	Oxidative wear and microfracture
Human teeth versus Al_2O_3	1 N, AS; 0.0005 m/s, 8,000 cycles	0.12–0.55	–	Fretting fatigue; adhesive wear
Dicor versus Al_2O_3	4.9 N, 0.0014 m/s, dry	0.7–0.077	$2.6*10^{-3}$	Microfracture
Dicor versus Al_2O_3	1 N, 0.0025 m/s, distilled water	0.4–0.6	10^{-3} to 10^{-4}	Localised fracture
$CaO–MgO–Al_2O_3–SiO_2$ (self-mated)	0.01–0.5 m/s; dry; contact pressure- 0.1–1.4 MPa	0.05–0.65	10^{-3} to 10^{-4}	Microcracking, abrasion
$MgO–CaO–SiO_2$ $P_2O_5–F$ versus ZrO_2	10 N, 0.025 m/s, dry	0.75	$0.7*10^{-4}$	Abrasive and adhesive wear
$K_2O–B_2O_3–Al_2O_3–SiO_2–MgO–F$ (M2 grade) versus steel	1 N, 0.0016 m/s, dry/AS; 100,000 cycles	0.88 (dry); 0.67 (AS)	$12*10^{-5}$ (dry); $2*10^{-5}$ (AS)	Tribomechanical wear (dry); tribochemical wear (AS)

5.5 Cell Proliferation and Differentiation, *in vitro*

In this study, Mouse fibroblast (L929) and Human osteoblast like cells (Saos-2) cell lines were used for cell-culture experiments. Figure 5.7 presents the scanning electron microscopic evidences of the cellular adhesion of SaSO2 cells on control M1, M2 and M3 GC after 72 h of culture. It can be noted that osteoblast-like cells (SaOS2) is differentiated and specialized cell line. From Fig. 5.7, it is clear that osteoblast-like cell SaOS2 shows the characteristic morphology and adhesion on GC samples, which support the cell proliferation (a coordinated process of growth and division) and cell–cell contacts. It has been observed that cell proliferation and cell-to-cell contacts are more in M2 and M3 GC sample as compared to M1 sample. However, cell fattening with very little actin cytoskeleton protrusions were noticed on the M1. In contrast, cells were fully spread on M2 and M3 samples with multiple lamellae extended in all the directions. Alkaline phosphatase, an ectoenzyme

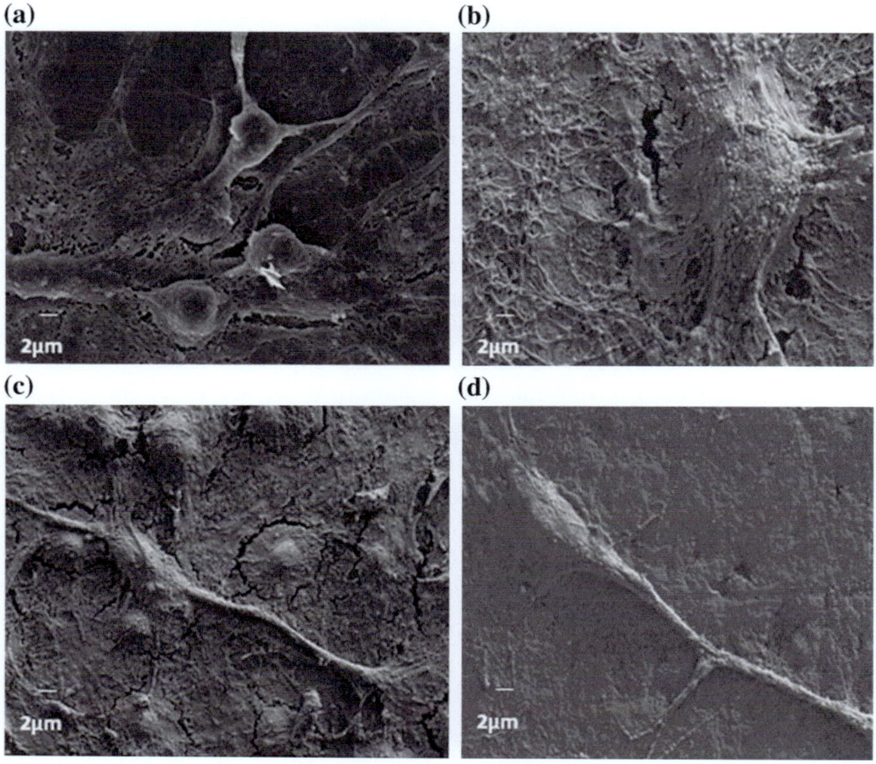

Fig. 5.7 SEM micrographs of adhesion and spreading behavior Saos-2 Osteosarcoma cell on **a** Control, **b** M1, **c** M2 and **d** M3 GC samples. The Dorsal ruffles, microspikes and filipodium are formed on the M1 surface, whereas M2 and M3 show Lamellipodia. The control material shows a normal osteoblast flattened morphology without any actin filament extension [10]

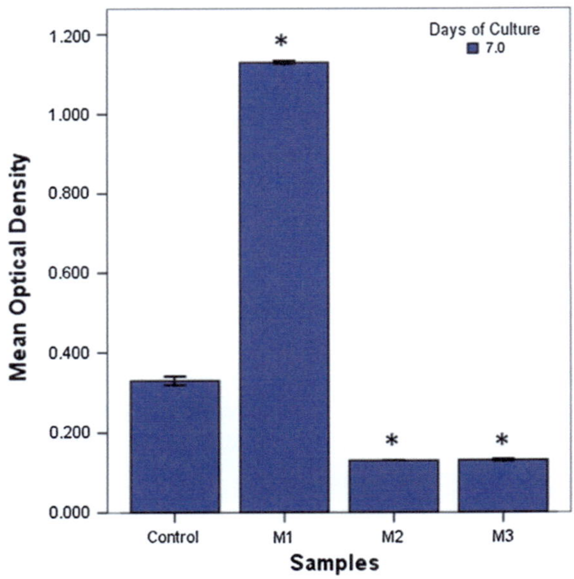

Fig. 5.8 Alkaline phosphatase (ALP) activity of the Saos2 cells on M1, M2 and M3 GC and control (Gelatin coated glass cover slip) after 7 days of culture in osteogenic medium. The cells were seeded initially at a density of $3*10^5$ cell/ml. * represents significant difference at $p < 0.05$ among sample sand error bars correspond to ±1.00 SE [10]

produced by osteoblast is involved in the degradation of inorganic pyrophosphate for mineralization to support the differentiation of cells.

The functionality of proliferated SaOS2 cells was further studied by examining the production of alkaline phosphatase. It needs to be mentioned here that Saos-2 cell line are already in differentiated stage and the purpose of carrying out ALP assay is to assess whether further differentiation of SaOS2 cell line is supported by investigated glass ceramics. It can be further noted that most of the SaOS2 cells of same passaging stage, having same differentiated stage, are seeded on all the three glass ceramic surfaces. Therefore, difference in ALP expression should primarily be attributed to difference in substrate composition or difference in ability of GC to promote further differentiation. The characteristics ALP activity of Saos-2 on control and the GC sample was assessed after 7 days of culture. M1 sample showed highest ALP activity and significant difference was noticed between the M2 and M3 sample with respect to control (see Fig. 5.8). The ALP activity decreased significantly in M2 and M3 sample, when compared to control sample. Based on the statistical significance analysis at $p < 0.05$, ALP activity on different GC can be ranked as M1 > M2 > M3.

5.6 Bactericidal Property, *in vitro*

The bacterial strain of *Staphylococcus epidermidis* and *Escherichia coli* were used to assess the *in vitro* antimicrobial properties of the glass ceramic samples. Since bacteria culture is not discussed in any details in a previous chapter related to biocompatibility, the bactericidal assay is discussed below in details. The above mentioned bacterial strains can be procured in the freeze dried condition from an authorized source, like the American Type Culture Collection (ATCC). Prior to seeding bacteria on the sample surface, they are to be inoculated in a Luria-Bertani broth, supplemented with yeast and beef extract, and culture was incubated overnight at 37 °C. In order to obtain pure culture, the streaking on agar plate was done with suspension culture to isolate the single colony. The single colony is to be picked from the agar plate and inoculates in the freshly prepared Luria-Bertani broth. The culture is then incubated over night for growth. The pure suspension culture of bacterial density of 1×10^7 CFU/ml/well can be used for seeding on the samples. The seeded samples are incubated and bacteria allowed to adhere up to 4 h. Subsequent to the incubation period, the surface is rinsed with PBS and then soaked in primary fixative of 3 % glutaraldehyde, 0.1 M sodium cacodylate, and 0.1 M sucrose for 45 min. The surface is washed twice with PBS buffer. The cells are dehydrated by replacing the buffer with ethanol series 30, 50, 70, 95 and 100 % for 10 min each and then, further dried by 100 % hexamethyldisilazane (HMDS) for 10 min. The dried samples are sputter coated with gold and examined under SEM. Turbidometric analysis is usually performed after 4 h of incubation. Followed by washing sample with PBS, each sample is placed in 3 ml volume of PBS solution and vortexed sufficiently to remove the adherent bacterial cell from the sample surface. Subsequently, the sample is removed from PBS and optical density (OD) of PBS suspensions is measured at 660 nm wavelength using UV-VIS spectrophotometer. The bacterial adhesion on the sample surface was expressed in terms of optical density (OD).

Figure 5.9 presents representative SEM images, revealing bacteria adhesion and antimicrobial properties against Gram-positive *S. epidermidis* bacteria after 4 h of incubation on the test samples *in vitro*. On the basis of the statistical analysis of the optical density results (Fig. 5.10) as well as SEM observation, M1 and M2 GC shows the bacteriostatic effect (adherence, but not killing of bacteria) against Gram positive bacteria, whereas M3 samples show bactericidal effect (killing of bacterial cells) compared to other two samples. SEM images also reveal noticeable difference with respect to control sample in terms of antibacterial property. It was clear that there was no colonization and biofilm formation of *S. epidemidis* on the surface of GC samples. After 4 h incubation of *S. epidemidis,* the number of bacterial colony forming unit adhered on control material was approximately $1*10^7$ CFU/ml. In contrast to control material, the number of CFU adhere on M1, M2 and M3 GC were approximately, $1.2*10^6$ CFU/ml, $1.38*10^6$ CFU/ml and $0.96*10^6$ CFU/ml, respectively. Overall, M3 sample showed highest antimicrobial activity with approximately 16 % of CFU/ml was observed as compared to the control material.

This observation can be attributed to the difference in composition of glass ceramics. From Table 5.1, it should be clear that M3 contain 2.53 % F^- and 42.57 % SiO_2; while M2 has 3.85 % F^- and 48.94 % SiO_2. Despite higher fluorine content, M2 has less quantified antimicrobial property (84 % CFU/ml in M3 vs. 77 % CFU/ml in M2) than M3. It appears that higher amount of SiO_2 in M2 suppresses beneficial effect of F^- as far as bactericidal effect is concerned.

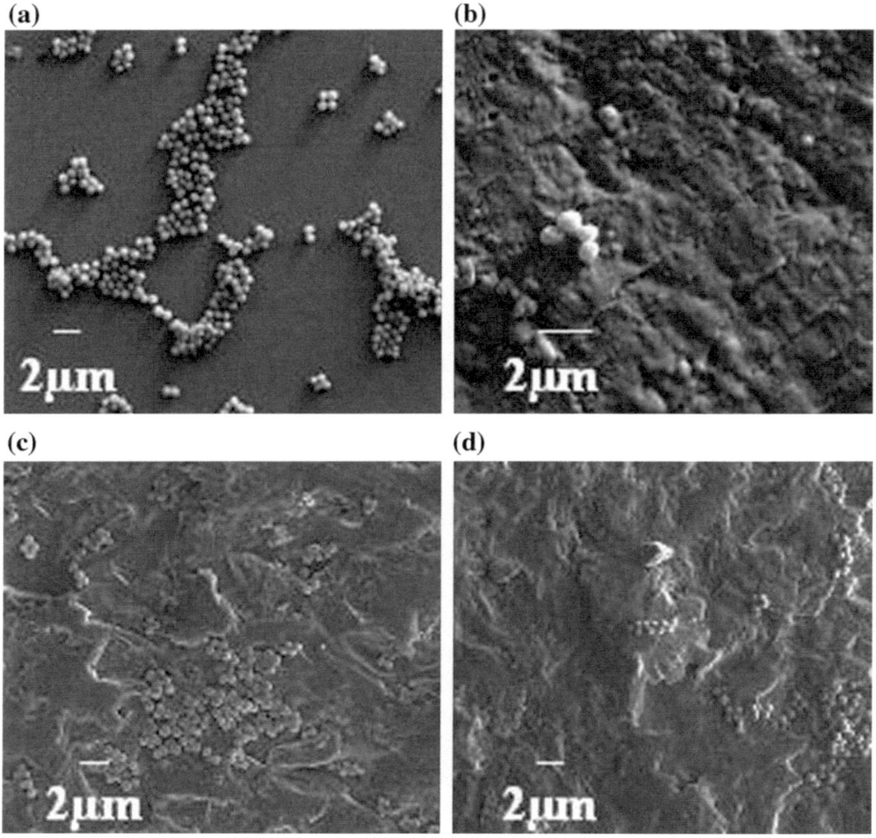

Fig. 5.9 SEM Electron Micrograph of the *Staphylococcus epidermidis* bacteria colony after 4 h of incubation on **a** Control, **b** M1, **c** M2 and **d** M3 glass ceramic samples [10]

It is reported that silicon oxide matrix containing any antimicrobial component demonstrates its antimicrobial activity against Gram-positive as well as Gram-negative bacterium. Coppelo and co-workers also observed that presence of 1 % of antimicrobial compound in silica matrix reduces 99 % colony-forming units (CFU) of Gram-negative bacterium [41]. However, a reduction in CFU/ml takes place for Gram-positive bacterium, when antimicrobial compound content was increased to 1.5 %.

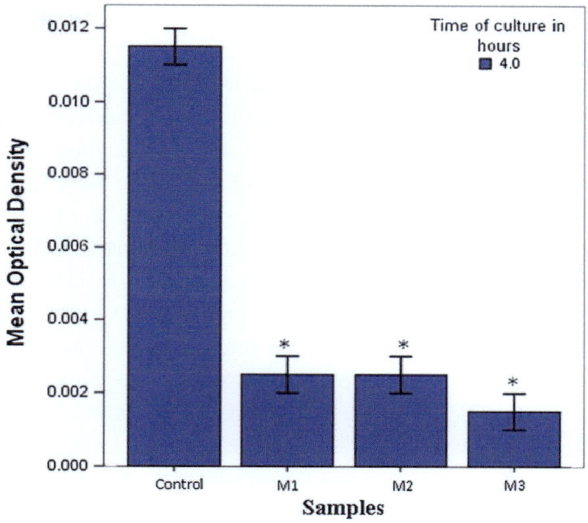

Fig. 5.10 Mean OD value after 4 h incubation of *Staphyloccous epidermidis* on control, M1, M2 and M3 GC. The Initial bacterial inoculum contains $1*10^7$ CFU/ml. * represents significant difference at among samples $p < 0.05$ with respect to control disc and Error bars correspond to ±1.00 SE [10]

5.7 Correlation Among Microstructure, Property and Cytocompatibility

As discussed in preceding sections, K_2O–B_2O_3–Al_2O_3–SiO_2–MgO–F based glass ceramic materials offer an excellent combination of hardness, E-modulus and strength property. It is known that the Dicor material which is being developed for dental restorative application exhibit a modest combination of hardness (350 MPa), E-modulus (66.2–91.1 GPa) and strength property (127 MPa). A comparison of presently developed material with Dicor and natural enamel is provided in Table 5.4. It should be understood that M1 or M2 glass ceramic can outperform Dicor.

In addition, the natural human cortical bone is characterized by a combination of mechanical property, depending on anatomical location. In particular, their E-modulus is 80 GPa or less and they have tensile strength values varying in the range of 60–160 MPa [42]. Although, the tensile strength is not measured in the present case, the tensile strength of the investigated glass ceramic should be less than the measured 3-point bending strength values (94 MPa). It is therefore, apparent that the developed glass ceramic can mimic the cortical bone properties.

More importantly, the *in vitro* dissolution experiment also proves the occurrence of Ca–P layer formation for M2 GC as well as modest level of metal ion dissolution at the ppm level. The biomineralisation capability *in vitro* is expected to enhance the bone formation capability of the M2 glass ceramic *in vivo*. From the above observations, the M2 glass ceramics can be considered as a potential bone replacement material, in particular for dental restorative application. However, further investigation needs to be carried out in future to characterise the optical translucency, machinability and long term durability property of the developed glass ceramics.

As far as the compositional aspect is concerned, the composition of base glass (not starting powder mix) was measured using ICP-AES and the results are presented in Table 5.1. It is clear that the base glass has SiO_2 as major constituent. As far as amount of B_2O_3 is concerned, M3 has little less than two times B_2O_3 than that of M1 and M2, which have almost similar B_2O_3 content. With respect to fluoride concentration, M1 has lowest and M2 has highest amount, while M3 has intermediate content. From the present experimental results, GC with lowest F^- content has the best cytocompatibility property.

As far as the cell-material interaction is concerned, initial attachment of cell is mediated by the integrins, a transmembrane protein which connect the extracellular matrix (ECM) to the cellular actin cytoskeleton through focal adhesions. It is known that the attachment of substrate-dependent cells, such as fibroblasts and osteoblast, to a substratum is a synchronized process, involving cytoskeleton reorganization, cell spreading, and formation of focal contact [43]. The cell adhesion of M1 samples shows extensive network of filopodium (Fig. 5.7). These actin cytoskeletons were helpful in sensing and exploring the local environment during the cell-biomaterial interaction and useful in maximizing the cell-cell contacts by polymerization of actin filaments. A maximum level of cell-cell contacts was observed in the M2 and M3 sample since long and directional lamellipodium were visible (see Fig. 5.7). The M2 and M3 glass ceramics exhibit intermediate level of actin filament extension and it helped in maximizing the cell-cell contacts by extending the 'dendritic' network of actin extension lamellipodium. In M1 sample, the cells are more flattened with low level of actin filament protrusion [44]. It is assumed that the actin cytoskeleton can be regulated by the ions released from the biomaterial that trigger signal transduction event and activate a family of protein that can help reorganization of cytoskeleton [45]. In particular, previous studies using NIH3T3 cells showed that the early spreading events are characterized by the extension of filipodium and cell-cell contacts by lamellipodium.

It has been reported that cytocompatlibilty property can be improved, when the B_2O_3 and F^- was added in an optimized concentration [46]. It has been observed that optimized concentration of B_2O_3, and F^- improve the osteoblast cell response and also exhibit bactericidal effect [47]. For instance, fluoride-containing HAp supports the cellular functionality. Alternatively, if B_2O_3 and F^- release is significantly higher, it shows the bactericidal effect [48]. The proliferating cell numbers obtained for the fluoridated GC samples (M1, M2 and M3) were greater than control glass cover slip, supporting the fact that fluoride has a stimulatory effect on cellular growth [49].

In our earlier study, we reported the *in vitro* dissolution characteristics of the presently investigated glass ceramics. One of the important results was the formation of CaP-like layer. It should be clear that such biomineralisation ability of the glass ceramic is one of the underlying reasons for exhibiting good cellular adhesion and supporting osteoconduction property of the investigated glass ceramic. As far as the release of different ions are concerned, *in vitro* dissolution results indicate that 0.16 ppm F^- ions are released from M2 glass ceramic over the period of one week (data not shown). The dissolution of K^+ and Mg^{2+} ions from M1 glass ceramic is much higher than M2 and M3, which have comparable metal ion leaching rate. For M1 sample, around 225 mg/l of K^+ ions and 4.75 mg/l of Mg^{2+} ions are leached over a period of 8 weeks. In view of the above information as well as considering that our cell culture experiments are conducted for less than a week, it is possible that metal ions/F^- ion are released although to lower extents, from M1, M2 or M3 glass ceramic during our *in vitro* testing. This is particularly true as leaching of ions from glass ceramic is kinetically driven process and thereby, is time dependent. From the above, it should be clear that lower F^- content and leaching of K^+ or Mg^{2+} ions is from M1 glass ceramic, both contribute to much better cytocompatibility.

Another factor is the difference in B_2O_3, a precursor for boric acid in cell culture media. The boron derivatives seem to be more active than boric acid with respect to the release of proteoglycans and proteins from fibroblasts, whereas boric acid is the most effective compound for release of proteoglycans from skeleton. In a different study, Fu et al. demonstrated that boron based bioactive glass ceramic scaffolds can exhibit significant proliferation of the MLO-A5 as well as on bone marrow stromal cells [50]. An important observation was that a threshold concentration of B_2O_3 is essential for the optimum cell growth and above a minimum threshold of B_2O_3 concentration, the cellular growth was inhibited. Based on our previous results of dissolution study, it can be said that M1 and M2 samples have good biocompatibility than those from M3 sample. It is also known that M1 and M2 samples contain 5.17 and 5.25 wt% B_2O_3, respectively. M3 sample with higher B_2O_3 content of 10.02 wt% shows minimum number of metabolically active cells (Fig. 5.8). Therefore, our results are in line with earlier observations that cellular functionality is inhibited at higher B_2O_3 concentration and the minimum threshold, as indicated by Fu et al. [35], should lie in the present case somewhere between 5 and 10 wt%.

It has been reported in literature that B_2O_3 on material surface or in solution stimulates the activity of intra and extracellular protease, which play a vital role in increasing the ECM and wound healing process. The ALP activity on GC is a phenotypic marker for early-stage differentiated osteoblasts. In M1 sample, the highest percent of viability of metabolically active cell is measured, when compared to other two compositions further corroborate highest ALP activity in the M1 sample. From MTT results of metabolically active cells, it is clear that after 2 days of culture, % viability decreased, due to leaching of F^- ions in culture media, which affects the cytocompatibility. The specific ALP enzyme activity was highest at the later stage of the cell growth. In view of the above observation as well as looking at Table 5.1, it appears that lower F^- content also favors cell viability and support osteoconduction property.

5.8 Summary

The influence of various ceramising treatments on the microstructure development, mechanical, biomineralisation, tribological and cytocompatibility of three different $K_2O–B_2O_3–Al_2O_3–SiO_2–MgO–F$ base glass compositions is discussed to illustrate how the concepts of material science can be integrated with relevant aspects of biological sciences to develop glass-ceramics for dental restorative applications. In particular, F^- content in base glass is varied in the range of 1.08–3.85 % and a set of heat treatment experiments were carried out at 1000–1120 °C for different time duration of 4–24 h. The variation in fluorine content has a marked influence on the crystallization kinetics as well as crystal morphology. The Avrami constant is highest (~ 1.4) and the activation energy of crystallization is around 272 kJ/mol for highly oriented and characteristic 'envelope' like crystals in glass ceramic with 1.08 % fluorine content, when heat treated at 1040 °C for 12 h. More importantly, an adherent Ca–P rich biomineralised layer, containing spherical particles of 2–3 μm size has been found to form *in vitro* on the glass ceramics surface (3.85 % fluorine), after dissolution in artificial saliva for 6 weeks. This observation indicates the biomineralisation capability of the investigated glass ceramic. Wear rate was experimentally measured to be varied in the order of 10^{-4}–10^{-5} mm^3/Nm and a systematic decrease in wear rate with test duration was recorded in both cases of testing medium. A comparison with the earlier developed glass-ceramics as well as commercial Dicor material reveals a better combination of the friction and wears resistance property of the investigated glass ceramic. One of the important results is that boron and fluorine component in the $SiO_2–MgO–Al_2O_3–K_2O–B_2O_3–F$ glass ceramics with fluorophologopite as major crystalline phase have significant influence on cytocompatibility, cell adhesion and cell attachment properties. The glass ceramic with lowest F^- content shows highest cell viability and also supports osteoconduction. The glass ceramics with highest B_2O_3 content (10.02 %) and moderate F^- (2.53 %) exhibits best antimicrobial property against both Gram negative (*E. coli*) and Gram positive (*S. epidermidis*) bacteria.

Based on the combination of mechanical and *in vitro* properties, the glass ceramics with highest amount of F^- content (3.85 %) has been proposed as a suitable material for dental restoration purpose. Also, the particular combination of E-modulus (69 GPa) and strength (80 MPa)/hardness (6.4 GPa) property indicates that this particular grade glass ceramic can be alternatively used for hard tissue replacement/bone analogue materials. However, further clinical trials are required to confirm such potential.

References

1. Kanchanarat N, Bandyopadhyay-Ghosh S, Reaney MI, Brook MI, Hatton VP. Microstructure and mechanical properties of fluorcanasite glass-ceramics for biomedical applications. J Mater Sci. 2008;43:759–65.

2. Gonzalez P, Serra J, Liste S, Chiuss S, Leon B, Perez-Amor M, Martinez-Fernandez J, de Arellano A, Lopez R, Varela-Feria FM. New biomorphic SiC ceramics coated with bioactive glass for biomedical applications. Biomaterials. 2003;24:4827–32.
3. Verne E, Ferraris S, Miola M, Fucale G, Maina G, Martinasso G, Canuto RA, Di-Nunzio S, Vitale-Brovarone C. Synthesis and characterisation of bioactive and antibacterial glass-ceramic part 1—microstructure, properties and biological behaviour. Adv Appl Ceram. 2008;107:234–44.
4. Uno T, Kasuga T, Nakayama S, Ikushima AJ. Microstructure of mica-based nanocomposite glass-ceramics. J Am Ceram Soc. 1993;76(2):539–41.
5. Baik DS, No KS, Chun JS. Mechanical properties of mica glass–ceramics. J Am Ceram Soc. 1995;78(5):1217–22.
6. Larry LH. Bioceramics: from concept to clinic. J Am Ceram Soc. 1991;74:1487–510.
7. Chen X, Hench LL, Greenspan D, Zhong J, Zhang X. Investigation on phase separation, nucleation and crystallization in bioactive glass-ceramics containing fluorophlogopite and fluorapatite. Ceram Int. 1998;24:401–10.
8. Roy S, Basu B. Hardness properties and microscopic investigation of crack-crystal interaction in SiO_2–MgO–Al_2O_3–K_2O–B_2O_3–F glass ceramic system. J Mater Sci Mater Med. 2010;21:109–22.
9. Molla AR, Basu B. Microstructure, mechanical and in vitro properties of Mica Glass-ceramics with varying fluorine content, J Mater Sci Mater Med. 2009; 4:869–882.
10. Kalmodia S, Molla AR Basu B. *In vitro* cellular adhesion and antimicrobial property of SiO_2-MgO-Al_2O_3-K_2O-B_2O_3-F glass ceramic, J Mater Sci Mater Med. 2010;21:1297–130.
11. Molla AR, Manoj Kumar BV, Basu B. Friction and wear mechanisms of K_2O-B_2O_3-Al_2O_3-SiO_2–MgO-F glass ceramics, J Eur Cer Soc. 2009;29:2481–2489.
12. Hench LL. Bioceramics: from concept to clinic. J Am Ceram Soc. 1991;74(7):1487–510.
13. Holland W, Vogel W. An introduction to bioceramics. Singapore: World Scientific Publishing Company Pvt. Ltd; 1993. p. 125–37.
14. Holand W, Beall G. Glass ceramic technology. Westerville: The American Ceramic Society; 2002.
15. Kokubo T, Ito S, Shigematsu M, Sakka S, Yamamuro T. Mechanical properties of a new type of apatite-containing glass-ceramic for prosthetic application. J Mater Sci. 1985;20:2001–4.
16. Kokubo T, Ito S, Sakka S, Yamamuro T. Formation of a high-strength bioactive glass-ceramic in the system MgO–CaO–SiO_2–P_2O_5. J Mater Sci. 1986;21:536–40.
17. Akao M, Aoki H, Kato K. Mechanical properties of sintered. hydroxyapatite for prosthetic application. J Mater Sci. 1981;16:809–12.
18. Dewith G, Vandijk HJA, Hattu N, Prijs K. Preparation, microstructure and mechanical-properties of dense polycrystalline hydroxy apatite. J Mater Sci. 1981;16:1592–8.
19. Liu D-M. Bioactive glass-ceramic: formation, characterization and bioactivity. Mater Chem Phys. 1994;36:294–303.
20. Vincenzini P, editor. Ceramics in clinical applications. Amsterdam: Elsevier; 1987.
21. Kitsugi T, Yamamuro T, Nakamura T, Kokubo T. Bone bonding behavior of MgO–CaO–SiO_2–P_2O_5–CaF_2 glass (mother glass of AW-glass-ceramics). J Biomed Mater Res. 1989;23:631–48.
22. Nakamuro T, Yamamuro T, Higashi S, Kokubo T, Ito S. A new glass-ceramic for bone replacement: evaluation of its bonding to bone tissue. J Biomed Mater Res. 1985;19:685–98.
23. Quinn JB, Sundar V, Lloyd IK. Influence of microstructure and chemistry on the fracture toughness of dental ceramics. Dent Mater. 2003;19:603–11.
24. Grossman DG. Processing of dental ceramic by casting method. Ceram Eng Sci Proc. 1985;6:19–40.
25. Radonjić LJ, Nikolić LJ. The effect of fluorine source and concentration on the crystallization of machinable glass-ceramics. J Euro Ceram Soc. 1991;7(1):11–6.

26. Cheng K, Wan J, Liang K. Crystallization of R_2O–MgO–Al_2O_3–B_2O_3–SiO_2–F (R = K^+, Na^+) glasses with different fluorine source. Mater Lett. 2001;47:1–6.
27. Park J, Ozturk A. Tribological properties of MgO–CaO–SiO_2–P_2O_5–F^- based glass-ceramic for dental applications. Mater Lett. 2007;61:1916–21.
28. Xiao H, Cheng Y, Yang Q, Senda T. Mechanical and tribological properties of calcia–magnesia–alumina–silica-based glass–ceramics prepared by in situ crystallization. Mater Sci Eng A. 2006;423:170–4.
29. Jahanmir S, Dong X. Wear mechanism of a dental glass-ceramic. Wear. 1995;181–183:821–5.
30. Bibby J, Bubb N, Wood DM. Fluorapatite-mullite glass sputter coated Ti6Al4 V for biomedical applications. J Mater Sci Mater Med. 2005;16:379–85.
31. Kumar R, Kalmodia S, Nath S, Sigh D, Basu B. Phase assemblage study and cytocompatibility property of heat treated potassium magnesium phosphate–silicate ceramics. J Mater Sci Mater Med. 2009;20:1689–95.
32. Xiang Q, Liu Y, Sheng X, Dan X. Preparation of mica-based glass-ceramics with needle-like fluorapatite. Dent Mater. 2007;23:251–8.
33. Alemany MI, Velasquez P, de la Casa-Lillo MA, De Aza PN. Effect of materials' processing methods on the '*in vitro*' bioactivity of wollastonite glass-ceramic materials. J Non-Cryst Solids. 2005;351:1716–26.
34. Salman MS, Salama NS, Darwish H, Mosallam Abo HA. HA forming ability of some glass-ceramics of the $CaMgSi_2O_6$–$Ca_5(PO_4)_3F$–$CaAl_2SiO_6$ system. Ceram Int 2006;32: 357–64.
35. Brook IM, Craig GT, Lamb DJ. In vitro interaction between primary bone organ cultures, glass-ionomer cements and hydroxyapatite/tricalcium phosphate ceramics. Biomaterials. 1991;12:179–86.
36. Soljanto P, Rehtijrvi P, Tuovinen HO. Ferrous iron oxidation by thiobacillus ferrooxidans: inhibition by finely ground particles. Geomicrobiol J. 1980;2:1–12.
37. Johnson WA, Mehl RF. Reaction Kinetics in process of nucleation and growth. Trans AIME. 1939;135:416–42.
38. Avrami M. Kinetics of phase change. J Chem Phys. 1939;7:1103–1112, 1941;9:177–84.
39. Goswami M, Sarkar A, Mirza T, Shrikhande VK, Sangeeta KR, Gurumurthy GP Kothiyal. Study of some thermal and mechanical properties of magnesium aluminium silicate glass ceramic. Ceram Int. 2002;28:585.
40. Sarkar D, Basu B, Chu MC, Cho SJ. Is glass infiltration beneficial to improve fretting wear properties for Alumina? J Am Ceram Soc. 2007;90:523–32.
41. Copello GJ, Teves S, Degrossi J, D'Aquino M, Desimone MF, Diaz LE. Antimicrobial activity on glass materials subject to disinfectant xerogel coating. J Ind Microbiol Biotechnol. 2006;33:343–8.
42. Ratner BD, Hoffman AS, Schoen FJ, Lemons JE. Biomaterials science: an introduction to materials in medicine. Published by Academic Press; 2004.
43. West KA, Zhang H, Brown MC, et al. The LD4 motif of paxillin regulates cell spreading and motility through an interaction with paxillin kinase linker (PKL). J Cell Biol. 2001;154:161–76.
44. Louise PC. Role of actin-filament disassembly in lamellipodium protrusion in motile cells revealed using the drug jasplakinolide. Curr Biol. 1999;9:1095–105.
45. Pavalko FM, Otey CA. Role of adhesion molecule cytoplasmic domains in mediating interactions with the cytoskeleton. Proc Soc Exp Biol Med 1994;205:282–93.
46. Issa Y, Brunton P, Waters CM, Watts DC. Cytotoxicity of metal ions to human oligodendroglial cells and human gingival fibroblasts assessed by mitochondrial dehydrogenase activity. Dent Mater. 2008;24:281–7.
47. Benderdour M, Hess K, Dzondo-Gadet M, Dousset B, Nabet P, Belleville E. Effect of boric acid solution on cartilage metabolism. Biochem Biophys Res Comm. 1997;234:263–8.
48. Farley JR, Wergedal JE, Baylink DJ. Fluoride directly stimulates proliferation and alkaline phosphatase activity of bone-forming cells. Science. 1983;22:330–2.

49. Campoccia D, Arciola CR, Cervellati M, Maltarello MC, Montanaro L. In vitro behaviour of bone marrow-derived mesenchymal cells cultured on fluorhydroxyapatite-coated substrata with different roughness. Biomaterials. 2003;24:587–96.
50. Hailuo F, et al. In vitro evaluation of borate-based bioactive glass scaffolds prepared by a polymer foam replication method. Mater Sci Eng C. 2009;29:2275–81.

Chapter 6
Processing, Tensile and Fracture Properties of Injection Molded HDPE–Al_2O_3–HAp Hybrid Composites

6.1 Introduction

The last few decades have witnessed significant research efforts in scientific community to integrate the concepts of materials science and biological sciences in order to develop biomaterials for orthopedic applications [1, 2]. Among various biomaterials, polymer-ceramic composites are investigated to a visible extent [3–6]. In last few decades, attention is directed towards the development of polymer-ceramic composites for load bearing orthopedic application. The concept of a ceramic reinforced polymer composite as a bone substitute was introduced in 1981 by Bonfield et al. [7]. They selected polyethylene (PE) in order to ensure the required ductility, while reinforcing the matrix with brittle HA particles. The bone-analogue concept by proposing composites comprising of a ductile polymer matrix [polyethylene (PE)] and a stiff ceramic phase (HA) was initially proposed by Bonfield et al. [8–15]. However, the lower strength, modulus and stiffness of those composites, compared to cortical bone, have limited their application as load bearing bone substitutes. Reis et al. [16] provided an alternative approach to the mechanical performance enhancement of HDPE/HA composites fabricated using shear controlled orientation in injection molding (SCORIM). Also, Pandey et al. [17, 18], reported that HDPE–HA composite could exhibit elastic modulus of 206–531 MPa, strength in the range of 20–24.3 MPa and % elongation at break of 8.3–163.

The main advantage of using such composites is the fact that by varying the type and amount of reinforcing phase, their mechanical as well as biological properties can be tailored for desired applications [19]. Polymer-ceramic composites, with tailored modulus and strength provide advantages in comparison with pure polymers, ceramics and metals for hard tissue replacement application. A myriad of polymers have been investigated for a number of different medical applications. Non-degradable polymers, used in orthopedic surgery, include high density polyethylene (HDPE), ultra-high molecular weight polyethylene (UHMWPE), poly (ethylene terephthalate) (PET), poly(propylene) (PP) and PTFE [20]. While

B. Basu and S. Ghosh, *Biomaterials for Musculoskeletal Regeneration*,
Indian Institute of Metals Series, DOI 10.1007/978-981-10-3017-8_6

HDPE/UHMWPE are used as articulating surfaces for total hip joint replacements (THR), PET/PP/PTFE are used to repair knee ligaments [21]. However, the release of HDPE particles following long term implantation, leading to chronic inflammation and aseptic loosening have led to attempts to strengthen HDPE.

Attempts to enhance the mechanical performance of thermoplastic polymers (e.g. high density polyethylene, HDPE) in order to allow for their use in hard tissue substitute have partially relied on the use of Hydroxyapatite (HA) reinforcement. HA with the molecular stoichiometric formula, $Ca_{10}(PO_4)_6(OH)_2$, attracted wider attention due to its excellent biocompatibility, bioactivity and osteoconductivity property. However, the poor compressive strength and toughness limits its applicability to low or non-load bearing application to human body [22]. These composites have been used successfully in clinical situations as orbital implants for conditions such as post-enucleation socket syndrome and orbital floor fractures [23, 24]. HAp–HDPE composites have also been used successfully as middle ear implants. Dornhoffer et al. [25] designed two ossicular replacement prostheses with a HAp head and a HAp–HDPE composite shaft. The properties of these HAp–HDPE composites have also been studied extensively [26–30]. A review of available literature reveals that the HDPE–HAp composites can have a combination of properties, including E-modulus of 200–530 MPa, strength of 20–24 MPa, and a large strain to failure of 200 % or more, when fraction of HAp is varied from 10–50 %, respectively [31]. It is recognized now that any improvement in mechanical properties of polymer-ceramic composites requires better interfacial adhesion between the reinforcement and the matrix. If load transfer is not effective across the interface, mechanical properties cannot reach expected values.

In an effort to further enhance the mechanical properties, recent research efforts are invested to develop HDPE–HAp composites with Al_2O_3 reinforcement [4]. Alumina has been chosen as the second reinforcing material because of its good biocompatibility, high wear resistance, and stability in physiological environments. Initial feasibility study was carried out using compression molding route, and the results demonstrated that improved physical properties (E-modulus, hardness), low COF, good wear resistance and, more importantly, that good cytocompatibility property (cell adhesion) can be favorably optimized by simultaneous addition of 20 vol.% Al_2O_3 and 20 vol.% HAp–HDPE matrix [3–6]. Therefore, it is believed that the HDPE–HAp–alumina could be a promising candidate material for orthopedic applications. Nevertheless, detailed mechanical properties in terms of tensile behavior and toughness measurements were not conducted in our earlier study and the application of such materials also demands the requirement of their processability to complex shapes using a commercially viable manufacturing route.

This chapter elucidates a systematic investigation of the role of polymer, ceramic and coupling agent on injection molding of HDPE–HA–Al_2O_3 hybrid biocomposites. The objective of this chapter is to emphasize the physico-mechanical behavior and biocompatibility of the composites, including the cell adhesion, cell viability, protein absorption and osseointegration.

6.2 Injection Moulding

Powder injection molding is a relatively new technology that extends the complex shaping capability of plastic injection molding into other high performance materials such as metals and ceramics [32, 33]. This process has found a niche in the area of manufacturing moderately high volume of small, extremely complex shaped components from various high performance materials. The process of powder injection molding typically consist of four major steps: feedstock preparation, injection molding, debinding, and sintering. In some special cases, the actual part is the as-molded material which is made by injection molding an intimate mixture of an organic component (generally a polymer) with some inorganic powder(s). The inorganic powder can be a metal, alloy, ceramic, or a mixture of these materials. In our case, the organic component is the HDPE and the inorganic particulate material is an equal-volume mixture of alumina and HAp.

In the context of HDPE–HA–Al_2O_3, the mechanical properties would critically depend on interfacial bonding between the reinforced material and the matrix [34, 35]. The adhesion of reinforcement to the matrix can have a significant influence on the characteristics and properties of the composites [36]. Polyethylene is a non-polar, hydrophobic polymer, and consequently, mechanical interlocking exists between the HA particles and the polyethylene matrix in a conventionally processed HA–HDPE composite [37]. The chemical interaction between the filler and the polymer will lead to much improved bonding, and hence, mechanical properties. The addition of coupling agents provides other beneficial effects besides improving composite blending and injection molding. The material interaction through polar coupling and hydrogen bonding also provides good adhesion between ceramic and polymer components.

It has been now widely recognized that the injection molding is advantageous for industrial scale production of net shaped ceramic and metal products of complex shapes. The major steps involved in injection molding process are powder mixing, moulding, debinding and sintering [19, 23]. Although not very popular among researchers, injection molding has been used in the past to shape biomaterials. Jaroslav and Martin [38] utilized the method for the preparation of HAp ceramics. Mondrinos et al. [39] developed a solid freeform fabrication (SFF)-based injection molding process for the fabrication of PCL and PCL–CaP scaffolds, that displayed *in vitro* cytocompatibility and suitable mechanical properties for hard tissue repair. Joseph et al. [40] addressed the effect of surface area and morphology of HAp on the rheology and processability of an injection-molding grade HAp–HDPE composite. Bakar et al. [41] reported mechanical properties of HAp–PEEK composites, produced by injection molding. Juang and Hon [42] prepared HAp ceramic bars by injection molding in an effort to study the effect of calcination temperature on the HAp powder and injection molding processability. Sahebian et al. investigated the influence of nanosized $CaCO_3$ addition on the toughness properties of HDPE nanocomposites, fabricated using injection molding [43].

6.3 HDPE Composites with 30 % Ceramic Fillers

6.3.1 Processing Related Challenges

Prior to discussing the processing results, it is important to mention the characteristics of the starting powders used in the injection molding process. An attempt made to determine the influence of ceramic loading on the processability of HDPE composites using Injection molding route. A torque rheometer was used to determine the maximum solid loading that is possible using a mixture of the alumina and HAp powder (in equal volume ratio). Torque rheometry measures the change in the mixing torque with various volume loading of solid. As shown in Fig. 6.1b, the addition of ceramic filler is associated with a concomitant drop in the mix temperature and an increase in mixing torque. Around 30 volume percent of solid addition, the mixing torque became quite erratic indicating that solid loading above this would result in an inhomogeneous mixture. It was decided that the maximum solids loading used will be 30 volume percent (15 v/o alumina and 15 v/o HAp).

The mixing temperature was 170 °C and the rotor speed used was 50 rpm. In order to systematically study the ceramic filler addition, equal amount of HAp and alumina powders were used along with HDPE to obtain a total of 10, 20 and 30 vol. % filler composites. Pure HDPE was also processed and used as a reference sample. The density of the pure HDPE, and the three other mixes with 10, 20, and 30 vol.% ceramics were determined using a gas pycnometer. The volume flow rate, viscosity, and the shear rate of the four materials were determined using an Extrusion Plastometer. The extrusion plastometer investigations were carried out at 160 °C using a load of 2.16 kg. Injection molding was carried out using a barrel and nozzle temperature of 160 °C, a mold temperature of 40 °C, pressure of 3000 psi, with a clamping force of 10 tons. During molding, the feedstock was pushed through the sprue into a runner and then into the die cavity (either a tensile bar or a flexural bar) through a gate. After injection molding, dog bone shaped tensile bars of dimension 21 mm (gage length) * 6.35 mm (width) * 3.84 mm (thickness) as well as flexural bars of dimension 15.06 mm (width) * 5.07 mm (thickness) * 60 mm (length) were obtained. During injection molding, the melt flow characteristics, like shear rate, volume flow rate, viscosity and melt density were measured using extrusion plastometer. A representative DSC run obtained with the used HDPE is provided in Fig. 6.1a. From the DSC plot, it is clear that the melting of HDPE starts at 126 °C with an endotherm peak around 136 °C. Total heat energy involved in the melting of HDPE is around 149.7 J/g. From the analysis of the area under the melting endotherm and knowing the total heat evolved for 100 % crystalline HDPE, it was found that SCLAIR grade HDPE polymer has around 51.2 % crystallinity. In Fig. 6.1c, both the experimental density at RT as well as melt density data at 160 °C are plotted against ceramic filler addition. An increase in the melt density with increasing ceramic filler additions is due to an increase in the amount of the powders, which have a higher density compared to the polymer (HDPE density of 0.95 g/cm^3) in the mixes. A closer look at the data in Fig. 6.1c reveal that the

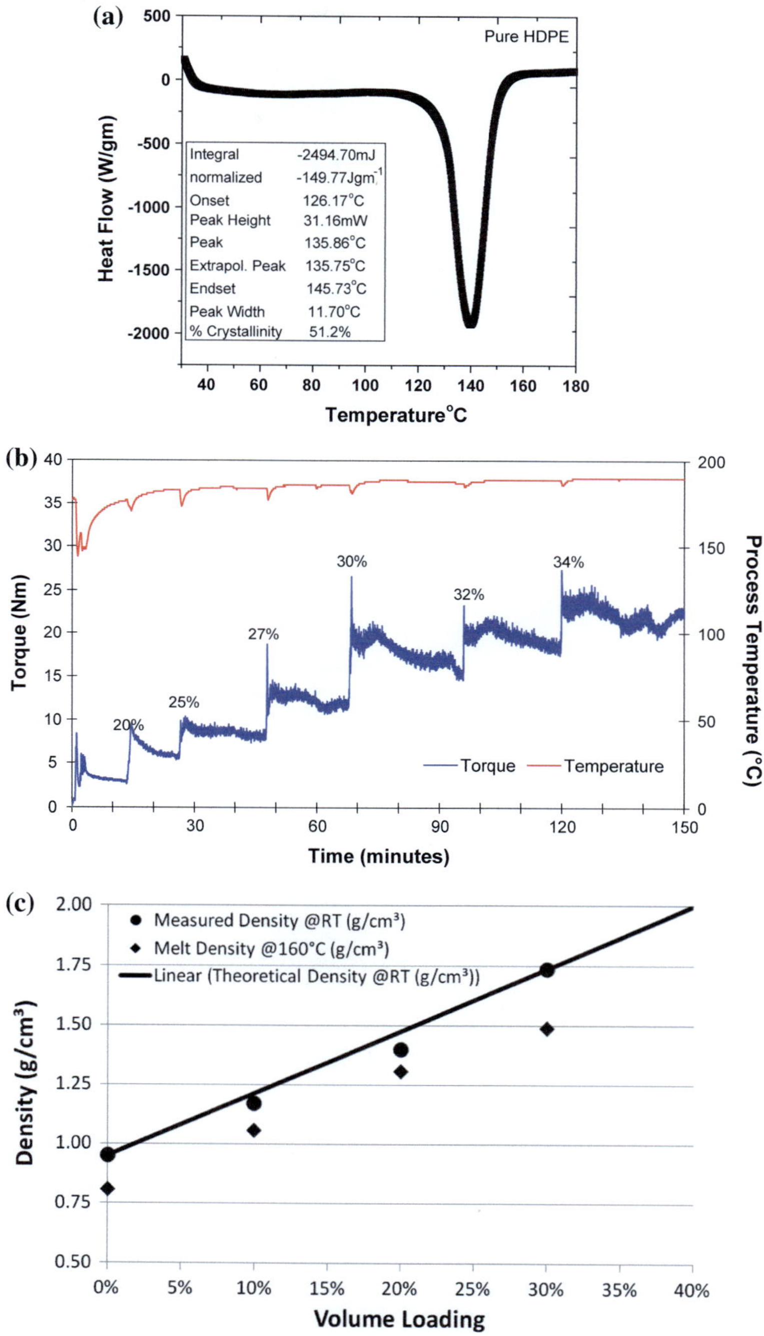

Fig. 6.1 DSC scan of pure HDPE showing the melting endotherm (**a**), change in torque with solids loading (**b**), theoretical and experimentally measured density at RT and measured melt density at 160 °C at different solids loading (**c**), variation of viscosity and volume flow rate with solids loading (**d**) and variation of shear rate with solids loading (**e**) [3]

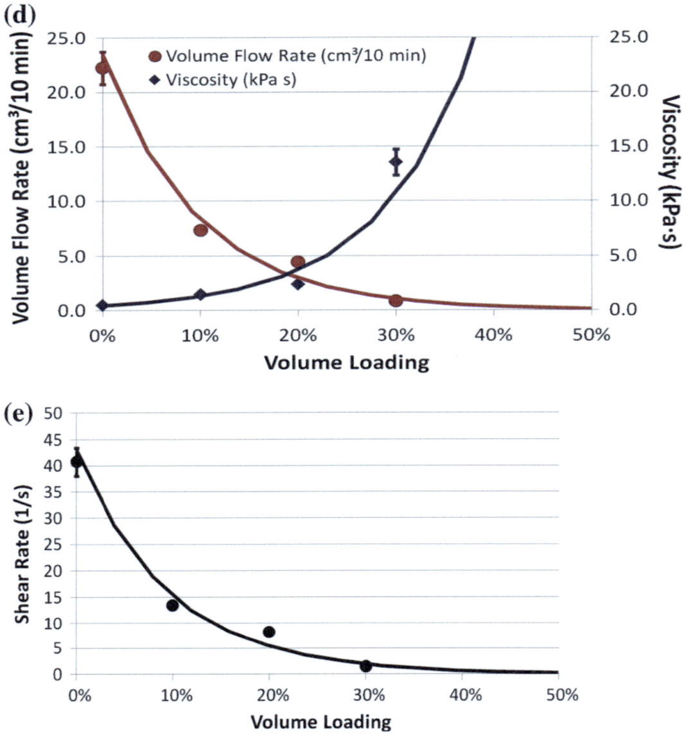

Fig. 6.1 (continued)

difference in density between experimentally measured values at RT and melt density is substantial at 30 vol.% ceramic addition; whereas such difference is comparable in case of pure HDPE and 10 or 20 vol.% filler addition.

In Figs. 6.1d–e, the melt flow characteristics for different composite compositions are summarized. The error bars in data of Figs. 6.1d–e represent the standard deviation for at least five set of experiments on each composition. It is clear from Figs. 6.1d–e that both shear rate and vol. flow rate consistently decrease with increasing ceramic filler addition. As the ceramic filler content increases, the ability of the material to flow (using the same temperature and pressure) is decreased. Thus, over the same period of time, the volume flow rate of the material is decreased with increasing additions of ceramic filler. In contrast, both the viscosity and melt density systematically increase with ceramic filler addition. The viscosity of a polymeric material is generally increased with addition of particulate material. Such increase or decrease in melt flow parameter is most significant when total ceramic loading is increased from 20 to 30 vol.%. For example, Fig. 6.1e illustrates how the shear rate of polymeric melt decreases in a non-linear manner with ceramic loading of up to 30 vol.%. Also, the observed trend in data is also consistent with our observations that any further addition in ceramic loading beyond 30 vol.% would make the melt so viscous that it would be difficult to injection mold.

6.3.2 Compositional Dependence of Viscoelastic Properties

Figure 6.2 plots the tensile test results in order to compare the behavior of three different composites. The stress–strain curve for pure HDPE, stretched in tension at 5 and 1 mm/min crosshead speeds, shows typical viscoelastic behavior of polymers (Fig. 6.2a), which is characterized by a yield point and large deformation at constant stress level at a strength value of lower than yield strength. Interestingly, a large strain to failure of around or more than 1000 % was measured for pure HDPE in the present case. Such large deformation in part can be related to lower crosshead speed used in our tests as well as large crystallinity (51.2 %) of the used HDPE. The tensile testing at a speed lower than 1 mm/min was not possible as the deformation exceeds the limit of the Instron machine used in the present investigation. Unlike metals, the neck does not continue shrinking until the specimen fails. Rather, the material in the neck stretches only up to a fixed drawing ratio, beyond which the material in the neck stops stretching. The neck propagates until it spans the full gage length of the specimen and this causes high fracture strain in pure HDPE.

In contrast, the non-linear viscoelastic deformation is highly restricted due to filler addition in HDPE and, as the amount of filler increases, the stress–strain behavior more closely resembles brittle fracture (Fig. 6.2). As discussed before, filler also influences the composite stiffness. Since the matrix is same for all the composites, the different behaviors may be attributed to the difference in filler amounts and more efficient load transfers between the filler and polymer. Typically, stiffer filler prevents the elongation of a highly ductile matrix due to efficient load transfer, ultimately decreasing the total elongation.

A general observation is that the ceramic filler addition of 10 % can increase the E-modulus of the composite to three times compared to that of pure HDPE. The tensile modulus was seen to consistently increase with HAp and alumina content in the composite. This is expected, considering that the modulus of HAp and alumina are ~85 and ~390 GPa, respectively [44, 45]. Importantly, the reinforcement effect was clearly evident even with the addition of 10 vol.% of ceramic fillers. At higher filler loading of 30 %, a significant increase in the modulus of about sixfold was observed as compared to pure HDPE. It may be noted that a maximum elastic modulus of close to 1 GPa has been measured with 30 % ceramic loaded HDPE composite, which is still lower than the lower bound elastic modulus of natural cortical bone (2–3 GPa) [16].

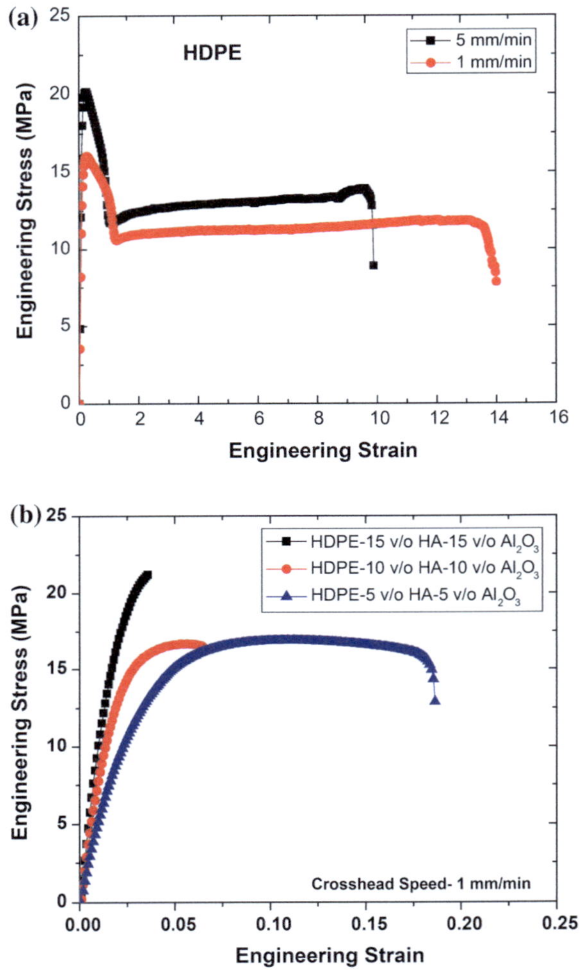

Fig. 6.2 Engineering stress–strain plot obtained using tensile testing at varying crosshead speeds for different materials: **a** pure HDPE, **b** HDPE–HAp–Al_2O_3 composites (composition in vol.%, abbreviated as v/o) [3]

6.3.3 *Notched Behavior in Flexure*

In the present case, long crack toughness method, i.e. SEVNB testing is adopted and the notched samples with a sharp precrack are fractured in 3-point flexure mode. Although the area under the stress–strain curves provides measures of fracture toughness, a more realistic measurement was obtained in the present study using SEVNB testing, as also followed in the work of Kim et al. [46]. For this, V-notch was created on the tensile face of the 3-point bend samples by machining and a sharp crack was introduced at the notch tip using a razor blade, following ASTM standard. All the notched samples were kept in an oven at 60 °C for 30 min to relieve the machining induced residual stress. The specimens were then fractured using three-point bending configuration with the span length of 50 mm and crosshead speed of 1 mm/min on Instron 1195 machine. The mode I fracture toughness, $K_{1c,}$ was determined using the following relationship [47].

$$K_{Ic} = \left(P_Q/BW^{1/2}\right)f(x) \tag{6.1}$$

where, W = 2B and x = a/W and $f(x) = 6x^{1/2}[1.99 - x(1 - x)(2.15 - 3.93x + 2.7x^2)]/(1 + 2x)(1 - x)^{3/2}$.

In the above expression, a = crack length (measured using SEM on the unfractured samples), P_Q = peak load, W = width and B = thickness.

The recorded load-displacement data are plotted in Fig. 6.3. Qualitatively, a general observation is that the initial linear mechanical response in flexure goes through a non-linear response prior to attaining the peak value, beyond which the notched samples can not sustain any increased load. Although the load drops in a non-linear manner, the entire flexure sample was not broken in case of pure HDPE and 10 or 20 vol.% filler addition. An important observation is that the peak load is attained at lower displacements, as the filler addition is increased and therefore, the load-displacement curve is shifted more towards left. Similar observations have been made in case of HDPE–$CaCO_3$ nanocomposites with the exception that the peak load is decreased with the second phase addition [43].

Based on Eq. (6.1), the mode I fracture toughness was determined and plotted in Fig. 6.4. The obtained fracture toughness was in the range of 0.55–1.0 MPa $m^{1/2}$, which is close to the lower bound of natural cortical bone fracture toughness (2–6 MPa $m^{1/2}$) [48]. It is clear from Fig. 6.4 that an enhancement in fracture toughness is only realised at the ceramic loading of 20 vol.%. An increase in the fracture toughness essentially indicates better energy dissipation at 20 vol.% solid loading. In general, a number of mechanisms contribute to the fracture toughness and it is often very difficult to determine the dominant mechanism [49]. In case of HDPE based composites, the major energy absorption mechanisms include crack

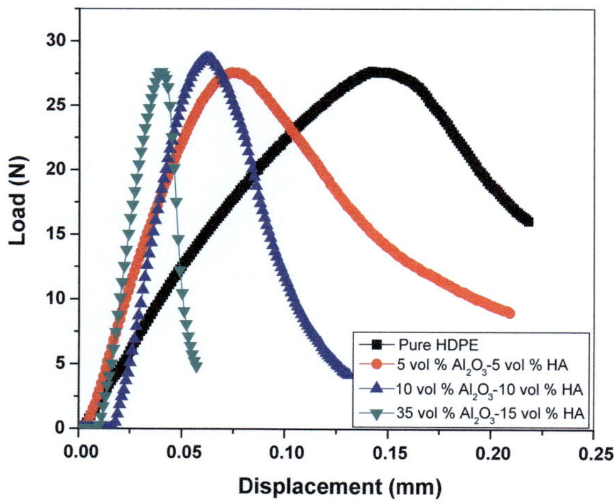

Fig. 6.3 Load-displacement plot obtained for 3-point bend test for fracture toughness measurement at crosshead speed of 1 mm/min for different materials: pure HDPE, HDPE–HAp–Al_2O_3 composites (10, 20 and 30 vol.%) [3]

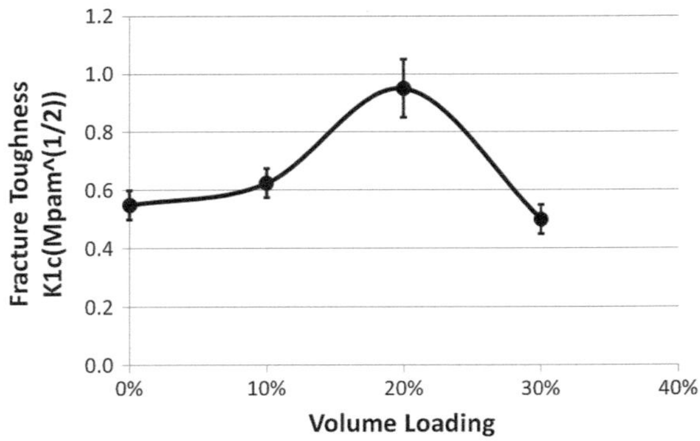

Fig. 6.4 Plot of fracture toughness, as measured using SEVNB testing, with vol.% ceramic loading to HDPE [3]

deflection, debonding between fiber and matrix, pull-out (extraction of fibers from the matrix), and fiber-bridging mechanism [50]. Some selected SEM images of the fracture surfaces after SEVNB testing are provided in Fig. 6.5. The alignment of HDPE fibers in the crack tip process zone is clearly observed. Also, the debonding at alumina particle/polymer interface can be clearly observed in Fig. 6.5. Similar, polymer-filler particle interfacial debonding is also noticed in other composite compositions.

6.4 HDPE Composites with 40 % Ceramic Fillers

In another set of HDPE composites with 40 vol.% filler loading injection molding was carried out using barrel temperature of 180 °C and mould temperature of 25 °C. The available cooling time was 30 s. The process was followed by the back pressure of ~3 MPa. Unreinforced HDPE is designated as IS1 sample. HA powders of 40 wt % were mixed with HDPE matrix and coupling agent to produce IS2, and IS3 is composed of 40 wt% Al_2O_3 with coupling agent. Compositions made from simultaneous addition (40 wt% fillers (1:1) of both HA and Al_2O_3) in the presence of coupling agent to 60 wt% HDPE powder was designated as IS4.

Conventionally sintered pure hydroxyapatite, densified at 1200 °C for 2 h in air is designated as HA. In all the composites, an optimal amount of 2 wt% titanium IV 2-propanolatotrisisooctadecanoato-O was used. Injection molding of all the HDPE materials was carried out using barrel temperature of 180 °C and mould temperature of 25 °C. The available cooling time was 30 s. The process was followed by the back pressure of ~3 MPa. More details on the processing can be found elsewhere. The elastic moduli of the investigated composites are provided in Table 6.1 and more details on property measurements can be found elsewhere.

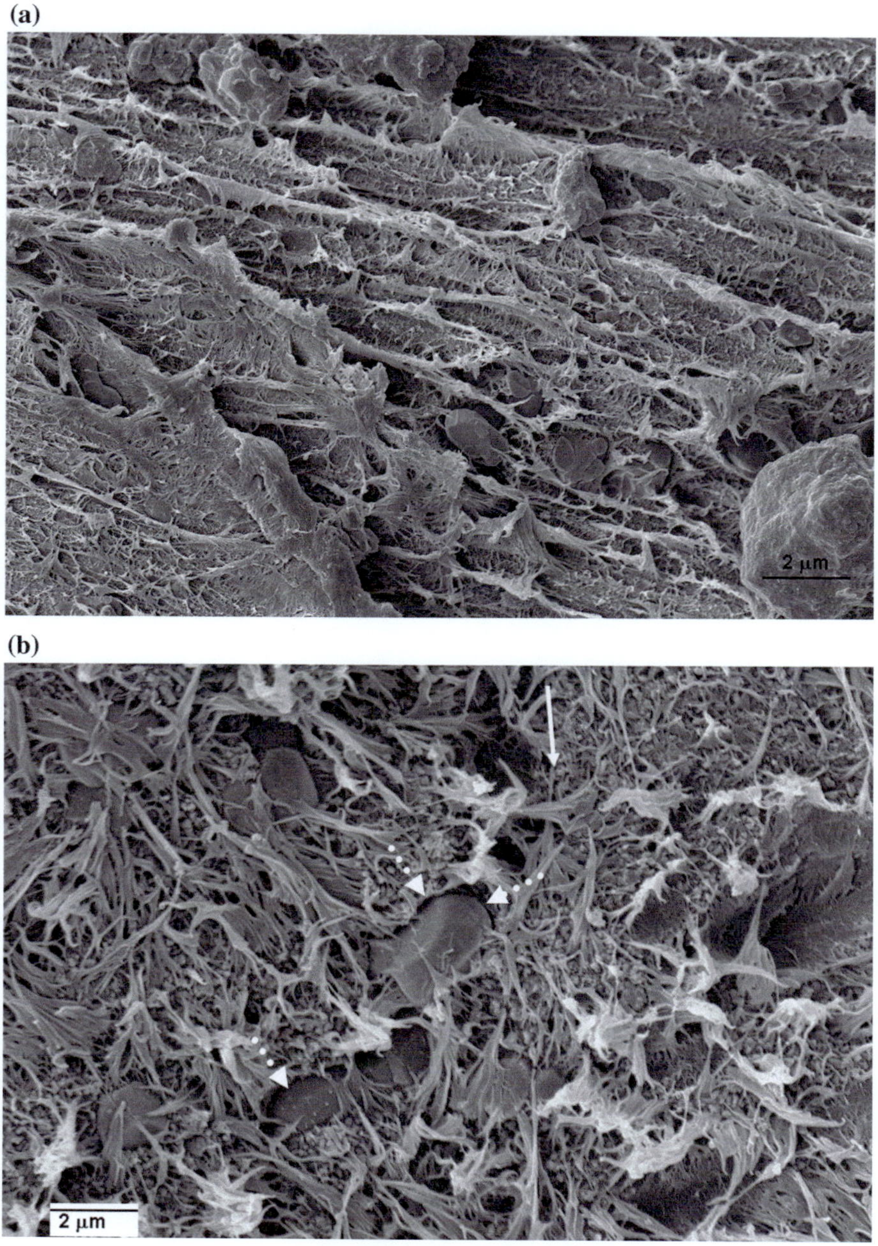

Fig. 6.5 Representative SEM images illustrating the observation of alignment of polymeric chains around the crack tip stress field adjacent to the V notch after SEVNB testing (**a**) and debonding of alumina platelets, indicated by *dotted arrow* during the propagation of the primary crack (bold crack). Both the images are obtained in case of 10 vol.% ceramic loaded composite [3]

Table 6.1 Summary of mechanical test data performed on injection molded samples [3]

Sample	Compositions	Compressive strength (MPa)	Compressive modulus (MPa)	Tensile strength (MPa)	Tensile modulus (MPa)	% elongation-at-break
IS1	HDPE	57.2 ± 1.4	469.1 ± 150.0	24.2 ± 0.9	564.3 ± 130.2	855.4
IS2	HDPE + 40 wt% HAp (2 % coupling agent)	67.0 ± 1.2	532.1 ± 148.8	14.3 ± 0.1	679.1 ± 110.4	47.9
IS3	HDPE + 40 wt% Al_2O_3 (2 % coupling agent)	71.6 ± 1.0	312.7 ± 108.2	16.4 ± 0.8	673.4 ± 102.7	95.0
IS4	HDPE + 20 wt% HAp + 20 wt% Al_2O_3 (2 % coupling agent)	70.2 ± 1.7	660.9 ± 118.0	18.7 ± 0.4	855.4 ± 108.4	6.8

6.4.1 Tensile Properties

It is well known that, most of the polymers, when loaded have a distinctive index to undergo viscoelastic deformation. This type of deformation is enhanced at indent contacts in the presence of complex stress conditions. In the present study, the indent impressions were not distinct, indicating the softness of the thermoplastic phase and a significant recovery due to viscoelastic nature. The mechanical behavior of the prepared hybrid composites (IS2–IS4) were investigated by means of uniaxial tensile testing of dog bone shaped samples and compression test of cylindrical shaped samples. For the comparative study, pure HDPE was also tested under identical testing conditions. The compression modulus and strength were obtained from stress versus strain response. The elastic modulus, tensile strength and elongation-at-break were calculated using tensile test data. All the experimental data are summarized in Table 6.1. The tensile modulus of the IS4 was about 1.5 times greater than the IS1. The results showed that IS1 composite had a tensile strength of 24.2 MPa ($\pm$0.9 MPa), whereas tensile strengths of IS2–IS4 were found to be 14.3 MPa ($\pm$0.1 MPa), 16.4 MPa ($\pm$0.8 MPa) and 18.7 MPa ($\pm$0.4 MPa), respectively. Tensile modulus of neat HDPE (IS1), was found to minimum i.e. 564.3 MPa ($\pm$130.2 MPa). It is clear from the results that the presence of filler and coupling agent increases the tensile modulus but decreases the strength. The ceramic fillers reduce the melt viscosity and increase plasticity of the overall composition. The IS4 composite shows highest tensile strength in comparison to IS2 and IS3 with a high tensile modulus. The presence of ceramic filler clearly increases stiffness of the composite with no significant effect towards the tensile properties. A small drop in tensile strength for composite samples is believed to be due to interfacial bonding between the HDPE matrix and HA/Al_2O_3 reinforcement.

Uniaxial compression tests were carried out on four cylindrical samples for each set of as-processed composites. The presence of Al_2O_3 as a filler provides better strength to the IS3 composite compared to IS2. Another important thing, which was noticed, is the E-modulus (elastic modulus (E-modulus)) of the composite, which can be easily correlated with the modulus of HA and Al_2O_3. The E-modulus of HA and Al_2O_3 are $\sim$85 and $\sim$390 GPa, respectively [22]. The maximum elongation was found in IS1 (pure HDPE) and minimum was measured in IS3, composite having 40 wt% Al_2O_3 (Table 6.1).

6.4.2 Cell Adhesion and Proliferation

It is known that bone marrow derived primary osteoblast cells repair cracks in bones, and help bones grow. They are the major cell type, comprising 90 % of cells in the mature skeleton and are important in the control of extracellular calcium deposition. It is also reported that the osteoblasts/osteogenic cells are the common primary mechanosensors in intact bone and are the key cells in mechanotransduction forces, adaptation of skeletal tissue, and in triggering of osteoblasts to form bone [51–53]. An earlier study demonstrates that loading of osteogenic cells inhibits osteoclast formation via direct contact effects as well as molecular signaling factors [54]. Also, Osteogenic cells are the most promising candidate to provide a cellular basis for mechanosensing and bone remodeling [55–58]. This is the prime motivation for the selection of primary osteoblast cells for in vitro biocompatibility assessment.

After 3 days of culture, the osteogenic cells reach confluency on the surface of all the investigated samples. Representative florescence microscopy images of cultured osteogenic cells are shown in Fig. 6.6. Most of the cells were flattened, polygonally shaped and spread, showing numerous highly extended filipodia and rough dorsal surfaces, as shown in Fig. 6.6. On higher magnification images (Fig. 6.6b, d and f), osteogenic cell proliferation can be seen to be intimately associated with an extensive network of actin filaments. Although the cell morphology on the composites is qualitatively similar, large filopodia extension together with cellular bridge formation are clearly observed on IS4 substrate (see Fig. 6.6f). Irrespective of the substrate composition, a number of brighter contrast nodules are widely spread around the cells and these nodules appear to be mineralized CaP crystals.

In order to quantify the cell proliferation and viability, MTT assay was carried out. The results of the quantitative measurements of the amounts of the formazan crystals are shown in Fig. 6.7. The results of the absorbance measurements revealed that a higher amount of the formazan extract was recovered from cells cultured on different substrates at all the timepoints. The following points emerge from the analysis of the results presented in Fig. 6.7 (a) for all the substrates, mean OD values increase with time in culture, reflecting that number of metabolically active osteogenic cells increases with time (cell proliferation), (b) among the composites, IS4 composite supports the growth of more number of metabolically active osteogenic cells than IS2 or IS3 composite and (c) at any given time point in culture, the number of metabolically active cells on IS4 is more than HA.

6.4.3 *in vitro Mineralization*

Similar to MTT results revealing statistically significant difference in terms of cell viability, the alizarin red staining results also provide significant difference among the substrate dependent mineralization behavior of cultured osteogenic cells (see Fig. 6.8). A critical analysis of Fig. 6.8 reveals a linear increase in absorbance values, indicating more Ca-deposition by cultured cells with incubation time and such trend is independent of the substrate composition. The increase in absorbance with culture time is by far more significant in case of IS4 composite than IS2 and IS3 (particularly for 14/21 days). Although absorbance measured in case of IS4 is higher than sintered HA for 7/14 days of culture, it is comparable at longer culture duration of 21 days.

The results summarized in preceding sections do establish that injection molded HDPE–20 % HA–20 % Al_2O_3 has HA-like cytocompatibility property.

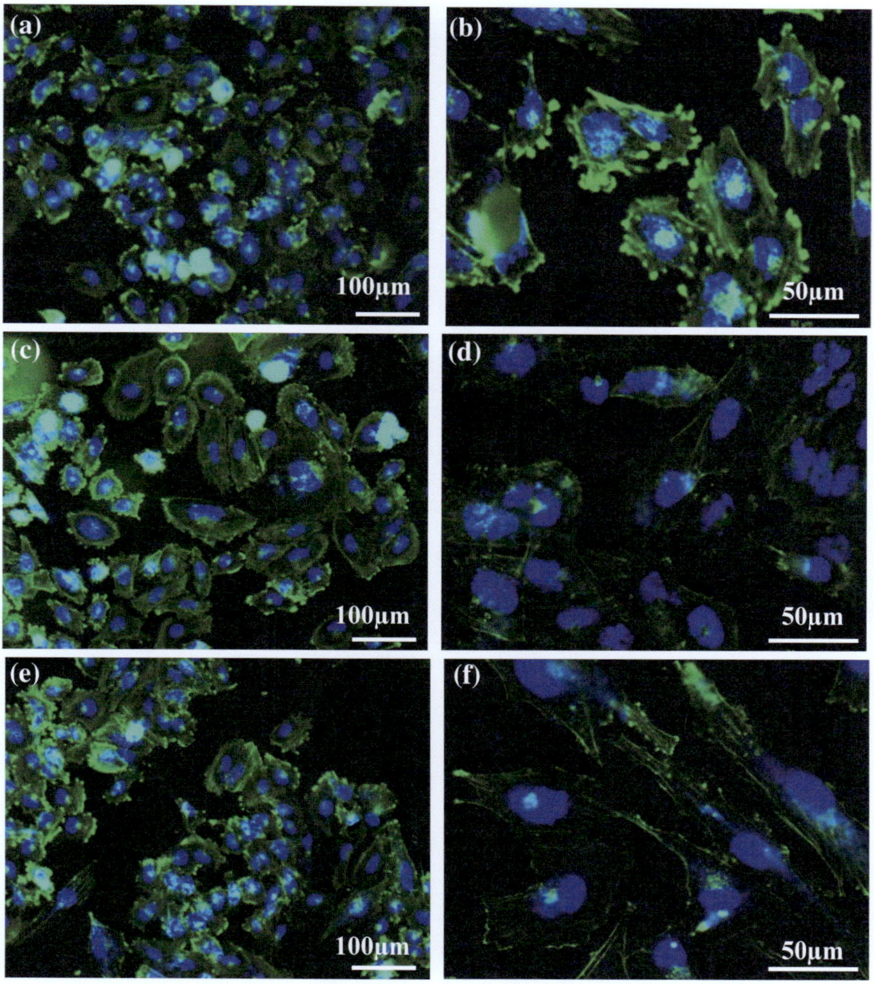

Fig. 6.6 Fluorescence microscopic images of osteogenic cells adhered on the investigated HDPE-based hybrid composites: **a**, **b** IM1, **c**, **d** IM2 and **e**, **f** IM3, after 72 h in culture. While the blue stained regions are nuclei, the green stained regions are actin filaments of the cytoskeleton [4]

6.4.4 Protein Adsorption and Cell Functionality

The difference in cell viability or cell proliferation among IS1–IS4 materials can be better explained in terms of difference in protein absorption behavior. Proteins are amphiphatic molecules, which typically adhere to the surface of a biomaterial in a nonspecific manner. In different cases, such nonspecific adhesion is sufficient to artificially immobilize proteins on the material surface and no surface modification is necessary. The high hydrophobicity of many proteins is reported to play an important role in their absorption on particles surface [59]. In the present study,

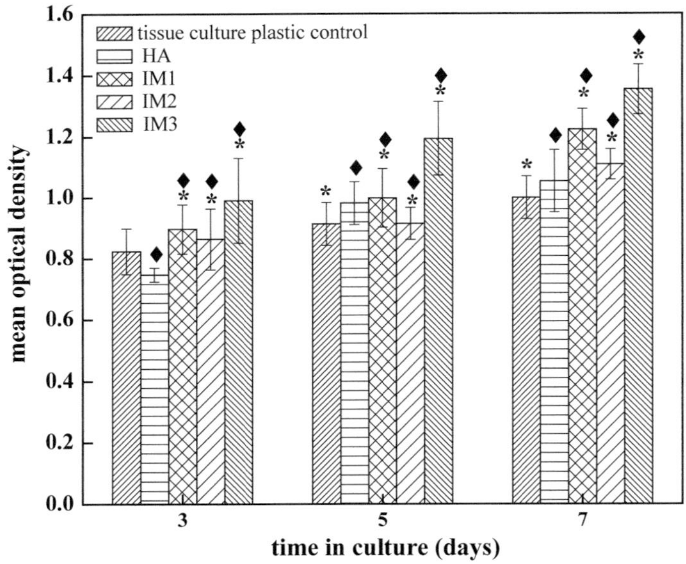

Fig. 6.7 Plot of MTT assay results showing the incubation time dependent cell viability on the investigated substrates. * represents the statistically significant difference among the samples with respect to time in culture and ♦ represents the statistically significant difference among the samples with respect to tissue culture plastic control [4]

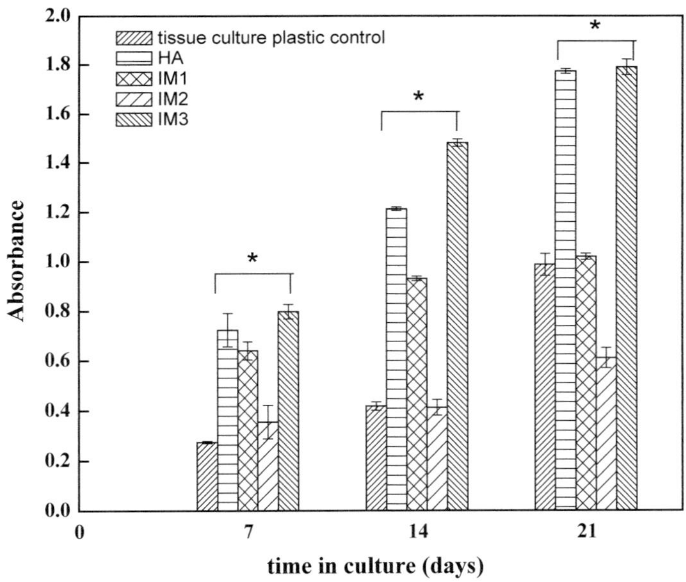

Fig. 6.8 The absorbance measured after alizarin red staining of primary cultured osteogenic cells for different composites at different timepoints over 21 days in culture (*$p < 0.05$). *represents the statistically significant difference among the samples with respect to control), irrespective of incubation periods (number of days) [4]

BSA (bovine serum albumin), a soft protein, is used. Soft protein is reported to be able to change their conformation better than the hard proteins. When absorbed on solid surfaces, soft proteins improve the efficacy of the absorption process, when compared to the hard proteins like fibrinogen, α-chymotrypsin, ribonuclease, lysozyme and β-lactoglobulin.

The protein absorption behavior of biomaterials is dependent on the presence of biocompatible phases in the composite surface. In the present case, the maximum absorption was measured on HA, IS2 and IS4, as shown in Fig. 6.9a. This might be due to the presence and homogeneous distribution of biocompatible HA filler. For example, a lower absorption was noticed in IS3, which is further reflected in cell viability measured by MTT assay. This might be correlated with the presence of Al_2O_3, which is comparatively less biocompatible than HA. The protein absorption behavior of the prepared composites shows statistically significant difference among IS2, IS3 and IS4, with respect to HA and control disc.

Apart from this, another study was performed to find out the incubation time, where protein absorption gets isolated (Fig. 6.9b). Along with control disc and sintered HA, all the samples were seeded with similar concentration of protein for identical timeframe. After every hour, one set of samples was studied for BCA assay. This experiment was performed continuously for number of hours to find the isolated stage. For the present study (up to 4 h), the protein density increased continuously, but no significant change was observed. After 5th and 6th h of incubation, no change is also observed in the absorbed protein concentration. The present experimental results reconfirm that the absorption mechanism is dependent on starting protein concentration and material composition.

6.4.5 *Influence of Surface Energy on in vitro Mineralization*

Three liquids i.e. MQ water, McCoy'S 5A and isopropyl alcohol (purity >99 %) were used to calculate surface tension and related properties. The surface free energy (SFE) of the substrates as well as the polar and dispersion components were determined using the Owens–Wendt–Rabel–Kaelble method [60]:

$$\sigma_S = \sigma_{S1} + \sigma_1 \cdot cos\theta \tag{6.2}$$

$$\sigma_1 = \sigma_1^d + \sigma_1^P \tag{6.3}$$

$$\sigma_S = \sigma_S^d + \sigma_S^P \tag{6.4}$$

$$\sigma_{S1} = \sigma_S + \sigma_1 - 2\left(\sqrt{\sigma_S^d \cdot \sigma_1^d} + \sqrt{\sigma_S^P \cdot \sigma_1^P}\right) \tag{6.5}$$

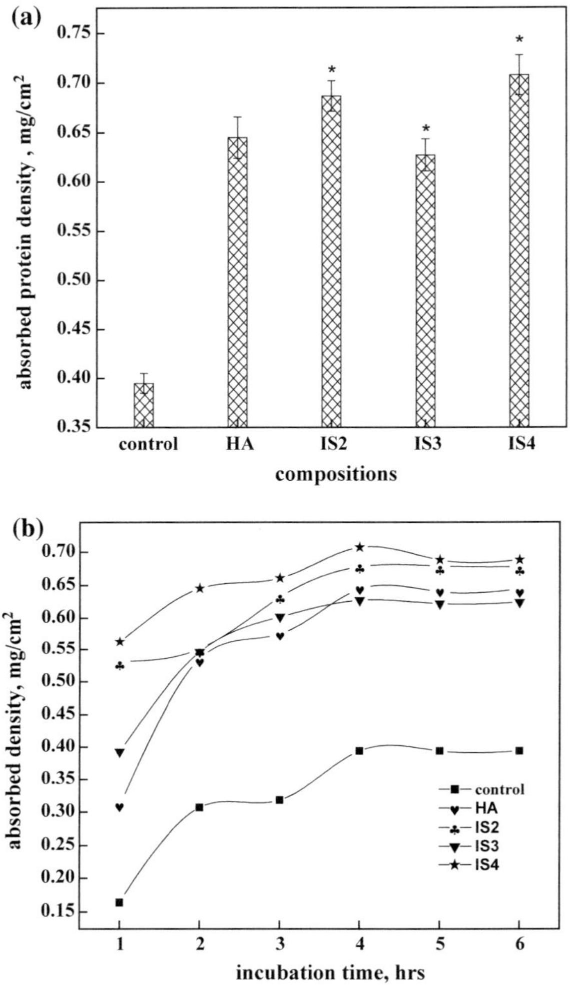

Fig. 6.9 **a** Protein absorption on the composites: incubation time versus absorbed protein density. (4 h incubation time), the asterisk mark (*) represent significant statistical difference at $p < 0.05$. **b** Study of protein absorption on the composites: incubation time versus absorbed protein density [4]

Here, σ_{S1} is the interfacial tension between the drop and the solid, θ is the CA on the solid surface, σ_s and σ_1 are the surface energy of the solid surface and surface tension of the liquid used, respectively. σ_S^d and σ_S^P stand for the respective dispersion and polar contributions of the solid, while σ_1^d and σ_1^P represent the dispersion and polar parts of the liquid. The values obtained from Eqs. 6.3–6.5 were substituted into Eq. 6.2. The linear regression method was used to fit the unknown quantities to obtain the surface energy of solid surfaces using the following equation:

$$\frac{1 + cos\theta}{2} \cdot \frac{\sigma_1}{\sqrt{\sigma_1^d}} = \sqrt{\sigma_S^P}\sqrt{\frac{\sigma_1^P}{\sigma_1^d}} + \sqrt{\sigma_S^d} \tag{6.6}$$

where θ is obtained from the CA measurements for various surface chemistries. The surface energy related data are summarized in Table 6.2.

The ability of the osteogenic cells to produce mineralized bone-like nodules, while cultured on the injection molded HDPE composites, is confirmed using alizarin red staining of the immobilized osteogenic cells. Mineralization is considered as the strongest indicator of true osteoblast differentiation and osteogenesis. IS4 composite supported better mineralization than that on control/HA at 7/14 days in culture. Here again, it can be said that the osteogenic properties of HA component is responsible for the mineralization of HDPE–HA–Al_2O_3 composites.

A significant increase in cell-mediated mineral deposition was measured with IM3 composites, that contained the largest amount of the osteoinductive HA particles. The significant increase in expression of alizarin red staining on IS2 and IS4 after 21 days *in vitro* culture is an encouraging result from the *in vitro* osteogenesis perspective (Fig. 6.8). Such calcium deposition is seen at earlier time points, but increased significantly at 21 days, particularly for IS4 composite. This result was promising as it illustrates the potential of the HDPE–20 wt% HA–20 wt% Al_2O_3 composites to support the development of bone tissue from an osteoblast proliferation stage through to extracellular matrix (ECM) deposition and further to cell-mediated early-stage mineralization.

Interestingly, this effect was not observed in case of IS2 or IS4 composite after 14 days in culture. Clearly, the addition of HA to the scaffolds results in a significantly increased level of calcium deposition and has a mild osteogenic effect. The results of this study suggested that this may be due to two distinct effects as a result of the inclusion of hydroxyapatite particles within the composite, namely the osteoinductive effect of the HA particles, when their inclusion does not significantly increase scaffold stiffness. The combination of the osteoinductive HA particles as well as a decrease in contact angle results in a dramatic increase in mineral deposition within 14/21 days.

The reduction in mineralization with the full replacement of HA by Al_2O_3 and a sharp increase in mineralization on IM3 with respect to IM2 substrates follows

Table 6.2 Composition of the various polymer-ceramic composites and their calculated surface free energy and its component [4]

Sample designation	Composite composition	Surface free energy (mN/m)	Polar component of SFE (SFED) (mN/m)	Dispersion component of SFE (SFEP) (mN/m)
IS1	HDPE + 40 wt% HA (2 % coupling agent)	21.7 ± 0.2	15.4 ± 1.1	6.3 ± 1.3
IS2	HDPE + 40 wt% Al_2O_3 (2 % coupling agent)	23.7 ± 0.6	15.8 ± 0.9	7.8 ± 0.3
IS#	HDPE + 20 wt% HA + 20 wt% Al_2O_3 (2 % coupling agent)	28.9 ± 0.9	18.3 ± 0.8	10.4 ± 0.3

similar correlation between cell viability and polar component of SFE (Table 6.2). Although such increase is rather modest after 7 days of culture, but mineralization is significantly increased after 14 and 21 days in culture.

The *in vitro* results of this study can be rationalized in the light of earlier literature reports. It is widely reported in literature that the surface topography, surface roughness and texture influences the adsorption of various proteins *in vitro* [61–63] which subsequently influences the cell attachment and adhesion [64–66]. Overall, the *in vitro* culture results clearly demonstrate that spatial cell growth and mineralization can be modulated on surfaces with higher SFE in case of HDPE-based composites.

6.4.6 Osseointegration Property

The HDPE–HA–Al_2O_3 composites (IS4) were implanted in the long bones of New Zealand rabbits weighing an average of 2 kg (±0.100 g). The experimental design composed of three groups of 4 rabbits in each group. HDPE implants were used as controls to assess whether the addition of HA in combination with Al_2O_3 degrades the osseointegration property. The injection molded plates were machined to obtain implants of 6 mm × 2 mm rod shape. The size and shape of the samples and defect size were prepared following the ISO-10993 guidelines.

All animals were recovered uneventfully from surgery. Their health status remained normal with no change in body weight. Implants were clearly identified in most cases. The implant margins were clearly identified at one week and indistinguishable at 12 weeks post implantation, indicating neobone deposition with good osseointegration with implant (Fig. 6.10). A closer look at the radiographs in Fig. 6.10 reveals the outline of implant, indicating stable placement at the cylindrical femoral defects (1 week). However, such visual difference between implant and neighboring bone is not much clear after 12 weeks of implantation. It is possible that the newly formed bone completely integrated with host bone and also underwent mineralization over 12 weeks of implantation. The mineralized neobone has similar density of host bone in the defect region, leading to the difficulty in distinguishing the control implant from host bone. This also suggests thickening of de novo cortical bone similar to natural femur during post surgery of 12 weeks.

Grossly, all implant sites showed good healing at all time periods for both control (HDPE) and test (HDPE–20 wt% HA–20 wt% Al_2O_3 with 2 % coupling agent) implant. Histologically, there was good healing at the implant/bone interface (Fig. 6.10). The orientation of host cortical bone was perpendicular to implant. The histological features were similar around both, HDPE and IS4 at all time periods.

One week post-implantation, most implants were identified in cortical bone with a moderate inflammatory infiltrate of macrophages (round to oval cells with a large nucleus) and fibroblasts (plump spindle shaped cells) at the host bone-implant interface. Trabeculae of osteoid with occasional osteoblasts were observed at the interface on both sides. Degenerative and necrotic changes were absent.

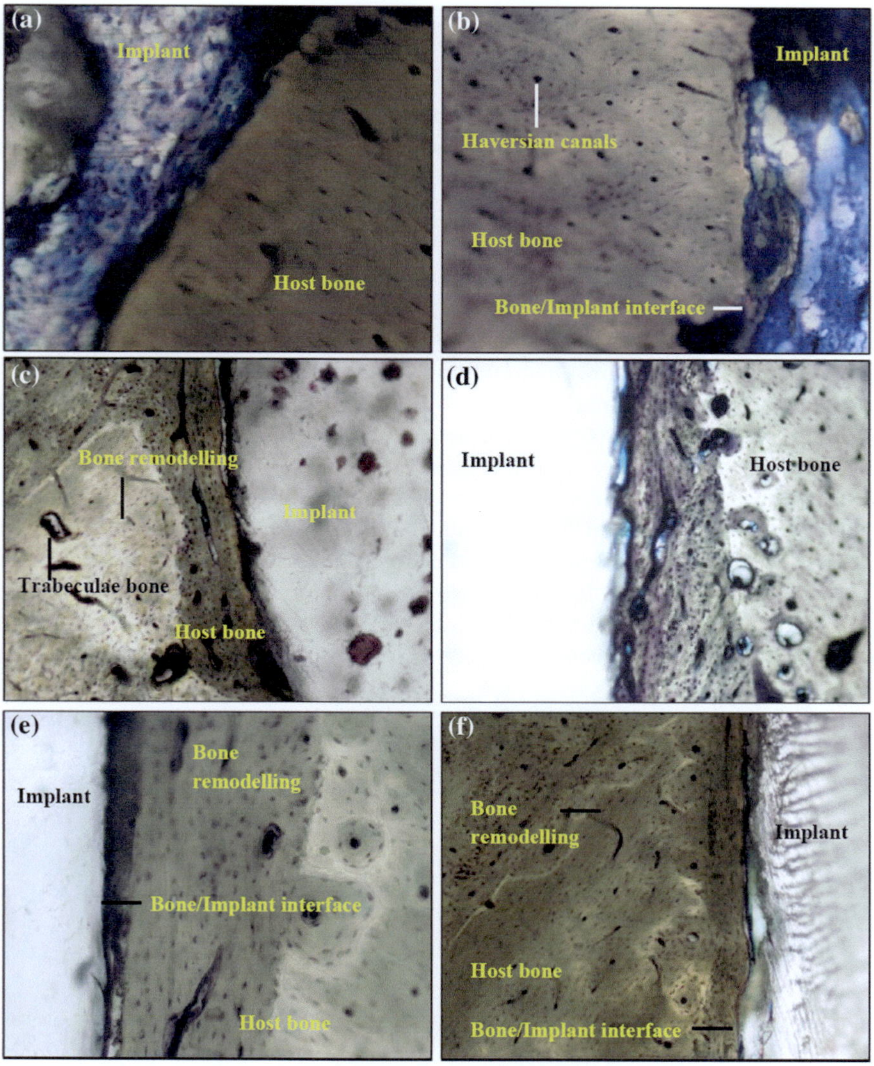

Fig. 6.10 Representative histology images revealing rabbit host bone-IM3 test implant composites interface (*left*) (**a**) and host bone-HDPE (control) (**b**) (*right*) interface at 1 week, **c** and **d** for 4 weeks and **e** and **f** for 12 weeks post implantation. The macrophages and fibroblasts are found to infiltrate at host bone interface with IM3 composite (**a**) and HDPE (**b**) at 1 week. Trabeculae of new bone are also observed in both, with osteoblasts identified in IM3 (**a**). Mature new bone deposition is present at 4 weeks at the interface of both implants with focal contact at the HDPE composite interface (**c**) and along the implant surface of both implants (control and IM3) at 12 weeks (**e** and **f**) Stevenel's blue X 10 (**a**–**d**) and X40 (**e**–**f**) [4]

Neovascularisation was not seen at this time period. New woven bone trabeculae were also noted at both edges of endosteal aspect of the cavity.

At four weeks of post implantation, new woven bone deposition was evident along both sides of host cortical bone of the cavity in both the groups with focal contact with implant and occasional macrophages, fibroblasts and fibrocytes at the interface. Degenerative and necrotic changes are absent at the bone-implant interface on both sides of the cavity in both groups. Neovascularisation was minimal in both control and test implant. Neobonetrabeculae were also noticed on periosteal and endosteal aspects in most cases.

At twelve weeks post-implantation, healing was complete with new mature bone interfacing mostly directly along both sides of the implant. In a few cases, fibrocytes persisted. Histomorphometrically, the neobone growth was further quantified from the multiple histology images. The neobone area at proximal/middle and distal sites around HDPE (control) and HDPE–20 wt% HA–20 wt% Al_2O_3 (IS4 grade-test implant) were measured and plotted in Fig. 6.11. The results reveal a considerable increase ($\sim$125 %) of neobone area at proximal and middle site for IS4 over control implants. However, such increase at distal site was not recorded.

A number of parameters derived from histopathological analysis have been qualitatively summarized in Table 6.3. In all the cases, the relative size of bone area around the implant was identical and therefore, the area of healing did not differ from implant to implant. In all the cases, mild inflammation as well as fibrous capsule formation was observed. More importantly, the necrotic cells or any sign of toxicity at tissue level could not be recorded even after 12 weeks of post implantation.

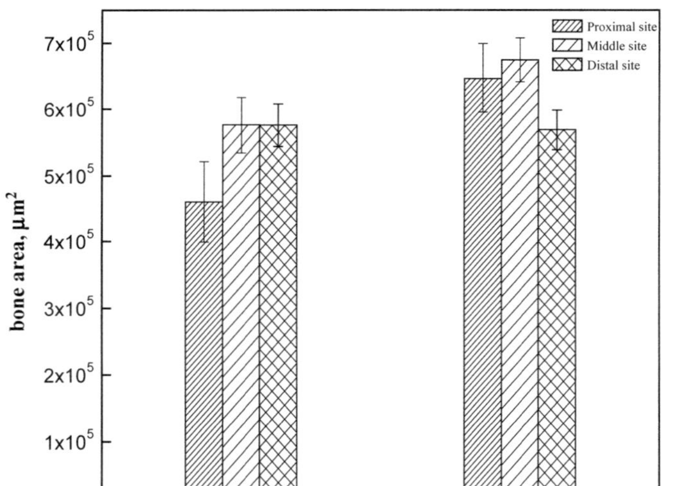

Fig. 6.11 Histomorphometrical evaluation of bone growth around control and implant after 12 weeks post-implantation in rabbit animal model [4]

Table 6.3 Summary of the qualitative histopathological evaluation of rabbit femur adjacent to IM3 composite at selected time intervals of implantation in rabbit animal model [4]

Materials	HDPE–20 wt% HA–20 wt% Al_2O_3, (IS4)		
Number of rabbits	4	4	4
Duration of implant (weeks)	1	4	12
Foreign body giant cells	Absent	Absent	Absent
Gross response	Positive	Positive	Positive
Degree of necrosis	Nil	Nil	Nil
Inflammation	Mild	Mild	Mild
Macrophages/fibrous capsule	Present	Present	Present
Foreign body debris	Absent	Absent	Absent
Fibrosis	Present	Present	Present
Relative size of implanted area (mm)	2–6	2–6	2–6
Toxicity rating	Nil	Nil	Nil

Summarising, Histomorphometrically, HDPE–20 wt% HA–20 wt% Al_2O_3 composite supports 125 % more neobone formation at proximally middle site of implant, when compared to HDPE (see Fig. 6.11). On the basis of the absence of mild inflammation and the presence of neobone formation at the interface, the suitability of the HDPE–20 wt% HA–20 wt% Al_2O_3 composite for the stabilization of bone defects or to bridge available gaps between existing host bone and another synthetic implant during orthopedic surgery is being established.

6.5 Summary

It is well known that the longevity of orthopedic implants depends upon implant fixation and interfacial stability. Loosening and failure of an implant are often reported due to instability and incomplete anchorage at the bone/implant interface [67]. There are different approaches to improve the stability/anchorage and these approaches include bioactive fixation by using compositionally tailored biomaterials, cemented fixation, mechanical bone implant interlocking, etc. To this end, the use of HDPE-based hybrid composites is established in this chapter.

At the closure, it can therefore be unambiguously confirmed that HDPE–20 wt % HA–20 wt% Al_2O_3 is clinically safe materials in rabbit animal model and such materials enhance bone formation both *in vitro* and *in vivo*. Also, the overall biocompatibility property is not compromised to any extent despite the presence of considerable amount of Al_2O_3 in the implant leading to an enhancement in mechanical properties. The clinical trials on HDPE–20 wt% HA–20 wt% Al_2O_3 are to conducted to further confirm its efficacy to stabilize bone defects or to bridge available gaps between existing host bone and another synthetic implant during orthopedic surgery in human patients.

References

1. Basu B, Katti D, Kumar A. Advanced Biomaterials: Fundamentals, Processing and Applications. USA: Wiley; 2009.
2. Ratner BD, Hoffman AS, Schoen FJ, Lemons JE. Biomaterials Science: An Introduction to Materials in Medicine. London: Elsevier Academic Press; 2004.
3. Basu B, Jain D, Kumar N, Choudhury P, Bose A, Bose S, Bose P. Processing, tensile and fracture properties of Injection Molded HDPE-Al_2O_3-HAp Hybrid Composites; J Appl Poly Sci. 2011;121:2500–2511.
4. Tripathi G, Basu B. *in vitro* osteogenic cell proliferation, mineralization and *in vivo* osseointegration of injection molded HDPE-based hybrid composites in rabbit animal model; J Biomater Appl. 2014;29:142–157.
5. Bodhak S, Nath S, Basu B. Fretting wear properties of hydroxyapatite, alumina containing high density polyethylene biocomposites against zirconia. J Biomed Mater Res, Part A. 2008;85:83–98.
6. Nath S, Bodhak S, Basu B. Tribological investigation of novel HDPE–HAp–Al_2O_3 hybrid biocomposites against steel under dry and simulated body fluid condition. J Biomed Mater Res Part A 2007;83A(1):191.
7. Bonfield W, Grynpas MD, Tully AE, Bowman J, Abram J. Hydroxyapatite reinforced polyethylene—a mechanically compatible implant material for bone replacement. Biomaterials. 1981;2:185–6.
8. Bonfield W, Grynpas MD, Tully AE, Bowman J, Abram J. Biomaterials. 1981;2:185.
9. Bonfield W. J Biomed Eng. 1988;10:522.
10. Huang J, Di Silvio L, Wang M, Tanner KE, Bonfield W. J Mater Sci Mater Med. 1997;8:775.
11. Huang J, Di Silvio L, Wang M, Rehman I, Ohtsuki IC, Bonfield W. J Mater Sci Mater Med. 1997;8:808.
12. Suwanprateeb J, Tanner KE, Turner S, Bonfield W. J Mater Sci Mater Med. 1997;8:469.
13. Guild FJ, Bonfield W. Biomaterials. 1993;13:985.
14. Ward IM, Bonfield W, Ladizesky NH. Polym Int. 1997;43:333.
15. Ladizesky NH, Pirhonen EM, Appleyard DB, Ward IM, Bonfield W. Comp Sci Technol. 1998;58:419.
16. Reis RL, Cunha AM, Oliveira MJ, Campos AR, Bevis M. J Mat Res Innovat. 2001;4:263.
17. Pandey A, Jan E, Aswath PB. J Mater Sci. 2006;41:3369.
18. Joseph R. Development of an injection moulding grade hydroxyapatite polyethylene composite. Ph.D. thesis, University of London, 2001.
19. Evans SL, Gregson PJ. Biomaterials. 1988;19:1329.
20. Laurencin CT, Ambrosio MA, Borden M, Cooper JA Jr. Tissue engineering: orthopedic applications. Annu Rev Biomed Eng. 1999;1:19–46.
21. Cooper SL, Visser SA, Hergenrother RW, Lamba NMK. Polymers. In: Ratner BD, Hoffman AS, Schoen FJ, Lemons JE, editors. Biomaterials Science. San Diego: Elsevier Academic Press; 2004. p. 67.
22. Pramanik S, Agarwal AK, Rai KN, Garg A. Ceram Int. 2007;33:419.
23. Downs RN, Vardy S, Tanner KE, Bonfield W. Hydroxyapatite polyethylene composite in orbital surgery. Bioceramics. 1991;4:239–46.
24. Bonfield W. Composite biomaterials: present and future. Bioceramics. 1998;11:37–40.
25. Dornhoffer JL. Hearing results with the Dornhoffer ossicular replacement prostheses. Laryngoscope 1998;108:531–6.
26. Huang J, Di Silvio L, Wang M, Tanner KE, Bonfield W. In vitro mechanical and biological assessment of hydroxyapatite rein-forced polyethylene composite. J Mater Sci. 1997;8:775–9.
27. Suwanprateeb J, Tanner KE, Turner S, Bonfield W. Influence of Ringer's solution on creep resistance of hydroxyapatite reinforced polyethylene composite. J Mater Sci. 1997;8:469–72.

28. Suwanprateeb J, Tanner KE, Turner S, Bonfield W. Influence of sterilization by gamma irradiation and of thermal annealing on creep of hydroxyapatite reinforced polyethylene composites. J Biomed Mater Res. 1998;39:16–22.
29. Wang M, Joseph W. Bonfield; Hydroxyapatite polyethylene composites for bone substitution: effect of ceramic particle size and morphology. Biomaterials. 1998;19:2357–66.
30. That T, Tanner KE, Bonfield W. Fatigue characterization of hydroxyapatite-reinforced polyethylene composite. I. Uniaxial fatigue. J Biomed Mater Res. 2000;51:453–60.
31. Pandey A, Jan E, Aawath PB. Physical and mechanical behavior of hot rolled HDPE/HA composites. J Mater Sci. 2006;41:3369–76.
32. German RM, Bose A. Injection molding of metals and ceramics. Princeton: Metal Powder Industries Federation; 1997.
33. Mutsuddy BC, Ford RG. Ceramic injection molding. New York: Chapman and Hall Inc.; 1995.
34. Wang M, Joseph R, Bonfield W. Biomaterials. 1998;19:2357.
35. Ton That PT, Tanner KE, Bonfield W. J Biomed Mater Res. 2000;51:461.
36. Wang M, Bonfield W. Biomaterials. 2001;22:1311.
37. Wang M, Porter D, Bonfield W. Br Ceram Trans. 1994;93:91.
38. Cihlar J, Trunec M. Injection moulded hydroxyapatite ceramics. Biomaterials. 1996;17: 1905–11.
39. Mondrinosa MJ, Dembzynskib R, Lub L, Byrapogub VKC, Woottonb DM, Lelkesa PI, Zhoub J. Porogen-based solid freeform fabrication of polycaprolactone–calcium phosphate scaffolds for tissue engineering. Biomaterials. 2006;27:4399–408.
40. Josepha R, McGregora WJ, Martynb MT, Tannera KE, Coatesb PD. Effect of hydroxyapatite morphology/surface area on the rheology and processability of hydroxyapatite filled polyethylene composites. Biomaterials. 2002;23:4295–302.
41. Abu Bakar MS, Cheang P, Khor KA. Mechanical properties of injection molded hydroxyapatite polyetheretherketone biocomposites. Compos Sci Technol. 2003;63:421–5.
42. Juang H, Hon M. The effect of calcination temperature on the behaviour of HA powder for injection moulding. Ceram Int. 1997;23:383–7.
43. Sahebian M, Gebarjad SM, Sajjadi SA, Sherafat Z, Lazzeri A. Effect of both uncoated and coated $CaCO_3$ on fracture toughness of HDPE/$CaCO_3$ nanocomposites. J Appl Polym Sci. 2007;104:3688–94.
44. Wang M, Porter D, Bonfield W. Processing, characterization, and evaluation of hydroxyapatite reinforced polyethylene composites. Br Ceram Trans. 1994;93:91–5.
45. Li GH, Hu ZX, Zhang LD, Zhang ZR. Elastic modulus of nano-alumina composite. J Mater Sci Lett. 1998;17:1185–1186.
46. Kim B, Park SW, Lee DG. Fracture toughness of the nano-particle reinforced epoxy composite. Compos Struct. 2008;86:69–77.
47. ASTM D5045-99. Standard test methods for plane-strain fracture toughness and strain energy release rate of plastic, American Society for Testing and Materials (ASTM); 2007.
48. Yan J, Clifton KB, Mecholsky J, John J, Reep RL. Fracture toughness of manatee rib and bovine femur using a chevron-notched beam test. J Biomech. 2006;39(6):1066–74.
49. Matthews FL, Rawlings RD. Composite materials: engineering and science. London: Chapman and Hall; 1994.
50. Silva RV, Spinelli D, Bose Filho WW, Claro Neto S, Chierice GO, Tarpani JR. Fracture toughness of natural fibers/castor oil polyurethane composites. Compos Sci Technol. 2006;66:1328–35.
51. Wolff JU. ber die Wechselbeziehungenzwischen der Form und der Function der EinzelnenGebilde des Organismus, 1901 (FCW Vogel, Leipzig, Germany).
52. Ehrlich PJ, Lanyon LE. Mechanical strain and bone cell function: a review. Osteoporos. Int. 2002;13(9):688–700.
53. Chen JH, Liu C, You L, Simmons CA. Boning up on Wolff's law: mechanical regulation of the cells that make and maintain bone. J Biomech. 2010;43(1):108–18.

54. You L, Temiyasathit S, Lee P, Kim CH, Tummala P, Yao W, et al. Osteocytes as mechanosensors in the inhibition of bone resorption due to mechanical loading. Bone. 2008;42(1):172–9.
55. Cowin SC, Moss-Salentijn L, Moss ML. Candidates for the, mechanosensory system in bone. J Biomech Engg. 1991;113(2):191–7.
56. Lanyon LE. Osteocytes, strain detection, bone modeling and remodeling. Calcif Tissue Int. 1993;53(Suppl 1):S102–7.
57. Klein-Nulend J, van der Plas A, Semeins CM, Ajubi NE, Frangos JA, Nijweide PJ, et al. Sensitivity of osteocytes to biomechanical stress *in vitro*. FASEB J. 1995;9(5):441–5.
58. Burger EH, Klein-Nulend J. Mechanotransduction in bone-role of the lacuno-canalicular network. FASEB J. 1999;13:S101–12.
59. Wang D, Douma M, Oleschuk RD, Horton JH. J Colloid Interface Sci. 2009;331:90.
60. Gupta A, Tripathi G, Basu B, Balani K. Dependence of protein adsorption on wetting behavior of UHMWPE–HA–Al_2O_3–CNT hybrid biocomposites. JOM. 2012;. doi:10.1007/s11837-012-0295-3.
61. Francois P, Vaudaux P, Taborelli M, Tonetti M, Lew DP, Descouts P. Influence of surface treatments developed for oral implants on the physical and biological properties of titanium. (II) Adsorption isotherms and biological activity of immobilized fibronectin. Clin Oral Implant Res. 1997;8:217–225.
62. Deligianni DD, Katsala N, Ladas S, Sotiropoulou D, Amedee J, Missirlis YF. Effect of surface roughness of the titanium alloy Ti-6Al-4 V on human bone marrow cell response and on protein adsorption. Biomaterials. 2001;22:1241–51.
63. Park JY, Davies JE. Red blood cell and platelet interactions with titanium implant surfaces. Clin Oral Implant Res. 2000;11:530–9.
64. Bowers KT, Keller JC, Randolph BA, Wick DG, Michaels CM. Optimization of surface micromorphology for enhanced osteoblast responses *in vitro*. Int J Oral Maxillofac Implant. 1992;7:302–310.
65. Keller JC, Schneider GB, Stanford CM, Kellogg B. Effects of implant microtopography on osteoblast cell attachment. Implant Dent. 2003;12:175–81.
66. Deligianni DD, Katsala ND, Koutsoukos PG, Missirlis YF. Effect of surface roughness of hydroxyapatite on human bone marrow cell adhesion, proliferation, differentiation and detachment strength. Biomaterials. 2001;22:87–96.
67. Sundfeldt M, Carlsson LV, Johansson CB, Thomsen P, Gretzer C. Aseptic loosening, not only a question of wear: a review of different theories. Acta Orthop. 2006;77:177–97.

Chapter 7
Case Study: Development of Acetabular Socket Prototype

7.1 Introduction

The natural hip joint in human body is generally called as a ball-and-socket joint, because the spherical head of the thighbone (femur) moves inside the cup-shaped hollow socket (acetabulum) of the pelvis. Healthy cartilage, which serves as a protective cushion allows smooth, low-friction movement of the joint. Additionally, a body fluid, composed of water, hyaluronic acid lubricates the joint area. During day-to-day activities, the average load on a hip joint is estimated to be up to three times body weight and sometimes it goes up to 10 times of the body weight during various strenuous activities such as jumping and exercise. Such activities cause severe wear of the prosthetic socket. If the cartilage is damaged by disease or injury, the tissues around the joint become inflamed, causing pain. With time, the cartilage wears away, causing the rough edges of bone to rub against each other, and this results in degenerative diseases like osteoarthritis causing hip pain and stiffness. An example of fractured hip joint has been shown in Fig. 7.1. In order to relieve pain, the clinical approach is to use total hip joint replacement with artificial biomaterials. The most common reason for having a hip or knee replaced is osteoarthritis, which mainly affects people over the age of 45.

Therefore, in an attempt to replicate natural hip joint, implantation of artificial hip joints is importantly required. Over the past half-century, there have been many advances in the design, construction and implantation of artificial hip joints. However, issues pertaining to the design of the prosthesis and the high cost of prosthesis limit the usage of such implants. The present chapter highlights the efforts to widen the base of this treatment and to increase its suitability in terms of design parameters. Before describing these developments in detail, the basic structure and components of an artificial hip prosthesis are discussed. A total hip replacement implant has three parts: the stem, which fits into the femur and provides stability; the ball, which replaces the spherical head of the femur and the cup, which replaces the worn-out hip socket. Figure 7.2 shows an X-ray radiograph of a

B. Basu and S. Ghosh, *Biomaterials for Musculoskeletal Regeneration*,
Indian Institute of Metals Series, DOI 10.1007/978-981-10-3017-8_7

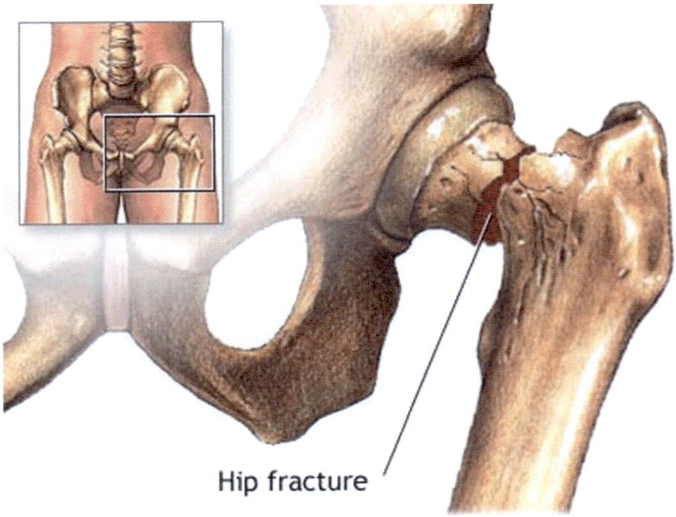

Fig. 7.1 A typical example showing the features of hip fracture [1]

natural implant in comparison to an artificial hip implant along with its different components. Generally, the stem part is made of metal with a metallic or ceramic femoral head attachment, whereas the Acetabular cup is usually made of polymers.

7.2 Clinical Perspective

From the clinical perspective, the inflammatory and degenerative joint diseases necessitate surgical joint replacement [2]. Total hip replacement is one of the most successful surgical procedures that have been developed so far, and is natural choice of treatment for long term function restoration and pain relief. Over the years there has been a change in the focus of total hip replacement implants. Earlier, in 1960s and 1970s, when this technique was still in its infancy, the main clinical focus was to reduce perioperative complications of infection, dislocation and implant failure. Since this process has now been well established, the next phase focused on improving the functionality of THR surgeries, in the form of pain reduction and improved quality of life. This was achieved mainly by improved surgical techniques as well as better implant design. Recently, the clinical need demands better long term sustainability of the implant to improve its acceptance and functionality even in a younger age group. Considering that the traumatic, inflammatory and degenerative diseases are increasingly affecting younger individuals, there is a need to look at newer materials with better wear resistance and biocompatibility properties to increase the life of the prosthetic joints. This involves a special emphasis on reduction of implant failure due to mechanical loosening and due to the ill effects of

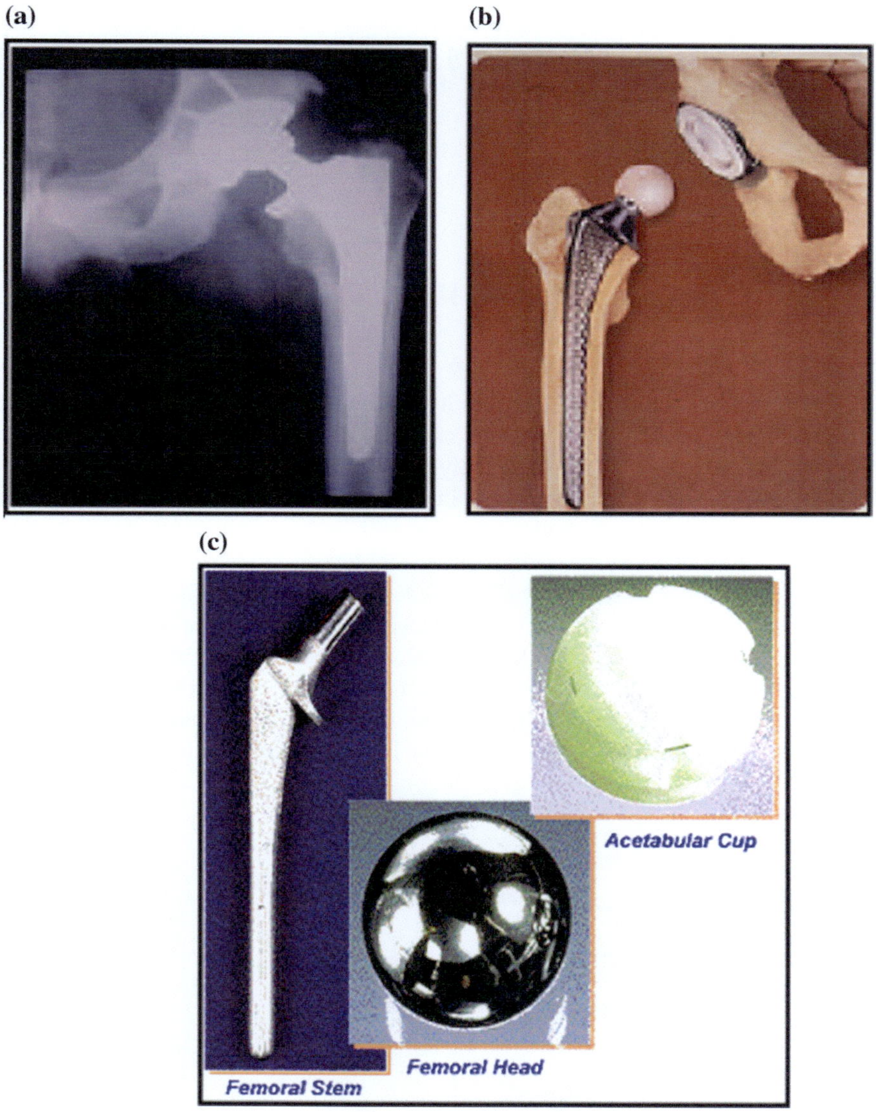

Fig. 7.2 An X-ray showing the natural hip joint (**a**); assembly of an artificial hip joint using biomaterials (**b**) and components of a total hip replacement (**c**) [45]

debris particles, which have been mentioned in the previous section [3]. This involves improvement of currently available biomaterials for artificial joints which have worked successfully for the older age group of patients. There is evidence that joints last 25–30 years with currently available bearing materials [4]. In addition, the wear particles of the currently available and most widely used UHMWPE bearing surfaces have shown to cause an inflammatory reaction at the bone—joint or bone cement interface. This results in painful loosening of the implant, warranting revision surgeries at a larger cost and patient trauma. Therefore, a bearing surface with high hardness and good wear resistant properties is urgently required. Moreover, effective processing of such composite in a cost-effective compression molding route would offer a long term solution towards the acetabular component development for THR application at reasonable cost.

7.3 Polymer Composites in THR

The first total hip joint replacement (THR) was introduced in the 1950s and over the last few decades, considerable progress has been achieved in the joint-replacement prostheses surgery. In US alone, the case of approximately 1,50,000 total hip replacement surgery are reported annually and significantly 20 % of these surgery is for replacement of the unsuccessful hip implants [5]. Figure 7.3 shows the different components of the total hip replacement as well as their post operation alignment along with other bones of the pelvis. From this figure, it is clearly visible that the femur head articulates with acetabular socket after assembly of the THR [6, 7].

It has been further reported that wear related mechanical failure has been a major concern as far as the life of hip implants are concerned. Therefore, there exists a need to develop new materials or to explore new material combination for THR application. In the early days of hip surgery, both femoral implants and acetabular cups made of stainless steel, Co–Cr–Mo or its alloy were considered suitable. In

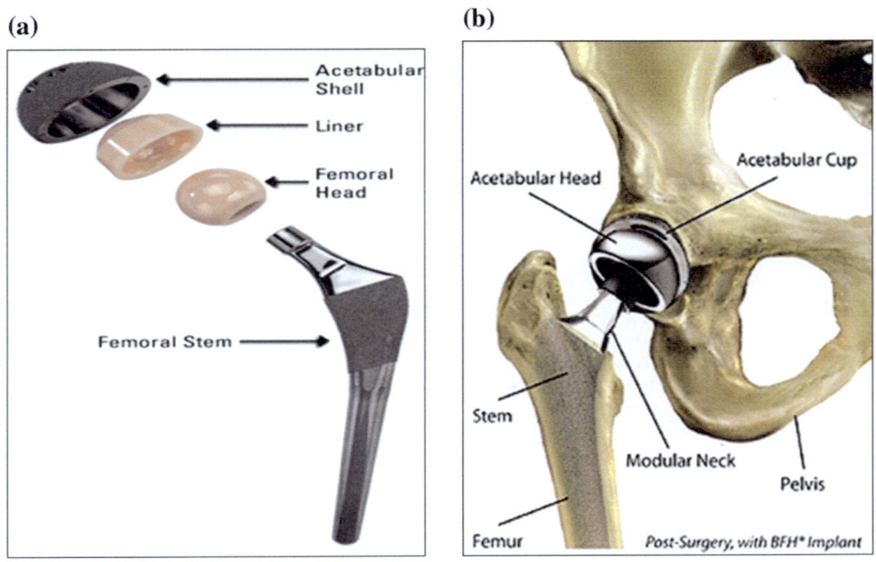

Fig. 7.3 Different components of a total hip replacement **a** before and **b** after implantation [45]

general, metallic implants exhibit good mechanical properties, but due to poor biocompatibility and the mismatch of elastic modulus with the surrounding host bone, they cause stress-shielding effect after implantation. This limits metal-to-metal pairing from extensive THR applications. Another important concern for the use of metallic implants is the metal ion release, which causes an increase in cobalt and chromium concentrations in blood and urine (in case of Co–Cr–Mo implants). The debris particles/lumps due to wear of metallic articulating surfaces lead to osteolysis, which ultimately results in implant loosening. This process has been outlined in Fig. 7.4, wherein the activation of the foreign body response due to wear debris generation has been described. These debris are phagocytosed by macrophages with particles of 500 nm or larger being transported to the lysosome. These metallic particles are corroded in the lysosome, leading to high concentration of ions such as Co and Cr in blood.

As far as the use of non-metallic THR components is concerned, low coefficient of friction, easy modulability, superior biocompatibility and low cost of making are important criteria for material selection for acetabular cup application in hip replacement surgery. However, it has been well established that the longevity of such joint implants depends on the wear performance of the polymeric acetabular cup. Charnley et al. [9] proposed 'low friction arthroplasty' and introduced PTFE

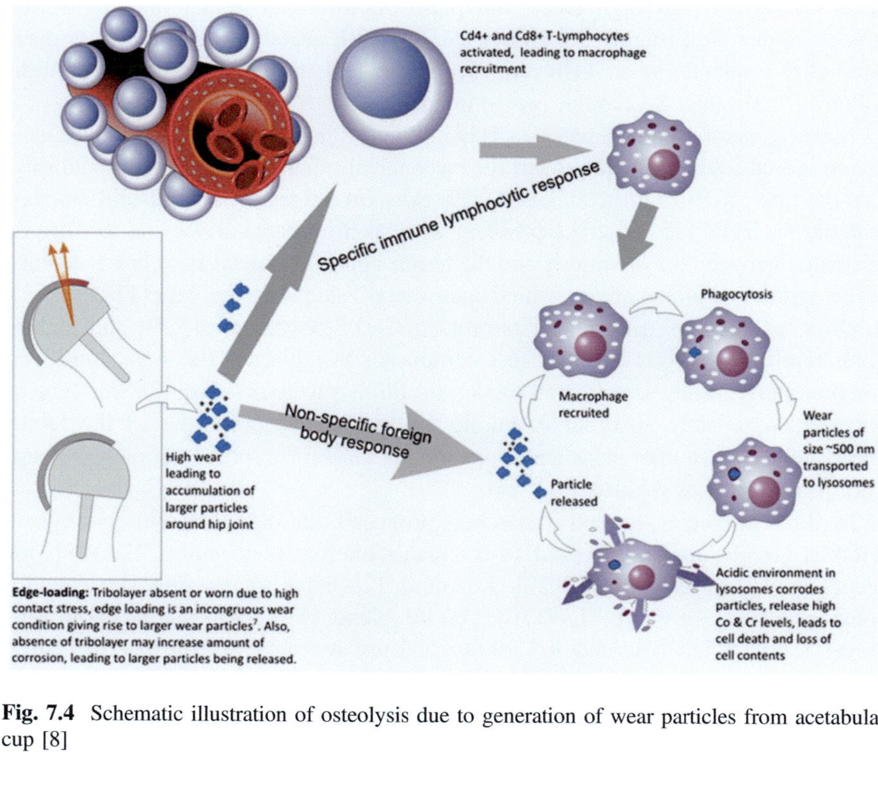

Fig. 7.4 Schematic illustration of osteolysis due to generation of wear particles from acetabular cup [8]

(polytetrafluroethylene) as acetabular cup in a metal to polymer pairing for total hip joint prosthesis. The major drawback of PTFE is its low wear resistance. It was experimentally observed that the stainless steel femoral head completely penetrated the PTFE acetabular cup within a few years of implantation. In an effort to solve the wear problem of metal-on-polyethylene in the hip joint, manufacturers have produced hip prostheses with ceramic-on-polyethylene. This accounts for the introduction of ceramic femoral heads, as an alternative to metallic heads. In 1969, Driskell et al. [10] introduced zirconia in the biomaterial application and later Christel et al. [11] was the first to illustrate the use of zirconia as femoral head in total hip replacement joints. Such orthopedic implants usually comprise a ceramic femoral head, mounted on metallic femoral stems, that articulates against a polymeric acetabular cup during in vivo use.

For more than 30 years, ultra-high-molecular-weight polyethylene (UHMWPE) has been used as a bearing material in total-joint-replacement prostheses [12]. UHMWPE has been successfully used as acetabular cup material due to its superior mechanical toughness, wear resistance and biocompatibility. However, the main concern about the use of UHMWPE is the adverse biological tissue response due to its wear debris [13, 14].

Although effective for short term use, UHMWPE has several other drawbacks besides generation of large amount of wear debris. Figure 7.5 shows an X-ray radiograph in which the wear debris particles are clearly visible outside the femoral head. UHMWPE has high creep and plastic deformation under long term load bearing application. Figure 7.6 shows a UHMWPE acetabular cup that was damaged after implantation in THR surgery. The extent of damage provides an indication of the progressive wear, post-implantation.

A new generation of biomaterials is being investigated over the last few decades, taking into account the requirement for mechanical reliability and biocompatibility. Carbon fiber (CF) reinforced UHMWPE was investigated as an useful implant material for THR [15, 16]. The problem with such implants arises due to stiffness mismatch between the prosthesis and the femur bone. The metal stem has 5–6 times higher stiffness than host bone, which causes stress shielding problem [17–19]. As a solution, a group of researchers introduced CF/PS(polystyrene) [20] and carbon fibres reinforced carbon [21] polymer composites to replace metal stem. Improved creep property, more strength and moderate stiffness were observed for this type of polymer composites. In order to mimic the mechanical properties of the femur bone, the use of another polymer composite i.e. CF/PEEK (polyether ether ketone) stem prosthesis was reported [22–24].

In different studies, several researchers proposed that high density polyethylene (HDPE) based composites could be a suitable alternative to pure UHMWPE for acetabular cup application [25–27]. Xue et al. [28] have successfully developed a composite by reinforcing HDPE/UHMWPE blend with multi-wall carbon nanotubes (MWCNTs). The detailed tribological test revealed that an extremely low wear rate was achieved ($\sim 12 \times 10^{-8}$ mm^3/Nm), when fretted against austenitic steel counter body. In another study, considerable improvement in the wear resistance property has been reported for a newly developed composite prepared by

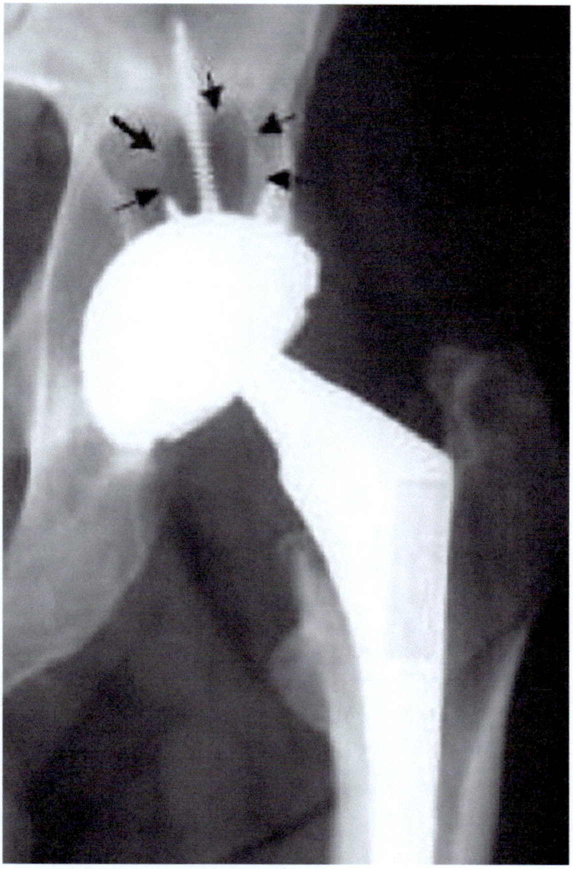

Fig. 7.5 Osteolysis taking place around the total hip replacements which is detected by X-ray diagnosis [45]

blending of UHMWPE and HDPE [29]. In another attempt to improve mechanical and tribological properties, HDPE-based biocomposites, reinforced by carbon and Kevlar fibers were developed and acetabular cup was prepared by compression moulding technique [30]. However, these materials are being developed on a lab scale and are yet to be tested for more realistic assessment of their suitability as socket materials.

In last few decades, substantial research efforts were also invested to develop bioactive composites as bone analogue replacement by reinforcing bioinert high density polyethylene matrix with bioactive hydroxyapatite (HAp) ceramic particulates [31–33]. Bonfield et al. [34–37] were the first to develop hydroxyapatite reinforced (up to 40 vol%) high density polyethylene (HDPE) biomaterial for skeletal applications and coined a trade name HAPEX™ for HA/HDPE composite [38]. In an early study Bonfield investigated the tribological behaviour of HA/HDPE composite against duplex stainless steel in dry and lubricated condition [39]. However, the result showed that beyond 10 vol.% HAp reinforcement, wear properties deteriorated in the presence of low viscous liquid. In spite of having

Fig. 7.6 Example of worn UHMWPE acetabular cup for total hip replacement [45]

excellent biocompatibility, the use of these composites in THR application is rather limited arising from the low wear resistance properties and strength mismatch of this composite with femur bone. The above discussion signifies that polymer-based new functional biocomposites for load bearing joint application are still awaited.

7.4 Acetabular Socket Prototype Fabrication

Natural bone is a polymer-ceramic composite based hard tissue, containing metabolically live cells, collagen and apatite mineral with an adequate balance of biochemical and mechanical properties. The strength, stiffness and toughness of natural bone are a result of its composite and anisotropic structure of organic (collagen) and inorganic (hydroxyapatite) materials. Without compromising on the required properties like mechanical strength and biocompatibility, a polymer-ceramic composite (HDPE–20 wt% HA–20 wt% Al_2O_3) implant has been developed. The different components of this composite were incorporated with specific aims. HDPE was used as a base material. It is important to mention here that the HDPE is a thermoplastic polymer, and hence was chosen as the base material. To HDPE, 20 wt% Hydroxyapatite was added to improve bioactivity and osseointegration and 20 wt% Al_2O_3 was added to improve stiffness and strength. In this section, the lab scale testing is first described and this will be followed by prototype fabrication.

7.4.1 Lab-Scale Testing of Coupon Samples

In order to investigate the performance of the developed optimal HDPE–HA–Al_2O_3 composition for the acetabular cup, this section summarizes the tribological, *in vitro* and *in vivo* biocompatibility results.

7.4.1.1 Fretting Wear Tests

This section contains a brief mention of the key results related to *in vitro* wear tests. In order to restore the function of body due to problems, like arthritis, joint infection, injury, and malignancy, the articulating joint should be repaired or replaced with suitable implantable devices. The biomaterial chemistry and wear resistance properties play a crucial role in the performance of such articulating devices. Concerning the mechanical properties of the developed material, higher hardness values were measured in the range of 190–256 MPa for chemically modified composites and such hardness values are near the hardness of human cortical bone (234–760 MPa). Compressive Modulus of around 700 MPa and tensile modulus of around 900 MPa were also recorded.

In order to investigate the wear properties, the developed hybrid composite was fretted against three mating counter bodies (spherical balls), including Al_2O_3, ZrO_2 and steel, *in vitro*. All the experiments were conducted at a load of 10 N for a duration of 100,000 cycles in both dry as well as simulated body fluid (SBF). Such planned set of experiments has been designed to address three important issues: (a) whether the improvement in physical properties (hardness, E-modulus) can lead to corresponding improvement in friction and wear properties; (b) whether the fretting in SBF can provide sufficient lubrication in order to considerably enhance the tribological properties, as compared to that in ambient conditions; and (c) whether the generation of wear debris particles can be reduced for various compositionally modified polymer composites, in comparison to unreinforced HDPE. The experimental results indicate the possibility of achieving extremely low coefficient of friction (see Fig. 7.7) as well as higher wear resistance (see Fig. 7.8) with the newly developed composites in SBF. A low wear depth of 3.5–4 μm is recorded in case of alumina counterface, irrespective of fretting environment. Similarly, in case of zirconia, the wear depth was 3–7 μm. Thus, on the basis of tribological results, one can conclude that the developed HDPE–HA–Al_2O_3 composite can be used either used as a liner for a metal back porous coated cemented cup or alone as an uncemented acetabular cup against different femoral ball head mating materials.

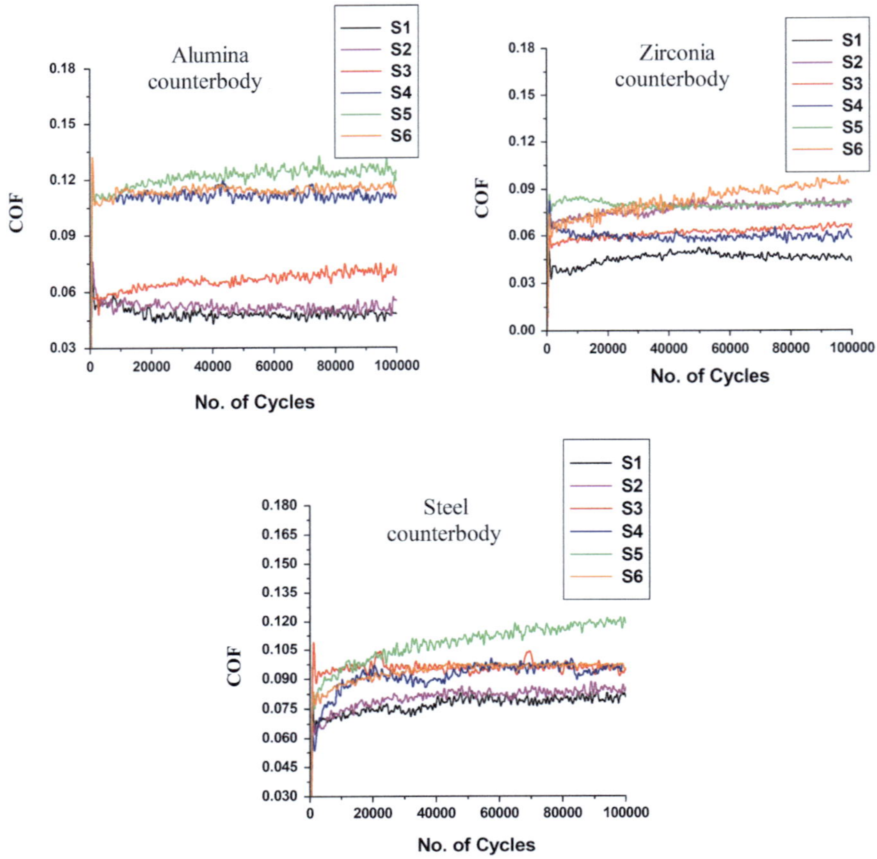

Fig. 7.7 Representative plots showing the variation in coefficient of friction w.r.t. number of cycles during fretting wear study. *S1* HDPE; *S2* HDPE–20HA; *S3* HDPE–20Al_2O_3; *S4* HDPE–20HA–20Al_2O_3; *S5* HDPE–40Al_2O_3; *S6* HDPE–40HA [40–42]

7.4.1.2 *in vitro* Cytocompatibility

in vitro studies on a material are a first step, before proceeding to in vivo evaluation. The biocompatibility tests as per ISO guidelines, using L929 mouse fibroblast cell line and $SaOS_2$ human osteoblast cells provide clear evidence in support of cell adhesion and cellular functionality of surfaces of chemically coupled composites. As shown in Fig. 7.9b, the number of metabolically active SaOS-2 cells increases systematically with an increase in incubation time for the HDPE–HA–Al_2O_3 composites. The results of the MTT assay with $SaOS_2$ cells therefore confirmed that the hybrid composite supported better cell viability as compared to the control. The cell-to-cell interaction and multidirectional extension of filopodia is also observed, as shown in Fig. 7.9a.

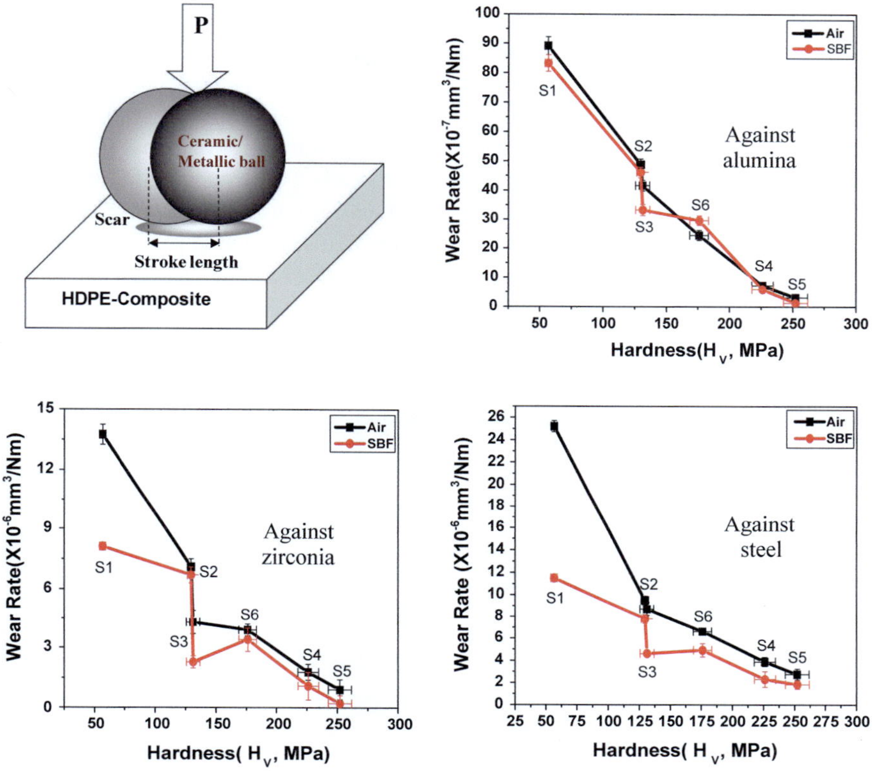

Fig. 7.8 Schematic of fretting mode (ball-on-flat) used for the ear resistance measurement (*upper left*). Representative plots (other three) showing the variation in wear rate with hardness of substrate during fretting wear study. *S1* HDPE; *S2* HDPE–20HA; *S3* HDPE–20Al_2O_3; *S4* HDPE–20HA–20Al_2O_3; *S5* HDPE–40Al_2O_3 [40–42]

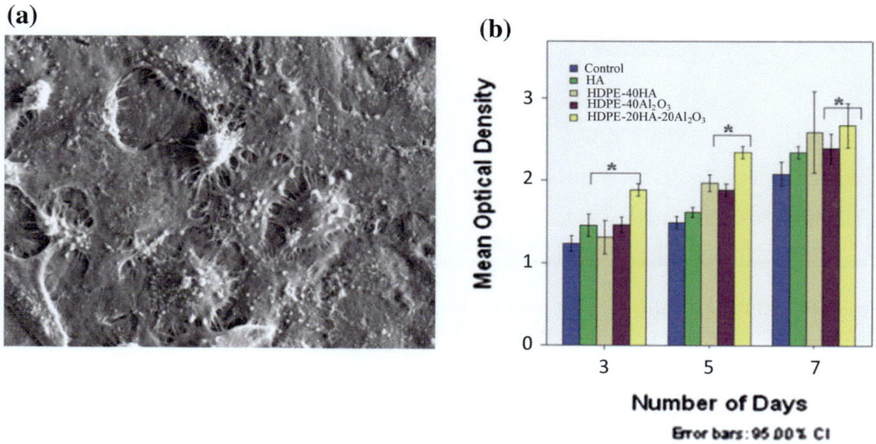

Fig. 7.9 **a** Representative SEM image showing the Saos-2 cells adhesion and growth on HDPE–20HA–20Al_2O_3. **b** MTT assay results show the cell viability/proliferation of various HDPE-composites surface during the different incubation period [43–44]

In order to investigate the effect of wear debris on the functionality of developed composite during *in vitro* conditions, eluate of HDPE–20 wt% HA–20 wt% Al_2O_3 particles was prepared and the noncytotoxic profiling of human osteoblast like cells, when treated with eluates of wear debris HDPE–20 wt% HA–20 wt% Al_2O_3 is shown in Fig. 7.10. The excellent osteoblast cell viability indicates that such debris particles from the investigated composite do not affect the functionality of human osteoblast cells. Since finer particles are generated *in vivo* due to friction and wear of the implants, the present study also indicates that such debris particles from the investigated composite would favourably support the cell functionality of human osteoblast cells. The *in vitro* results are summarized in Fig. 7.10.

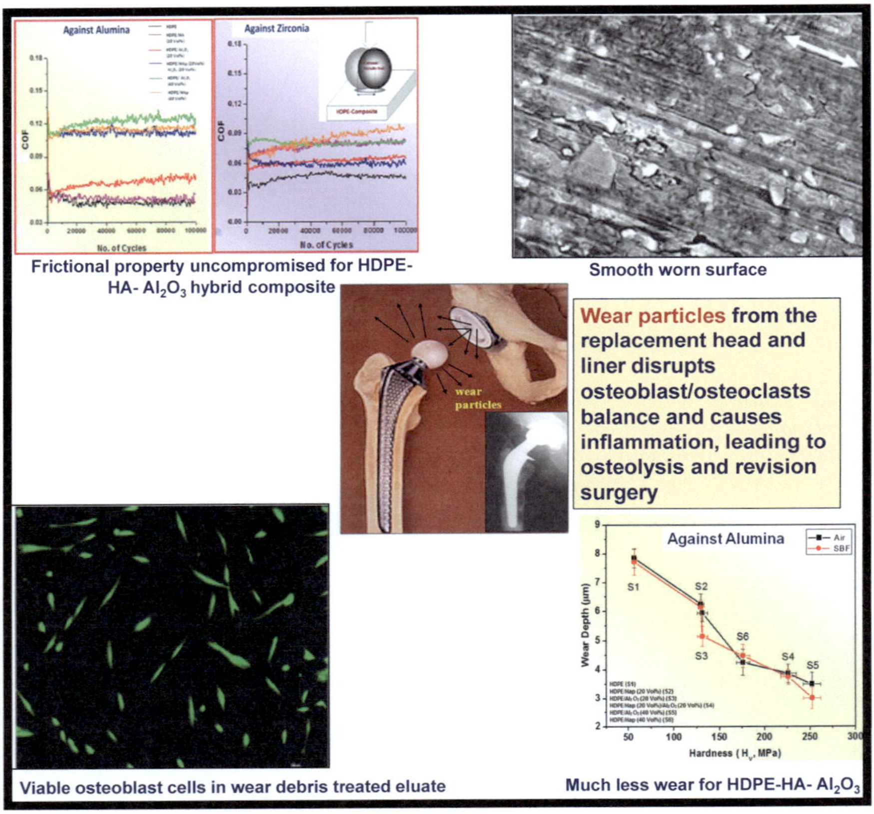

Fig. 7.10 Schematic illustration showing the central theme of wear particle induced osteolysis as the major problem driving the development of new wear resistant acetabular sockets (*center*). The frictional behaviour during fretting of several HDPE-based composites against Al_2O_3 (*top left*), typical worn surface (SEM image in *top right*), osteoblast cell morphology in wear debris treated eluate (*bottom left*) and wear resistance data (*bottom right*). The entire combinations of results establish good tribological properties of HDPE–20 % Al_2O_3 composite and non-cytotoxic nature of wear debris [40–44]

7.4.1.3 *in vivo* Osseointegration

The frictional behaviour and wear resistance data established good tribological properties of HDPE–20 % HA–20 % Al_2O_3 composites. As a next step, the biocompatibility property of this composite is established in rabbit animal model using both segmental defect and cylindrical femoral defect. The in vivo tests were carried out after approval from animal ethics committee. As shown in Fig. 7.11, a segmental defect model in rabbit animal was chosen. The in vivo biological response unambiguously confirmed good in vivo osseointegration property of the composite materials after implantation for 14 weeks as long segmental bone defect in rabbit

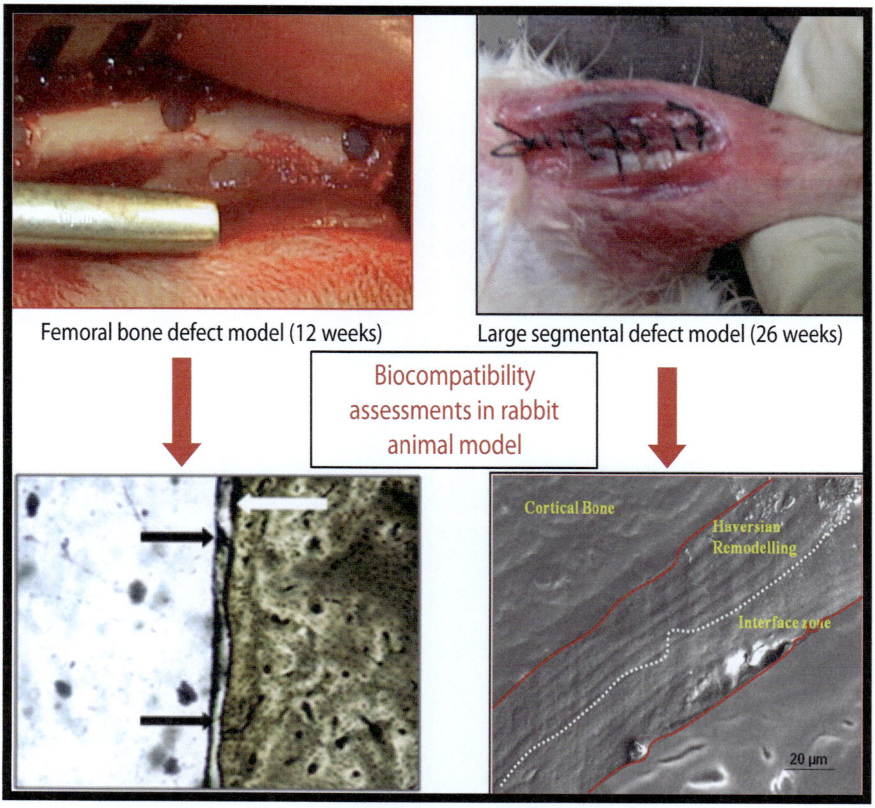

Fig. 7.11 Illustration of good osseointegration property of HDPE–20 % HA–20 % Al_2O_3 hybrid composite in animal model-cylindrical femoral bone defect (*left panel*) and large segmental bone defect (*right panel*) [44]

model. Some illustrative results are summarized in Fig. 7.11. The neo-bone formation at the implant/host bone interface in both the defect models confirms good osseointegration properties of the new hybrid composite.

The histological observations of bone cell activity as well as other observations of the bone/implant interface indicates that HDPE–20 wt% HA–20 wt% Al_2O_3 composite can support neobone deposition at the interface with cortical bone in rabbit model (see Fig. 7.12). The inflammation of any of the experimental animal (rabbit) during post implantation period was not observed. On the basis of histological analysis and radiographic observations of the bone/implant interface after short term implantation (1/4/12 weeks) in rabbit animal model, it can be stated that the investigated HDPE–20 wt% HA–20 wt% Al_2O_3 composite exhibits a consistent pattern in deposition of neobone, which appears to remodel over survival period. Minimal fibrosis and occasional macrophages separate the neobone/material interface.

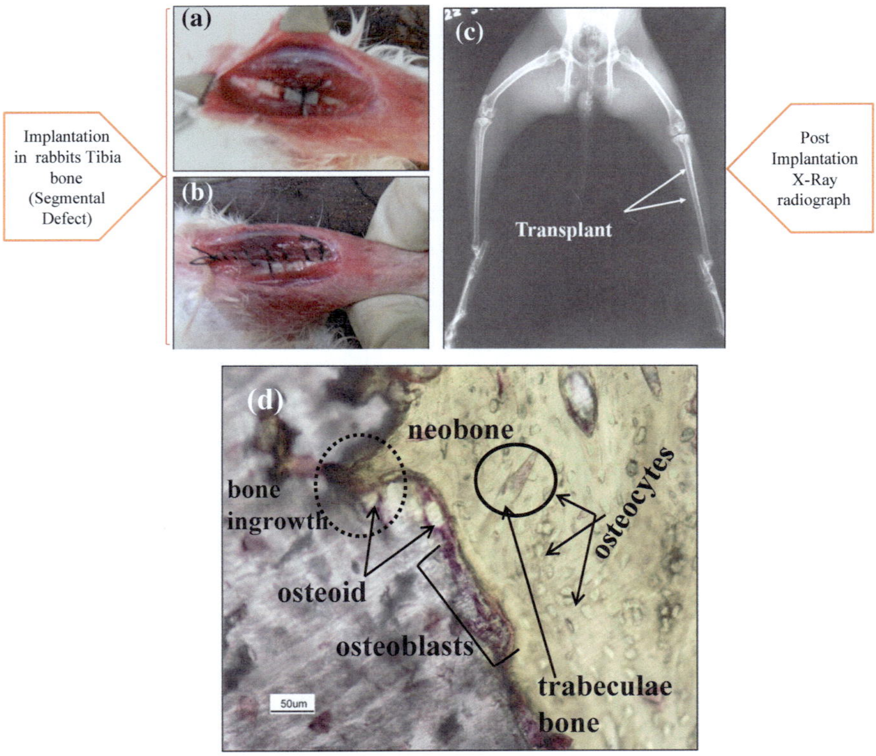

Fig. 7.12 Digital camera images showing, **a** rabbit's tebia zone with cylindrical long segmental defect along with the biocomposite material before implantation, **b** rabbit's tebia zone after implantation, **c** X-ray radiographs of post implant rabbit's tebia after 14 weeks of implantation of CS7 implant, **d** new bone formation, easily differentiated from host bone [44]

7.4.2 *Prototype Fabrication*

7.4.2.1 Engineering Design and Mold Preparation

Once all the lab-scale and pre-clinical study establishes acceptable biocompatibility property of a given biomaterial, (e.g. HDPE–20 % Al_2O_3) the planning for device fabrication has to be initiated as part of translation research [45]. As a first step, the engineering design of device prototype is shown in Fig. 7.13. Various engineering design views along with the dimensions are used in fabricating molds for compression molding experiments. One has to consider typical processing induced shrinkages at this stage.

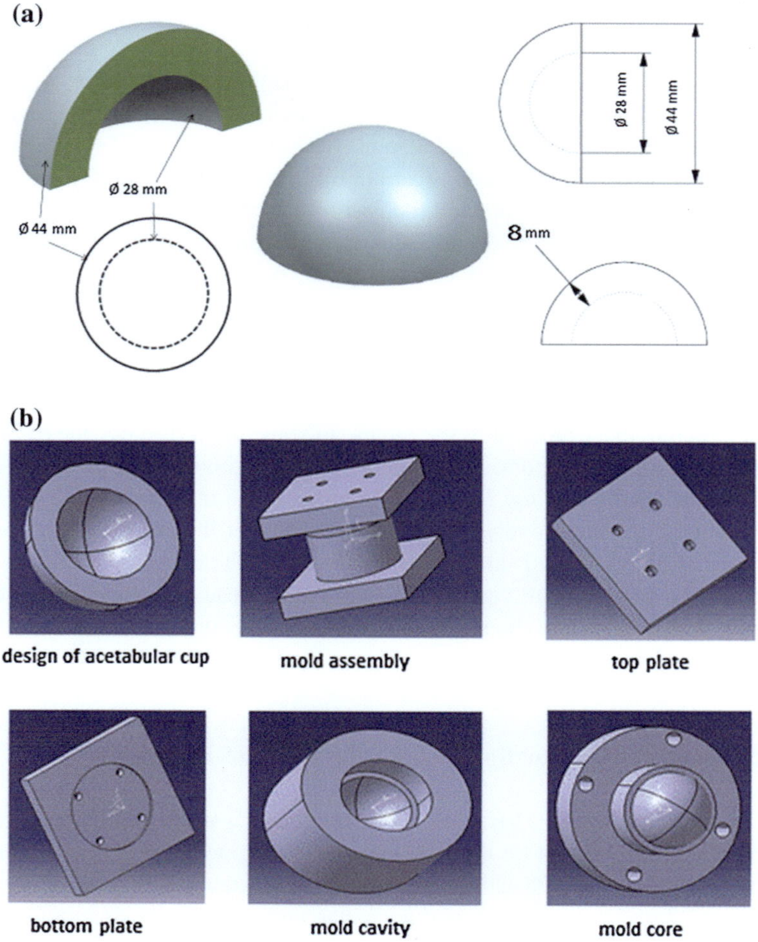

Fig. 7.13 The engineering drawing and 3D sketch of the acetabular cup (**a**) and CAD model of the product as well as different components of compression mold assembly are shown (**b**) [45]

Typically, the shrinkages are less in compression molding. If the flow materials with good flowability property are used, they do not shrink as much as softer flow materials. As a result of lower shrinkages, warpage is lessened and dimensional accuracy, is better in compression molding route. Also, the cycle times for compression molded parts using preheated preforms can be less than for injection molded parts. In view of the anticipated negligible shrinkage, no dimensional allowance was given while fabricating molds from the engineering design of Fig. 7.13a.

The fabrication of near net shape acetabular socket in compression molding route required the use of custom-made mold cavity and core. The shrinkage was limited to 2 % in single cavity mode, while both cavity and core was fabricated using semi-automatic CNC lathe for smooth finish operation. The heat treatment was carried out for the cavity material, OHNS (Oil Hardened Non Shrinkage steel) to avoid bulging due to heavy load and full hardening leads to the attainment of 45–50 HRC hardness. Additionally, retainer plates and band heater have also been specially designed for the compression molding.

For fabrication of the acetabular cup using compression molding, a specially designed mold consisting of one or more cavities and a force side is used. CAD design of various compression mold assembly is shown in Fig. 7.13b. The mold is heated by electric cartridge heaters, steam or oil to a certain temperature range. The molds used for compression molding consist of a cavity and a force or plunger, as shown in Fig. 7.14a. The guide pins maintain the proper relation between these members. The mold cavity forms one surface of the molded part and the molding compound is generally loaded in this member. The mold plunger forms the outer surface of the piece being closed. The molding compound is thus confined to the open space between plunger and cavity, while it hardens. The provision for heating is often included in the mold design. Some presses are equipped with heating platens, which transfer the heat to the molds. In some mold designs, the cavity and plunger or the top and bottom retainer shoes are drilled out to permit steam or hot water to circulate the heating. Figure 7.14b, c show the inner and outer surface of the acetabular cup, respectively. The smooth surface of the cup that was obtained after compression molding is also evident from these photographs. While the inner surface of the cup will articulate with the femur head, the outer surface will articulate with the pelvic bone. The smoothness of the surface is of utmost importance to minimize wear.

7.4.2.2 Compression Molding Based Fabrication of Prototypes

One of the important aspects for prototype development was to optimize compression molding conditions. Based on the labscale experiments to obtain 10 mm diameter, cylindrical samples of HDPE–20 wt% HA–20 % Al_2O_3 composites, we initially used 130–140 °C. After several trial experiments, the optimal compression molding conditions include the pre-heating temperature of mold as 190 °C with mold pressure varying in the range of 3–7 MPa. Another innovative step was to add 2 wt% Titanium IV, 2-propanolato, trisisooctadecanoato-O coupling agent during dry

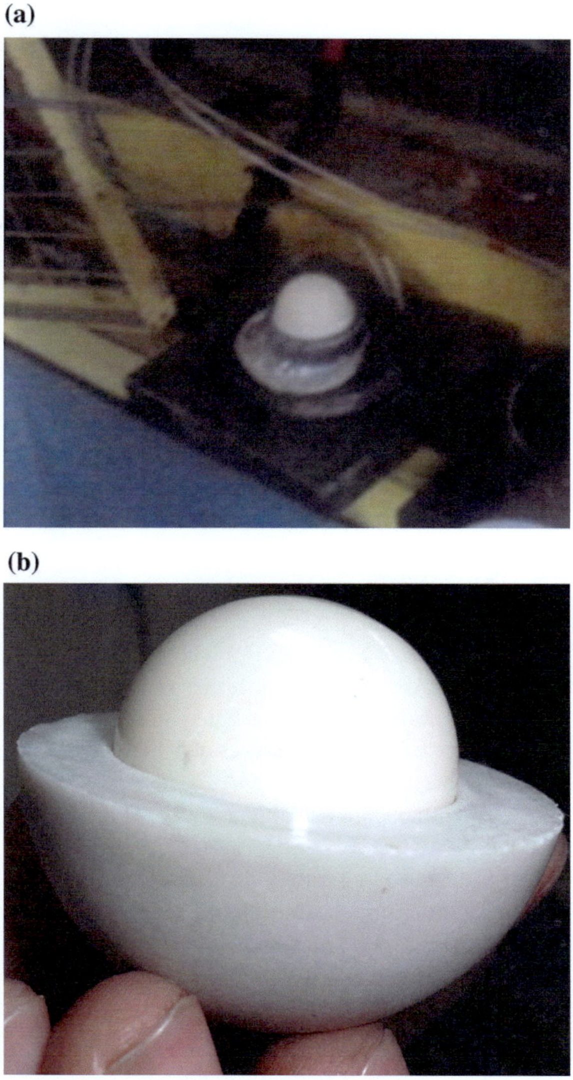

Fig. 7.14 **a** The acetabular cup on the mold after molding **b** the conformal contact of as-fabricated acetabular cup with a commercial alumina femoral ball head [45]

mixing of HDPE with HA and Al_2O_3. The process parameter optimization was a key step in developing acetabular sockets with smooth surface profile ($R_a \sim$ 1–2 μm).

Compression molding process consists of three stages namely melting, compression and cooling. A number of experiments are conducted to optimize the ratio of polymer and ceramic phase as well as coupling agent. The chosen coupling agent (titanate) provided better bonding between polymer and ceramic phase without contributing any adverse effect of the original property of the individuals. As a follow up work, a prototype of acetabular cup was developed after active discussion with clinicians, which helped to design and fabricate an acetabular cup, suitable for the Indian population. The design adopted during the present prototype development is the one that is most commonly used with the Indian patients undergoing total hip joint replacement surgery. Based on an adult person's acetabular cup dimensions, we have arrived at the following dimensions for design and development of cup to be used in total hip joint replacement which consist of an outer diameter—of 44 mm, an inner diameter of 28 mm and a cup thickness of 8 mm (see Fig. 7.3). The molding was carried out at constant pressure of 10 MPa and at a temperature of 130 °C for 30 min in case of monolithic HDPE and at 140 °C for 45 min for the composites. The CMP 30 Compression molding machine used for prototype development is a less tonnage machine enough for this purpose (Approx. 8 tonnes). The processing temperature used for the developed material was ~140 °C, so the pre-heating temperature of the mold was set to 190 °C (during compression, measured temperature of the mold was 185 °C) and mold pressure varying from 3 to 7 MPa. It is important to mention here that the temperature inside the mold (~140 °C) will be lower than the mold temperature (185 °C).

7.5 Summary

It can be reiterated that there is a whole range of prostheses available for conditions ranging from degenerative arthritis, rheumatoid arthritis, previous failed surgery, bone and joint tumors, and unstable painful joints, where supporting ligaments have been damaged. One of the key result is that the hybrid composites with up to 40 wt% ceramic fillers (HDPE–20 % HA–20 % Al_2O_3) loading can be compression molded into a patient-specific acetabular socket shape using near-net-shape manufacturing protocol.

Despite such large addition of ceramics, both the tribological and biocompatibility properties are not compromised to a significant extent. In particular, the novelty in this product design and development can be summarized in reference to, (a) unique composition of hybrid composite with proven biocompatibility property in rabbit model, (b) non-cytotoxic profiling of wear debris particulates, (c) better friction and wear resistance in fretting mode and (d) excellent wear resistance in HIP simulator tests. Figure 7.10 shows a schematic describing the novel, unique properties that were observed for the composite developed in this work. In brief, extremely low co-efficient of friction was observed along with low wear depth in wear and fretting experiments that were carried out for this hybrid composite. The surface roughness of the obtained prototype is around 0.6–0.7 μm. Such smooth roughness without additional machining makes the adopted manufacturing process a cheaper and easily scalable for mass scale production feasible. Subsequent machining operations are not required. The *in vitro* cell culture studies also depicted good proliferation of $SaOS_2$ cells on the hybrid composite as well as in the presence of the eluates. The use of tailored addition of titanate coupling agent is another novel step in the manufacturing of the prototype. Another important novel aspect is the near-net shape manufacturing as the prototype obtained at the end of

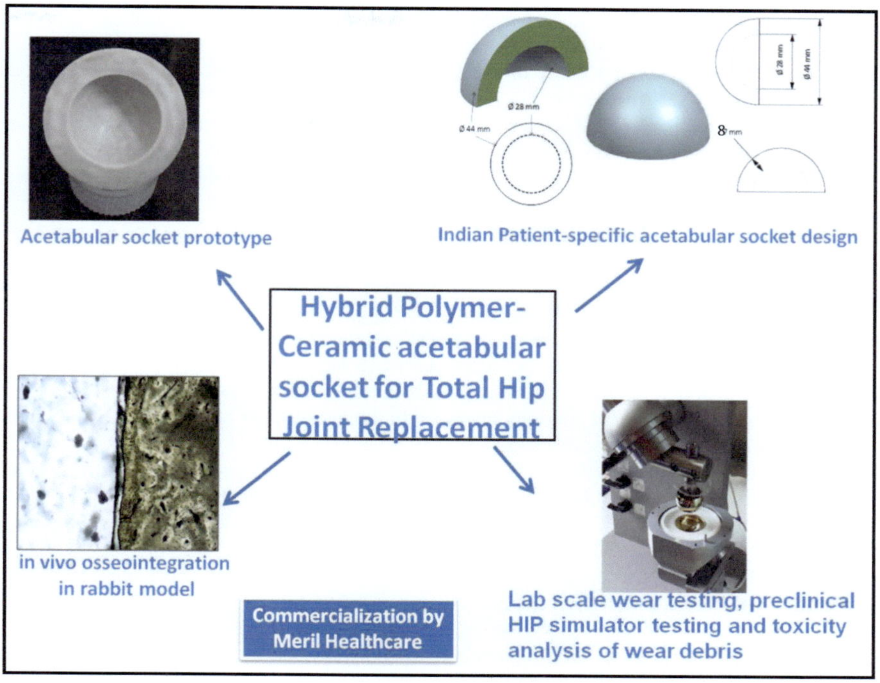

Fig. 7.15 A schematic figure showing the development of acetabular cup in its various stages, including prototype design, in vivo studies and hip simulator studies [45]

compression molding cycle replicates exactly the mold shape without any visible warpage or porosity.

To conclude, this chapter describes in detail the actual process of development of an acetabular socket for total hip prosthesis. The chapter describes the tedious process of development of a suitable composite as well as optimization of processing parameters. Figure 7.15 depicts the entire study performed in this chapter in a nutshell. This chapter not only presents compression molding-based approach for the fabrication and manufacture of implants but also provides a guideline to developing such new materials with enhanced functionality for orthopedic applications.

References

1. www.azom.com.
2. Kanis JA, Johnell O. Requirements for DXA for the management of osteoporosis in Europe. Osteoporos Int. 2005;16:229–38.
3. Katz JN, et al. Failures of total hip replacement: a population-based perspective. Orthop J Harvard Med Sch. 9:101–6.

4. Lichte P, Pape HC, Pufe T, Kobbe P, Fischer H. Scaffolds for bone healing: concepts, materials and evidence. Injury-Int J Care Inj. 2011;42:569–73.
5. Dowson D. A comparative study of the performance of metallic and ceramic femoral head components in total replacement hip joints. Wear. 1995;190:171–83.
6. Davidso JA, Georgette FS. State of art materials for orthopedic prosthetic devices: on implant manufacturing and materials technology. Proc Soc Manufact Eng. (Itasca, IL).
7. Long M, Rack HJ. Titanium alloys in total joint replacement—a materials science perspective. Biomaterial. 1998;19:1621–39.
8. Grammatopoulos G, Pandit H, Kamali A, Maggiani F, Glyn-Jones S, Gill HS, Murray DW, Athanasou N. J Bone Joint Surg Am. 2013;95:e81.
9. Charnley J. Arthroplasty of the hip, discussion South African orthopaedic association (1955). Jour Bone Joint Surg. 1956;38:592.
10. Helmer JD, Driskell TD. Research on bioceramics, in: proceedings of the symposium on use of ceramics as surgical implants. SC, USA: Clemson University; 1969.
11. Christel P, Meunier A, Dorlot JM. Ann NY Acad Sci. 1988;534:234.
12. Li S, Burstein AH. Current concepts review: ultra-high molecular weight polyethylene. Jour Bone Joint Surg. 1994;76-A:1080–90.
13. Amstutz HC, Campbell P, Kossovsky N, Clarke IC. Mechanism and clinical significance of wear debris-induced osteolysis. Clin Orthop Relat Res. 1992;276:18.
14. Schmalzried TP, Jasty M, Harris WH. Prosthetic bone loss in total hip joint arthroplasty. J Bone Joint Surg A. 1992;74:849–63.
15. Rushton N, Rac T. The intra-articular response to particulate carbon fiber reinforced high density polyethylene and its constituents: an experimental study in mice. Biomaterials. 1984;5:352.
16. Sclippa E, Piekarski K. Carbon fiber reinforced polyethylene for possible orthopedic uses. Jour Biomed Mater Res. 1973;7:559.
17. Amstutz H. Mechanism and clinical significance of wear debris-induced osteolysis. Clin Orthop Relat Res. 1992;276:7.
18. Blumenthal N. A new technique for quantitation of metal particulates and metal reaction products in tissue near implants. Jour Appl Biomater. 1994;5:191.
19. Schneider E. Comparative study of the initial stability of cementless hip prostheses. Clin Othop. 1989;248:200.
20. St John K. Applications of advanced composites in orthopedic implants. In: Szycher M, editor. Biocompatible polymers, metals and composites. Lancaster: Technomic Publishing; 1983. p. 861.
21. Christel P, Meunier A, Leclercq S, Bouquet P, Buttazzoni B. Development of carbon-carbon hip prostheses. J Biomed Mater Res Appl Biomet. 1987;21:191.
22. Akay M, Aslan N. Numerical and experimental stress analysis of a polymeric composite hip joint prostheses. J Biomed Mater Res. 1996;31:167.
23. Peter T, Tognini R, Mayer J, Wintermantel E. Homoelastic, anisotropic osteosynthesis system by net-shape processing of endless carbon fiber reinforced polyetheretherketone (PEEK). In: Goh J, Nather A, editors. Proceedings of 9th conference on biomedical engineering, Singapore, National University of Singapore; 1997. p. 317.
24. Wintermantel E, Bruinink A, Ruffiex K, Petitmermet M, Mayer J. Tissue engineering supported with structured biocompatible materials: goals and achievements. In: Speidal M, Uggowitzer, editors. Materials in medicine. Switzerland: ETH zurich; 1998. p. 1.
25. Guo Q, Luo W. Mechanisms of fretting wear resistance in terms of material structures for unfilled engineering polymers. Wear 2002;249:924–31.
26. Kehler BA, Baker NP, Lee DH, Maggiore CJ, Nastasi M, Tesmer JR, Walter KC, Nakamura Y, Ullrich BM. Tribological behavior of high-density polyethylene in dry sliding contact with ion-implanted Co–Cr–Mo. Surf Coat Techol. 1999;114:19–28.
27. Anderson JC. High density and ultra-high molecular weight polyethenes: their wear properties and bearing applications. Trib Int. 1982:43–7.

28. Xue Y, Wu W, Jacobs O, Schäde B. Tribological behaviour of UHMWPE/HDPE blends reinforced with multi-wall carbon nanotubes. Polym Test. 2006;25:221–9.
29. Saha D, Roy Chowdhury SK, Banthia A. Characterization of particulate debris obtained from hip joint prosthesis. In: Proceedings of the national conference on biomaterials, IIT, Kharagpur, India; 2000.
30. Roy Chowdhury SK, Mishra A, Pradhan B, Saha D. Wear characteristic and biocompatibility of some polymer composite acetabular cups. Wear. 2004;256:1026–36.
31. Bonfield W. Hydroxyapatite-reinforced polyethylene as an analogous material for bone replacement. In: Ducheyne P, Lemons JE, editors. Bioceramics: materials characteristics versus in vivo behavior. Annals of the New York Academy of Science; 1988. 253:173–77.
32. Bonfield W, Grynpas MD, Tully AE, Bowman J, Abram J. Hydroxyapatite reinforced polyethylene—a mechanically compatible implant. Biomaterials. 1981;2:185–9.
33. Huang J, Silvio LD, Wang M, Tanner KE, Bonfield W. In vitro mechanical and biological assessment of hydroxyapatite reinforced polyethylene composite. J Mater Sci Mater Med. 1997;8:775–9.
34. Wang M, Porter D, Bonfield W. Processing, characterisation, and evaluation of Hydroxyapatite reinforced polyethylene composites. Brit Ceram Trans. 1994;93:91–5.
35. Guild FJ, Bonfield W. Predictive modelling of hydroxyapatite–polyethylene composites. Biomaterials. 1993;14:985–93.
36. Wang M, Deb S, Bonfield W. Chemically coupled hydroxyapatite-polyethylene composite: processing and characterization. Mater Lett. 2000;44:119–24.
37. Wang M, Bonfield W. Chemically coupled hydroxyapatite-polyethylene composite: structure and properties. Biomaterials. 2001;22:1311–20.
38. Bonfield W. From concept to patient—engineering solutions to medical problems, in Engineers and society: the 1997 CSE international lecture. London: The Royal Academy of Engineering; 1997. p. 5–11.
39. Wang M, Chandrasekaran M, Bonfield W. Friction and wear of hydroxyapatite reinforced high density polyethylene against the stainless steel counterface, J Mater Sci Mater Med. 2002;13:607–11.
40. Bodhak S, Nath S, Basu B. Friction and Wear Properties of Novel HDPE-HAp-Al_2O_3 Biocomposites against Alumina counterface, J Biomaterials Applications 2009;23:407–433.
41. Bodhak S, Nath S, Basu B. Fretting wear properties of hydroxyapatite, alumina containing high density polyethylene biocomposites against zirconia; J Biomed Mater Res: Part A 2008;85:83–98.
42. Nath S, Bodhak S, Basu B. Tribological investigation of Novel HDPE-HAp-Al_2O_3 hybrid biocomposites against Steel under Dry and Simulated Body Fluid Condition; J Biomed Mater Res: Part A 83 A 2007;1:191–208.
43. Tripathi G, Dough JE, Dinda A, Basu B. *In vitro* cytotoxicity and *in vivo* osseointergration property of compression molded HDPE-HA-Al_2O_3 hybrid biocomposites; J Biomed Mater Res. Part A 2013;101(6):1539–1549.
44. Tripathi G, Dubey A, Basu B. Evaluation of physico-mechanical properties and in vitro biocompatibility of compression molded HDPE Based Biocomposites with HA/Al_2O_3 Ceramic Fillers and titanate coupling agents; J Appl Polym Sci. 2012:124:3051–3063.
45. Basu B, Kumar A, Sundaresh DC, Sreedhar P, Singh JB: Polymer-Ceramic Hybrid Composite Material for Orthopedic Prostheses and Method of Producing the Same; Indian patent filed in 2015(File No.: 3011/CHE/2015).

Chapter 8
Case Study: 3D Printed Cartilage

8.1 Introduction

With increase in life expectancy, the motivation of our society is shifting towards living longer and with a good quality of life. Majority of the world's aging population is now willing to take advantage of medical technology to solve their health problems. Osteoarthritis, Rheumatoid arthritis and lower back pain are most common ailments of aging population, in which cartilage tissues get damaged or degenerated. Once damaged, cartilage tissue has poor propensity to regenerate.

Tissue engineering and Regenerative medicine offer fascinating promises for developing bioengineered tissue constructs for organ replacement. In 1980s, Tissue engineering field was initiated with attempt to engineer cartilage-like tissues, as it is an avascular tissue, without neuron, consisting of only one cell type. Since last 30 years researchers focused on developing cartilage-like constructs to replicate structure-function properties native cartilage by culturing various types of cells on different types of polymeric scaffolds [1]. However despite sincere efforts of brilliant tissue engineers around the world, development of clinically relevant, functionally equivalent load bearing, engineered, three dimensional (3D) cartilaginous tissue constructs remains elusive. Clinical success was achieved in few exceptions where non-load bearing cartilage constructs (cartilaginous tissue for airway and nasal reconstruction) were implanted in restricted patient cohorts [2, 3].

Orientation of cells, alignment of collagenous fibrous proteins and spatial gradient extracellular matrix is highly important to replicate anatomical features of fibrocartilaginous tissues of mandibular condyle [4], intervertebral disc [5] or meniscus tissues [6]. Under physiological dynamic compression loading condition, collagen fibers dominantly contribute in reinforcing these tissues. However, currently practiced Tissue engineering strategies still could not coherently simulate the complexity of these anatomical features and cellular or ECM alignment present in human cartilage, hence fail to replicate mechanical properties of these cartilage tissues.

B. Basu and S. Ghosh, *Biomaterials for Musculoskeletal Regeneration*,
Indian Institute of Metals Series, DOI 10.1007/978-981-10-3017-8_8

From this translational point of view, 3D bioprinting has fascinating potential to offer a paradigm shift in cartilage tissue engineering research, as this will allow deposition of cells in a specific alignment, control spatial orientation of matrix proteins, in order to engineer anatomically relevant sized cartilage grafts in arbitrary shapes as required for the particular defect in a patient-specific manner.

Silk fibroin protein from *Bombyx mori* silkworm has been widely used in the biomedical research sector for Tissue engineering. This case study will highlight development of silk-based bioink and its application for cartilage 3D bioprinting.

8.2 Limitation of Scaffold-Based Cartilage Tissue Engineering

Cartilage tissue consists of only one cell type, chondrocytes, which is embedded within ECM network, which is composed of primarily water (80 % wet weight), collagen type II (50–60 % dry weight), large (aggrecan, versican) and small molecular weight (biglycan, lumican, decorin, fibromodulin) aggregating proteoglycans such as chondroitin sulfate, heparin sulfate (around 25–35 % dry weight) etc. These ECM components impart mechanical property of cartilage tissue, where collagen fibrous network mainly governs the tensile strength, and the proteoglycans (e.g., aggrecan) along with glycosaminoglycan (GAG) impart compressive modulus, aggregate modulus of the cartilage tissue. In order to replicate such impressive mechanical properties, various kind of scaffold-based culture systems have been explored for cartilage tissue engineering, for example nonwoven [7], woven [8], hydrogel, porogen-leached 3D porous scaffolds [9] etc. Depending upon mechanical properties and architecture of those scaffold-based culture systems researchers attempted to recapitulate the nonlinear, anisotropic and viscoelastic nature of articular cartilage, albeit unsuccessfully. There are several unsolved challenges.

First step in cartilage engineering is expansion of primary chondrocytes *in vitro*, followed by culturing cells over scaffolds. During this *in vitro* proliferation phase, with cumulative population doublings chondrocytes quickly lose their native phenotype and potential for chondrogenic ECM production. In Petri dishes articular chondrocytes gradually lose their round morphology to assume an elongated fibroblast-like shape. Synthesis of Type II collagen and aggrecan get reduced and type I collagen and versican synthesis get increased [10]. Research is still ongoing to explore various strategies to preserve native phenotype of chondrocytes during re-differentiation.

Native articular cartilage has distinct zonal structures, namely superficial zone, the middle zone and the deep zone. But scaffold-based strategies invariably develop uniform distribution of cells due to random distribution of pores in scaffolds, as a result eventually develop uniform cartilaginous construct. Attempts are being made to generate engineered cartilage with the inhomogeneity of adult articular cartilage, zonal ECM features and depth-dependent cellular phenotypes [11–13]. But despite extensive endeavors to replicate depth-dependent cellular and mechanical inhomogeneity, engineered constructs at the best depict zonal characteristics akin to immature articular cartilage [11, 12]. Moreover, as these engineered constructs do not recapitulate a collagen fibrous architecture (Benninghoff-like pattern) and cellular alignment, their region-specific compressive properties are several magnitudes inferior compared to that of adult articular cartilage. 3D bioprinting strategies may offer unique strategies in achieving anatomically depth-dependent cellular and mechanical inhomogeneity.

So far cartilage tissue engineers assumed that "chondrocyte" represents a single cell population originated from a single differentiation process. But recent developmental biology studies demonstrated that indeed, in human articular cartilage each chondrocyte subpopulation possess distinct morphology as well as distinct gene expression profile. Each of these subpopulations have specific functions [14]. Only 3D bioprinting strategies may enable us to deposit distinct subpopulation of chondrocytes at defined location.

Similarly, specific types of ECMs are present at different regions of cartilage tissues, which also change within the tissue during embryonic development and aging process. For example, chondroitin-4 sulphate is the most predominant type of GAG in immature (embryonic) articular cartilage, but chondroitin-6 sulphate is predominantly present in adult articular cartilage [15]. In addition, the sulphation patterns of chondroitin-4-sulfate change in human cartilage tissue with ageing, along with an increase in 6- to 4-sulphated N-acetyl glucosamine residues ratio [16]. Traditional scaffold-based strategies fail to replicate such intricate variation in ECM composition. 3D bioprinting may enable the researchers to use different type of bioink to recapitulate different ECM composition to engineer cartilage tissues of different developmental or degeneration stages [1].

Architectural mismatch between the pre-fabricated 3D scaffold and the defect site in arthritis lesions often cause lack of integration between the neo-engineered construct and the surrounding host cartilage [17]. This problem often causes failure in clinical trials and hence an alternative approach is needed. 3D bioprinting has the potential to develop arbitrary shaped grafts designed according to the dimension of particular lesion in a patient-specific manner.

Another major problem associated to chondrogenic differentiation of bone marrow-derived or adipose derived progenitor cells (hMSCs) is that cells rapidly undergo hypertrophic differentiation [18]. In early time points of culture hMSCs express chondrogenic markers (sox-9, aggrecan, collagen type II), leading toward formation of "transient cartilage". But with increasing culture time upregulation of hypertrophic (collagen type X and MMP13) and osteogenic (cbfa-1, osteocalcin, bone sialoprotein) markers are commonly seen [19]. It is already known that articular cartilage does not support bone growth while transient cartilage does. The tissue engineered construct is capable of responding to its surroundings the same way as transient cartilage cells do *in vivo*. It would be interesting if 3D printed cartilage constructs can be used to reveal the underlying molecular mechanisms of hypertropic differentiation.

8.3 3D Bioprinting

3D bioprinting offers potential to replicate complex tissue architectures by programmable fabrication of cell-laden architectures, which allows precise deposition of cells, biomaterials and matrix components. This technique is unique as the patterns, pitch, macroporosity, cellular orientations, and biochemical composition can each be independently controlled. Further the compositional and mechanical cues of the bioink are expected to govern cellular proliferation, synthesis and deposition of ECM. Such designer approach for cartilage tissue engineering could potentially lead to unprecedented reproduction of spatially controlled cell-instructive microenvironments, targeted zonal-specific signaling, to develop functionally relevant cartilage.

8.3.1 Bioink

A bioink should fulfill several stringent requirements for 3D bioprinting, in terms of tailorable chemistry, precisely optimized window of rheology to ensure ease of printability, ability to pico- to nanoliter level bioink droplet formation, as well as instant solidification. Hydrogel bioinks recapitulate some features of the natural ECM. Cell encapsulation in a highly hydrated polymeric bioink helps to deposit newly synthesized matrix in pericellular region. Bioinks should exhibit shear-thinning behaviour (viscosity should reduce with increasing shear rate) so that the bioink can smoothly pass through small diameter nozzles of the printer without choking. At the same time they should exhibit instant solidification and shape retention soon as they exit the nozzle. It should also have predictable and tunable mechanical properties such as tensile or compressive modulus, as well as controllable biodegradation. Bioink composition should be such that the cells embedded in bioink should be able to degrade and remodel pericellular matrix, as well as it should support multilineage differentiation of progenitor cells.

Keeping all such goals in mind, silk protein (either isolated from silkworm cocoon or recombinant source) might be considered as a potential bioink for 3D bioprinting. Similar to many other biopolymers, silk fibroin protein inherently does not meet many of these requisites due to following reasons. At low concentrations (i.e. <20 wt%) regenerated fibroin protein inks show shear thinning behavior [20]; but the concentration is too dilute for printing. Beyond this 20 wt% concentration regenerated fibroin protein inks display Newtonian fluid-like behaviour [20], where viscosity is not affected by increasing shear rate. At the same time we should keep it in mind that, during printing fibroin macromolecular chains undergo shear induced ($<100\ s^{-1}$) conformational changes from random coil to β-sheet crystal, resulting from thermodynamic and kinetic processes [21]. This crystallite formation can choke the micrometer diameter nozzles. Hence, it is highly challenging to print silk fibroin protein solution without additives.

At the same time it would be interesting to use silk for 3D bioprinting, as it is composed of amino acids. The amphiphilic nature of silk protein chains can be exploited to form precise volume of fibroin protein droplets in the pico- to nanoliter level, or continuous hydrogel filaments by optimizing the rheological characteristics of the inks (e.g., viscosity, hydrophilicity, surface tension) at wide range of pH, ionic strength [20]. The biodegradation of silk-bioink based constructs could be governed either by modulating the secondary conformations (α-helix, β-sheet, β-turn content) and/or by controlling length of fibroin chain during fibroin protein isolation process.

We solved the above mentioned bottlenecks of using silk fibroin protein as bioink, and reported the formation of complex 3D microperiodic architectures by direct-writing into a coagulating bath [20] (Fig. 8.1a). The aqueous solution of silk fibroin ink was extruded in a coagulation reservoir comprising of 86 % methanol. Following principles of wet spinning technique, a continuous filament-like fibroin protein was dispensed from the nozzle that retained its shape after coagulation. 86 % methanol was found to be optimal, as it produced a coagulated ink filament with optimum elasticity required to maintain the shape of the structure while spanning unsupported gaps of the underlying layers, while being flexible enough to retain smooth flow through the nozzle. Higher concentration of alcohol bath resulted in instant solidification and lower concentration led to non-uniform coagulation or partial dissolution, resulting in broken filaments. After fabricating one layer (x-y direction), the nozzle was uplifted in the z-direction by programming to print the consecutive layers until the desired 3D pattern was generated. That was the first report on 3D printing of silk protein. Further, we could also successfully culture mesenchymal progenitor cells (hMSCs) over such patterns (Fig. 8.1b), and induced chondrogenic differentiation [20]. hMSCs changed shapes from a fibroblastic spread morphology to a roundish morphology, along with gradual modulation in expression of actin cytoskeleton during chondrogenic differentiation.

But this strategy cannot be followed for 3D bioprinting of cartilage, as cells would die instantly within the alcohol based coagulating baths. Moreover, addition of another biopolymer may act as dopant, which and may impart shear thinning behavior, hence eventually improve flowability. With that logic, gelatin was mixed with silk fibroin protein at various ratios to develop silk fibroin-gelatin interpenetrating hydrogel based ink [21]. Fibroin-gelatin ink demonstrated shear thinning behavior across wide range of concentrations. Moreover elastic behavior of the blend ink was dramatically increased (<2000 fold) compared to only silk ink. For 3D printing, the elastic modulus of the ink should be higher than the viscous modulus, particularly at low strain %. But that is not the case in most commonly used polymer solutions, in the form of viscous sol state.

Immediately after printing the bioink should undergo sol to gel transition. In the process, the whole constructs should also be stabilized as fast as possible in order to preserve 3D morphology. Hence there is a need to develop rapid, cytocompatible gelation strategy. An interesting strategy for physical cross-linking of self-curing silk bioinks was reported by Jose et al. [22] which utilized the evaporation-induced modulation of silk protein conformations during curing when blended with the

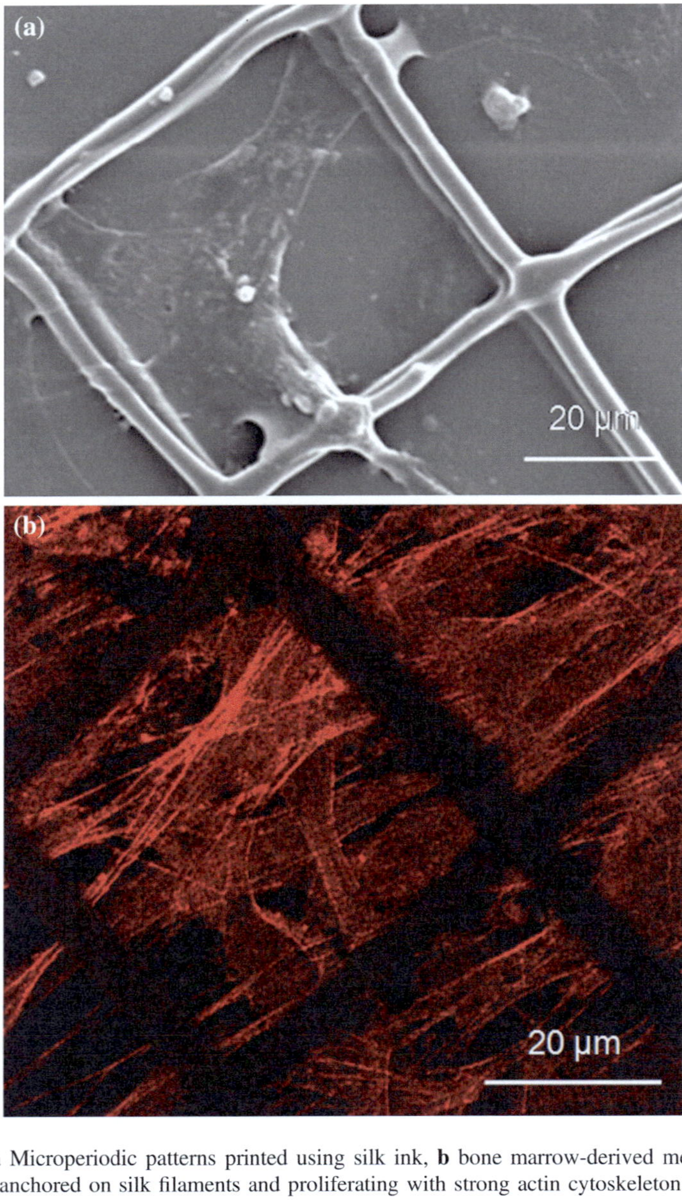

Fig. 8.1 **a** Microperiodic patterns printed using silk ink, **b** bone marrow-derived mesenchymal stem cells anchored on silk filaments and proliferating with strong actin cytoskeleton expression

nontoxic polyol additives. This strategy might allow bypassing of deleterious curing mechanisms, such as high temperatures, UV light, enzymatic or chemical cross-linking, which are prevalent in alternative 3D rapid prototyping techniques and have been required in prior silk bioprinting approaches. We explored two different in situ crosslinking strategies, namely sonication and enzymatic

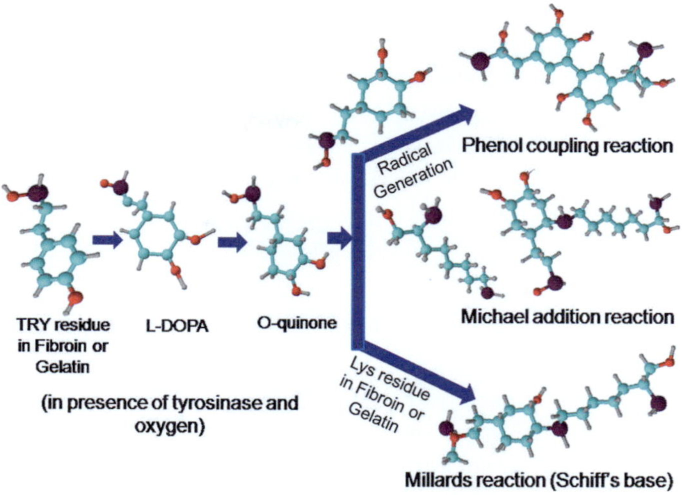

Fig. 8.2 Schematic showing the mechanism of in situ crosslinking of fibroin–gelatin bioink by tyrosinase enzyme to form 3D bioprinted cell-laden constructs

crosslinking [23]. Tyrosinase enzyme oxidizes the accessible tyrosine moieties of proteins into reactive o-quinone units without cleaving the peptide backbones of the protein macromolecules. Tyrosinase can induce oxidation to 10–11 and 20 % tyrosine residues of silk and gelatin respectively. In next step, oxidized quinone groups either get combined to neighbouring quinone groups or undergo nonenzymatic reactions with amines of both gelatin and silk (Fig. 8.2). In depth investigation of such gelation strategy is not only crucial for successful development of 3D printed constructs, but it is important to generate insight about various chemical groups developed during the gelation and how that will affect cellular behavior.

8.3.2 Printers

Three major classes of 3D printers are commonly used, based on their working principles; (i) droplet based bioprinting (electrohydrodynamic jetting, inkjet-based, acoustic-based ejection), (ii) laser-assisted (stereolithography or its modified versions, laser guidance direct-writing and laser-induced forward transfer), (iii) extrusion-based. Although first generation of 3D Bioprinting commonly used inkjet-based or laser-based strategies, but currently extrusion-based techniques are gaining popularity. Various factors are responsible for this selection, for example affordability, versatility, ease of dispensing diverse types of ink.

Typically an extrusion-based printing technique, for example the Direct-write 3D printer present in our laboratory, are equipped with a combination of controlled ink-dispensing unit nozzle, a computer controlled automated robotic translation system which can move along x-, y-, and z-directions for bioprinting, an air compressor unit. Using Direct-write assembly, we could print wide range of biologics for printing (single cells, aggregates, organoids etc.) as cell-cell junction plays crucial role in forming N-cadherin mediated gap junction, which is prime requirement for chondrogenesis [24]. Various advanced level 3D printers with multi-head nozzles and capacity to deposit multiple materials are being offered by various companies.

8.3.3 Printing Process and Parameters

Success of 3D bioprinting is still a challenging proposition as optimization of printing parameters (deposition speed, shear force applied, movement of nozzle, temperature, humidity), design aspects (orientation of cell-laden filaments, pore size and porosity, interconnectivity), post-printing stabilization methods (thermal, chemical, UV-based method or enzymatic cross-linking, sonication), the concentration and gradient of growth factors, morphogens, viscosity and surface wetting characteristics of ink can have a drastic effect on the fate of cells and tissues. Moreover, bioprinting needs software algorithms which can convert clinical images (MRI and CT-scans on a patient-specific basis) to working CAD models to make high resolution 3D patterns. Our understanding is still rather limited for fabricating 'off the shelf' engineered cartilage tissues by optimizing all these parameters.

> Clinical imaging techniques, such as magnetic resonance imaging (MRI) and/or computed tomography (CT) are used to develop computer-aided design (CAD) and computer-aided manufacturing (CAM) technology based models. Then the CAD/CAM models are transformed into the manufacturing output file, which are transmitted to the 3D bioprinters for biofabrication of target tissue.

8.4 Characterization of 3D Printed Constructs

8.4.1 Cell Viability

Two major concerns in 3D bioprinting are achievement of high percentage of cell viability as well as maintaining long term cell viability. Cell viability has been demonstrated to drastically vary with dispensing micronozzle diameter and the applied pressure [25, 26]. During extrusion, cells may face shear forces within the bioink/hydrogel system due to applied pressure, which in turn could potentially induce apoptosis or rupture cell membranes [27]. Nozzle diameter and viscosity of the bioink predominantly determine the level of shear force on the encapsulated cells, as well as the maximal time required for developing a clinically relevant sized constructs [28]. Thus, optimizing these parameters would be critically important to achieve maximum cell viability.

Some studies reported initial drop in cell viability [29] immediately after printing, followed by gradual recovery by cell proliferation. This finding highlights the requirement for a high cell seeding density to start with. However, in most of the 3D bioprinting studies show viability only up to 3–7 days [30]. There is curious paucity of long term cell viability results. Bioink's hydrated network should allow diffusion of nutrients, gases and metabolites and metabolic waste removal for increased cell viability.

For the first time, we demonstrated the use of silk-gelatin bioink that could support 82–87 % of viability of human nasal turbinate tissue-derived Mesenchymal stem cells (hTMSCs) for at least a month [23]. Live and dead staining (Fig. 8.3) results should be validated by other relevant techniques such as DNA quantification, Cell Counting Kit-8 (CCK-8) to extrapolate data on cell proliferation.

8.4.2 *Printing of Chondrogenic Construct and Characterization*

The nature of bioink and composition as well as matrix stiffness are key determining factors for regulating proliferation of cells confined within the hydrogel bioink. In this context, Das et al. [23] encapsulated hTMSCs silk-gelatin matrix prepared by different cross-linking methods (tyrosinase crosslinking or sonication).

Sonication-induced fibroin-gelatin bioink exhibiting around 2 times higher β-sheet content, compared to enzyme-induced bioink. Difference in β-sheet-crystal content resulted in several interesting observations. Firstly, the lesser stiffness of bioink hydrogel was more suited for cell proliferation. Secondly, sonication-induced fibroin-gelatin bioink resulted in relatively lower degradation profile. Most interestingly, when we explored multilineage differentiation of hTMSCs, we noticed that due to higher order of matrix stiffness cells had higher propensity to express osteogenic markers with time dependent upregulation. The stiffer matrices are known to initiate specific cell-mediated contractions within the confined 3D microenvironment hence preferentially regulating osteogenic commitment [31]. While tyrosinase crosslinked matrices with lesser β-sheet content, favoured chondrogenic differentiation of encapsulated hTMSCs, as evidenced by strong alcian blue staining, as well as immunofluorence staining for aggrecan and collagen type II (Fig. 8.4). In presence of chondrogenic differentiation factors at day 14, the level of early chondrogenic transcription factor SOX9 mRNA expression was 5.5-fold upregulated in sonication-induced silk-gelatin bioink compared to alginate. At day 21, hTMSCs embedded in tyrosinase crosslinked construct showed statistically

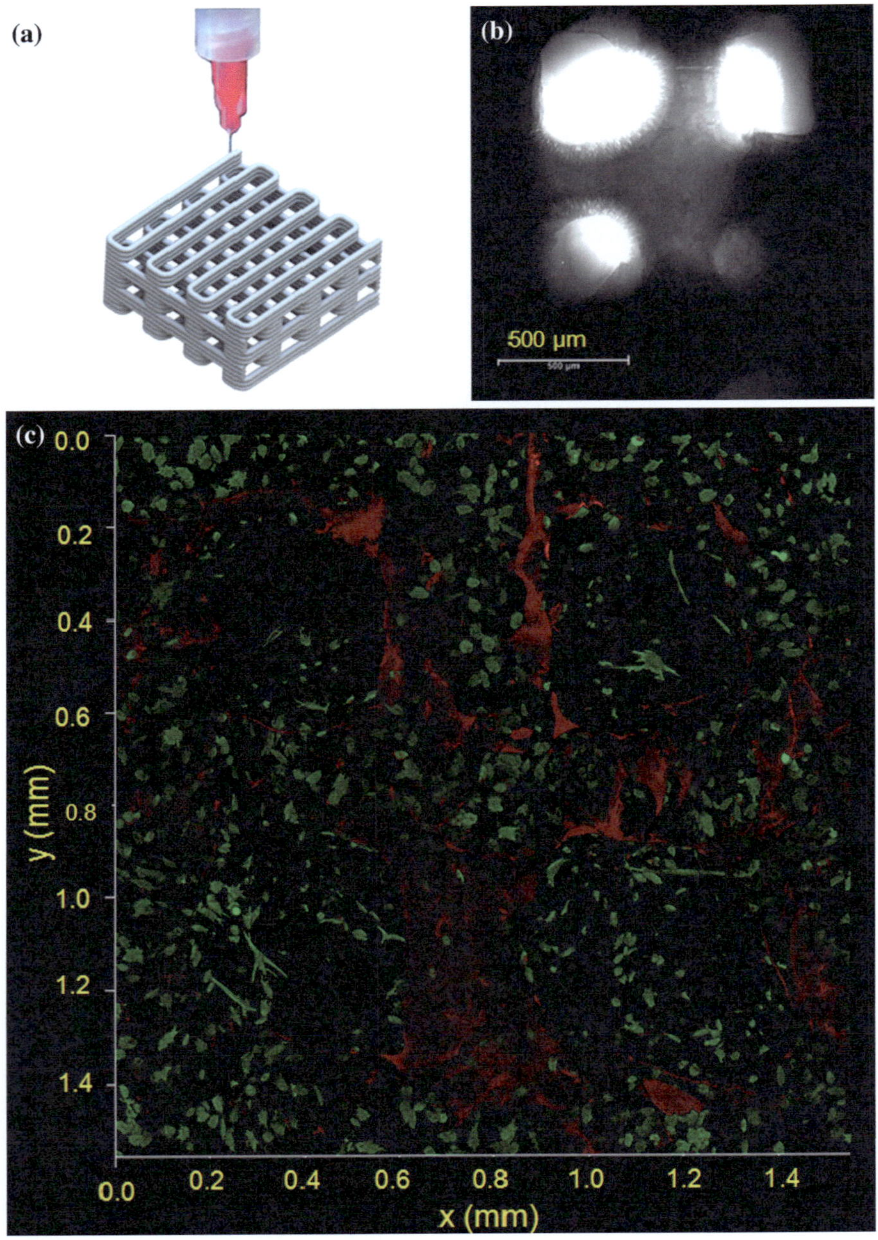

significant expression of SOX9 than that of cells in sonication-induced bioink (twofold higher) and alginate (fivefold higher). Aggrecan gene expression was higher in tyrosinase crosslinked constructs at all the time points. At day 21 aggrecan expression was 13-fold higher compared to that of alginate construct. Collagen

◀ **Fig. 8.3** **a** 3D bioprinting of anatomically relevant sized cartilaginous construct using silk fibroin-gelatin ink, in which dispersed cells are distributed within filamentous pattern, and filament-filament distance can be pre-decided **b** 3D microperiodic patterns showing filament structures and porous architecture of the printed construct, **c** live and dead viability staining showed high cell viability of chondrocytes encapsulated within printed constructs. Green fluorescence is representing the live cells and red is generated by the autofluorescence from silk-gelatin filaments. We could successfully identify cells deposited in multiple layers within the 3D printed pattern designed to replicate anatomical 3D organisation of tissue

type-II mRNA expression level was higher in tyrosinase crosslinked silk-gelatin construct compared to that of sonication-induced bioink and alginate at all the time points, with 2.5- and 5-fold changes at days 7 and 14, respectively. However, Collagen type-II expression was not statistically significantly different between two types of silk-gelatin constructs at day 21. Taken together, silk fibroin-gelatin bioink based 3D printed constructs supported chondrogenic differentiation of progenitor cells.

In another study based on silk-gelatin bioink, we deposited bone marrow-derived hMSCs and primary articular chondrocytes, either as dispersed cells or aggregates, and we investigated the effect of matrix on regulating the chondrogenic signaling mechanisms of deposited cells [32]. Our results indicated that the silk-gelatin matrix facilitated the proliferation (measured DNA content), migration, matrix synthesis and remodelling of the deposited hMSCs into phenotypically stable cartilaginous tissue with suppressed hypertrophy. The committed chondrogenic differentiation along with reduced hypertrophy was found to be a combinatorial effect of; bioink composition, matrix diffusivity and remodelling, cell type and seeding modality, proteoglycan production. The role of two major signaling pathways, hypoxia mediated HDAC4 and TGF-β mediated SMAD4 signaling was evidently noticed [32].

Hypertropic differentiation of chondro-progenitor cells are a common problem to all the polymeric matrix explored so far. For example, cells in Gelatin methacrylamide ink (gel-MA), which is commonly used for 3D Bioprinting [33], also reported to undergo hypertropic differentiation. gelMA-encapsulated hMSCs were cultured *in vitro* for 2 weeks to provide a cartilage template that was subsequently remodeled *in vivo* into mineralized bone tissue harboring bone marrow cavities. So cells in gelatin methacrylamide bioink underwent hypertropic differentiation, similar to endochondral ossification *in vivo* [34]. Hence, our findings strongly indicate that silk fibroin-gelatin bioink seems to be advantageous compared to other bioinks in providing signals for minimizing hypertropic differentiation.

Taken together, we have developed innovative silk fibroin-gelatin hydrogel-based bioink formulations to biofabricate intricate complex and phenotypically stable 3D cartilaginous structures. Silk-gelatin hydrogel ink has provided the cells with a biologically relevant microenvironment for matrix remodeling and accumulation of newly synthesized matrix. Despite these promising results significant challenge are still remaining to be solved in the bioprinting of articular cartilage constructs, to develop a load-bearing mechanical environment, strategies

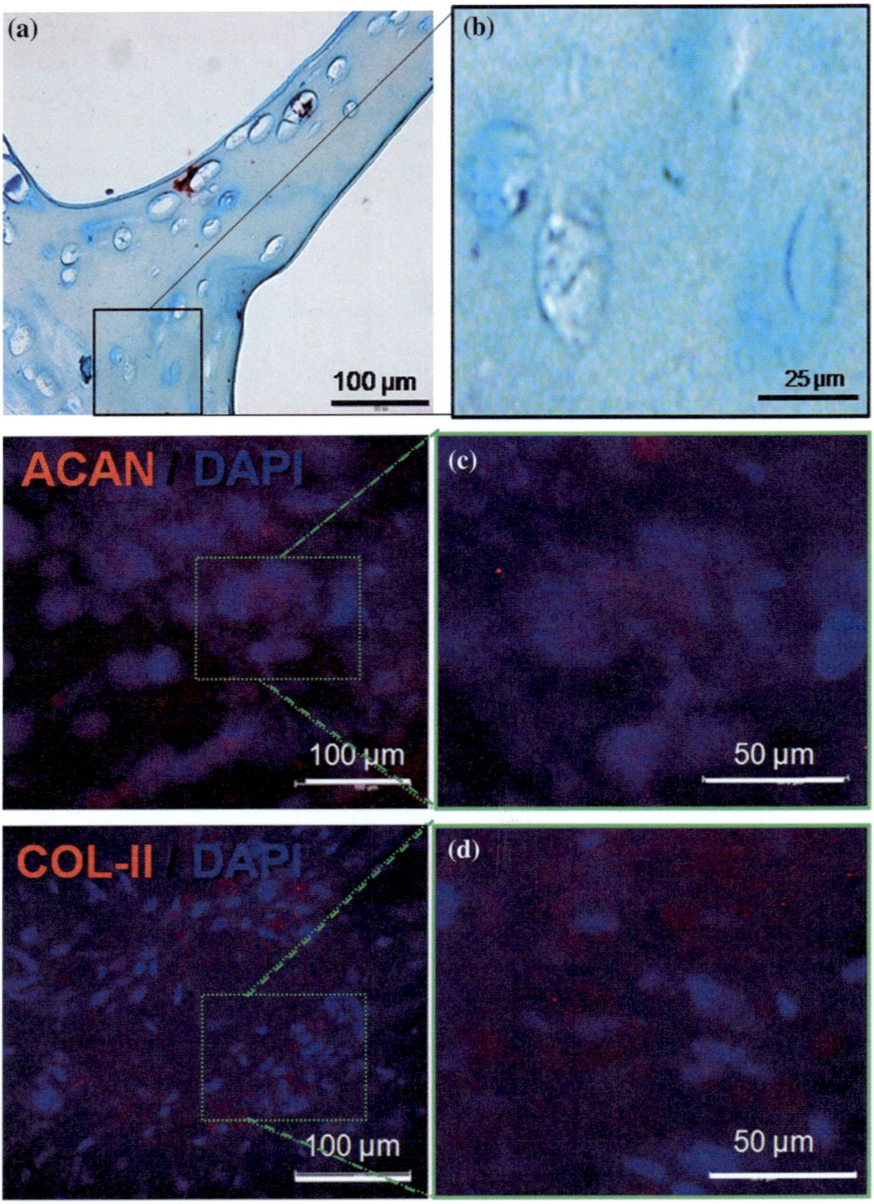

Fig. 8.4 Chondrogenic differentiation of mesenchymal stem cells in 3D printed constructs, showing **a** alcian blue staining, **b** alcian blue staining around the cells, **c** aggrecan staining and **d** collagen type II staining at 4th week. Intensity of staining increased with culture time

to minimize hypertropic differentiation etc. Hopefully, with the versatility of 3D bioprinting technologies it will soon be possible to develop more complex functional cartilaginous tissue interfaces in patient-specific manner. Various other researchers around the world are also exploring development of innovative bioinks [35] for cartilage bioprinting, as well as strategies to minimize hypertropic differentiation by probing role of microenvironment created by such bioinks [36].

8.4.3 Bioprinting of Complex Cartilaginous Tissue Constructs

Attempts are being made to develop more complex cartilaginous tissues, for example heterogeneous hydrogel based constructs for 3D printing of osteochondral grafts. Both human chondrocytes and osteogenic progenitor cells were bioprinted and cultured *in vitro* for 3 weeks, followed by 6 weeks implantation subcutaneously in immunodeficient mice. This study showed that it was possible to develop different types of tissues in same construct [37].

In another study, polycaprolactone fibres and chondrocytes were suspended in a fibrin-collagen hydrogel, which have been printed to develop a cartilage construct. The study demonstrated that the constructs formed cartilage-like tissues both *in vitro* and *in vivo*, and with enhanced mechanical properties [38].

Congenitally malformed external ears (microtia) offer significant psychological as well as social burden. Moreover many patients demand auricular prosthesis or synthetic implant (silicone, nylon, Teflon based structures) after trauma, where external ear is damaged. Although auricular cartilage is a fibrocartilage, but complex shape, and patient-specific intricate architecture make it a huge challenge to engineer external ear. So far attempts have been made to replicate complex 3D shape of human auricle by using nonwoven fabrics, hand sculpted matrix, impression molds, which fail to reproduce patient-specific curvature and architecture, which is exactly symmetrical to the other ear. In collaboration with clinicians from Safdarjung hospital, New Delhi, we tried to develop a proof of concept prototype using MRI data from a patient, and develop a mirror image CAD model. Finally 3D printed auricular cartilage was prepared (Fig. 8.5). Next steps would involve coculturing multiple cell populations, in order to achieve covering of auricular cartilage with perichondrium and skin (by culturing fibroblast, keratinocytes and melanocytes) [39]. Advanced biofabrication techniques involving simultaneous or sequential deposition of multiple cell types will allow us to achieve this target. Tensile modulus of native human auricular tissue is around 16 MPa, whereas the ultimate tensile strength is 2.18 MPa. Auricular cartilage also possesses remarkable elasticity and bending modulus, which is still difficult to achieve. Commonly used hydrogels cannot achieve such mechanical characteristics. Development of fibre-reinforced or particle-reinforced printed constructs may enable to progress toward developing functional, clinically relevant auricular construct.

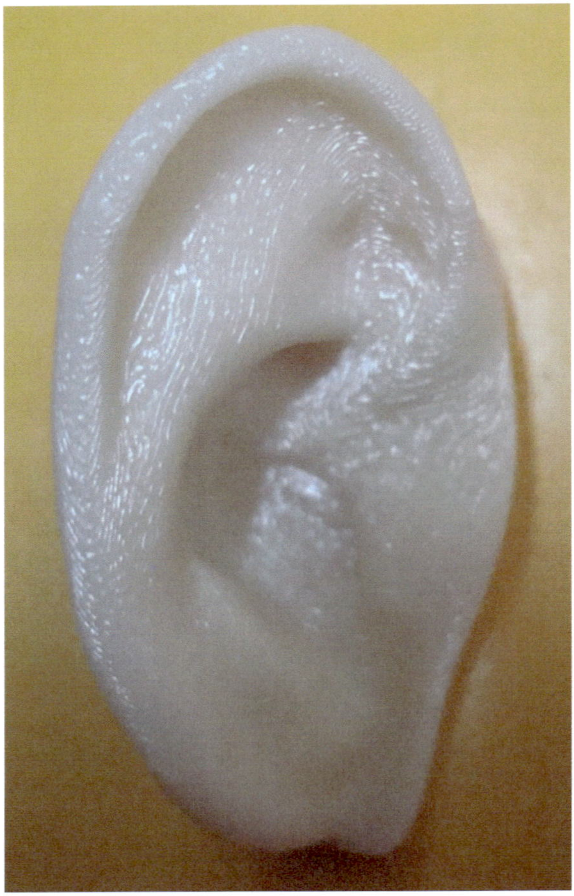

Fig. 8.5 Auricular cartilage formation by 3D bioprinting

8.5 Future Perspective

During human tissue development in embryonic stage, cellular orientation, differentiation and matrix synthesis are tightly regulated by time-dependent delivery of morphogens or soluble factors in spatially controlled 3D microenvironment. Conventional tissue engineering strategies employ culturing cells on 3D porous scaffolds, leading to uniform localisation of cells and those factors and eventually resulting in unwanted non-physiological cellular responses. Moreover, these strategies invariable fail to recapitulate the cellular organization and complex anatomical architecture of human organs. Next generation of tissue engineering should have personalized approach. Bioprinting offers fascinating potential to generate clinically relevant sized cartilage constructs (centimeter scale). Moreover, understanding of heterogeneous tissue microenvironment using hydrogel-based bioinks possessing differential stiffness, chemical compositions may pave the way to develop complex cartilage tissues such as, osteochondral tissue, intervertebral disc, auricular tissue etc.

References

1. Bhattacharjee M, Coburn J, Centola M, Murab S, Barbero A, Kaplan DL, Martin I, Ghosh S. Tissue engineering strategies to study cartilage development, degeneration and regeneration. Adv Drug Deliv Rev. 2015;84:107–22.
2. Macchiarini P, Jungebluth P, Go T, Asnaghi MA, Rees LE, Cogan TA, Dodson A, Martorell J, Bellini S, Parnigotto PP, Dickinson SC, Hollander AP, Mantero S, Conconi MT, Birchall MA. Clinical transplantation of a tissue-engineered airway. Lancet. 2008;372 (9655):2023–30.
3. Fulco I, Miot S, Haug MD, Barbero A, Wixmerten A, Feliciano S, Wolf F, Jundt G, Marsano A, Farhadi J, Heberer M, Jakob M, Schaefer DJ, Martin I. Engineered autologous cartilage tissue for nasal reconstruction after tumour resection: an observational first-in-human trial. Lancet. 2014;384(9940):337–46.
4. Fazaeli S, Ghazanfari S, Everts V, Smit TH, Koolstra JH. The contribution of collagen fibers to the mechanical compressive properties of the temporomandibular joint disc. Osteoarthr Cartil. 2016;24(7):1292–301.
5. Bhattacharjee M, Chameettachal S, Pahwa S, Ray AR, Ghosh S. Strategies for replicating anatomical cartilaginous tissue gradient in engineered intervertebral disc. ACS Appl Mater Interfaces. 2014;6(1):183–93.
6. Puetzer JL. Koo E2, Bonassar LJ. Induction of fiber alignment and mechanical anisotropy in tissue engineered menisci with mechanical anchoring. J Biomech. 2015;48(8):1436–43.
7. Bhattacharjee M, Miot S, Gorecka A, Singha K, Loparic M, Dickinson S, Das A, Bhavesh NS, Ray AR, Martin I, Ghosh S. Oriented lamellar silk fibrous scaffolds to drive cartilage matrix orientation: towards annulus fibrous tissue engineering. Acta Biomater. 2012;8(9):3313–25.
8. Moutos FT, Estes BT, Guilak F. Multifunctional hybrid three-dimensionally woven scaffolds for cartilage tissue engineering. Macromol Biosci. 2010;10(11):1355–64.
9. Chameettachal S, Murab S, Vaid R, Midha S, Ghosh S. Effect of visco-elastic silk-chitosan microcomposite scaffolds on matrix deposition and biomechanical functionality for cartilage tissue engineering. J Tissue Eng Regen Med. 2015.
10. von der Mark K, Gauss V, von der Mark H, Müller P. Relationship between cell shape and type of collagen synthesised as chondrocytes lose their cartilage phenotype in culture. Nature. 1977;267(5611):531–2.
11. Sharma B, Williams CG, Kim TK, Sun D, Malik A, Khan M, Leong K, Elisseeff JH. Designing zonal organization into tissue-engineered cartilage. Tissue Eng. 2007;13(2): 405–14.
12. Klein TJ, Schumacher BL, Schmidt TA, Li KW, Voegtline MS, Masuda K, Thonar EJ, Sah RL. Tissue engineering of stratified articular cartilage from chondrocyte subpopulations. Osteoarthr Cartil. 2003;11(8):595–602.
13. Ng KW, Ateshian GA, Hung CT. Zonal chondrocytes seeded in a layered agarose hydrogel create engineered cartilage with depth-dependent cellular and mechanical inhomogeneity. Tissue Eng Part A. 2009;15(9):2315–24.
14. Karsenty G, Wagner EF. Reaching a genetic and molecular understanding of skeletal development. Dev Cell. 2002;2(4):389–406.
15. Torzilli PA, Grande DA, Arduino JM. Diffusive properties of immature articular cartilage. J Biomed Mater Res. 1998;40(1):132–8.

16. Maeda S, Miyabayashi T, Yamamoto JK, Roberts GD, Lepine AJ, Clemmons RM. Quantitative analysis of chondroitin sulfate isomers in intervertebral disk chondrocyte culture using capillary electrophoresis. J Vet Med Sci. 2001;63(9):1039–43.
17. Theodoropoulos JS, De Croos JN, Park SS, Pilliar R, Kandel RA. Integration of tissue-engineered cartilage with host cartilage: an in vitro model. Clin Orthop Relat Res. 2011;469(10):2785–95.
18. Pelttari K, Winter A, Steck E, Goetzke K, Hennig T, Ochs BG, Aigner T, Richter W. Premature induction of hypertrophy during in vitro chondrogenesis of human mesenchymal stem cells correlates with calcification and vascular invasion after ectopic transplantation in SCID mice. Arthritis Rheum. 2006;54(10):3254–66.
19. Seda Tigli R, Ghosh S, Laha MM, Shevde NK, Daheron L, Gimble J, Gümüşderelioglu M, Kaplan DL. Comparative chondrogenesis of human cell sources in 3D scaffolds. J Tissue Eng Regen Med. 2009;3(5):348–60.
20. Ghosh S, Parker ST, Wang X, Kaplan DL, Lewis JA. Direct-write assembly of micro-periodic silk fibroin scaffolds for tissue engineering applications. Adv Funct Mater. 2008;18(13): 1883–9.
21. Das S, Pati F, Chameettachal S, Pahwa S, Ray AR, Dhara S, Ghosh S. Enhanced redifferentiation of chondrocytes on microperiodic silk/gelatin scaffolds: toward tailor-made tissue engineering. Biomacromolecules. 2013;14(2):311–21.
22. Jose RR, Brown JE, Polido KE, Omenetto FG, Kaplan DL. Polyol-silk bioink formulations as two-part room-temperature curable materials for 3D printing. ACS Biomater. Sci. Eng. 2015;1(9):780–8.
23. Das S, Pati F, Choi YJ, Rijal G, Shim JH, Kim SW, Ray AR, Cho DW, Ghosh S. Bioprintable, cell-laden silk fibroin-gelatin hydrogel supporting multilineage differentiation of stem cells for fabrication of three-dimensional tissue constructs. Acta Biomater. 2015;11: 233–46.
24. Klumpers DD, Mooney DJ, Smit TH. From skeletal development to tissue engineering: lessons from the micromass assay. Tissue Eng Part B Rev. 2015;21(5):427–37.
25. Nair K, Gandhi M, Khalil S, Yan KC, Marcolongo M, Barbee K, Sun W. Characterization of cell viability during bioprinting processes. Biotechnol J. 2009;4(8):1168–77.
26. Kundu J, Shim JH, Jang J, Kim SW, Cho DW. An additive manufacturing-based PCL-alginate-chondrocyte bioprinted scaffold for cartilage tissue engineering. J Tissue Eng Regen Med. 2015;9(11):1286–97.
27. Tasoglu S, Demirci U. Bioprinting for stem cell research. Trends Biotechnol. 2013;31(1): 10–9.
28. Malda J, Visser J, Melchels FP, Jüngst T, Hennink WE, Dhert WJA, Groll J, Hutmacher DW. Engineering hydrogels for biofabrication. Adv Mater. 2013;25(36):5011–28.
29. Chang R, Nam J, Sun W. Effects of dispensing pressure and nozzle diameter on cell survival from solid freeform fabrication-based direct cell writing. Tissue Eng Part A. 2008;14(1):41–8.
30. Fedorovich NE, De Wijn JR, Verbout AJ, Alblas J, Dhert WJ. Three-dimensional fiber deposition of cell-laden, viable, patterned constructs for bone tissue printing. Tissue Eng Part A. 2008;14(1):127–33.
31. Das RK, Zouani OF. A review of the effects of the cell environment physicochemical nanoarchitecture on stem cell commitment. Biomaterials. 2014;35(20):5278–93.
32. Regulation of chondrogenesis and hypertrophy in 3D bioprinted constructs, in communication, ACS Biomaterials).
33. Yue K, Trujillo-de Santiago G, Alvarez MM, Tamayol A, Annabi N, Khademhosseini A. Synthesis, properties, and biomedical applications of gelatin methacryloyl (GelMA) hydrogels. Biomaterials. 2015;73:254–71.
34. Visser J, Gawlitta D, Benders KE, Toma SM, Pouran B, van Weeren PR, Dhert WJ, Malda J. Endochondral bone formation in gelatin methacrylamide hydrogel with embedded cartilage-derived matrix particles. Biomaterials. 2015;37:174–82.

35. Kesti M, Eberhardt C, Pagliccia G, Kenkel D, Grande D, Boss A, Zenobi-Wong M. A versatile bioink for three-dimensional printing of cellular scaffolds based on thermally and photo-triggered tandem gelation. Acta Biomater. 2015;11:162–72.
36. Studer D, Millan C, Öztürk E, Maniura-Weber K, Zenobi-Wong M. Molecular and biophysical mechanisms regulating hypertrophic differentiation in chondrocytes and mesenchymal stem cells. Eur Cell Mater. 2012;24(24):118–35.
37. Fedorovich NE, Schuurman W, Wijnberg HM, Prins HJ, van Weeren PR, Malda J, Alblas J, Dhert WJ. Biofabrication of osteochondral tissue equivalents by printing topologically defined, cell-laden hydrogel scaffolds. Tissue Eng Part C Methods. 2012;18(1):33–44.
38. Xu T, Binder KW, Albanna MZ, Dice D, Zhao W, Yoo JJ, Atala A. Hybrid printing of mechanically and biologically improved constructs for cartilage tissue engineering applications. Biofabrication. 2013;5(1):015001.
39. Otto IA, Melchels FPW, Zhao X, Randolph MA, Kon M, Breugem CC, Malda J. Auricular reconstruction using biofabrication-based tissue engineering strategies. Biofabrication. 2015;7:032001.

Chapter 9
Clinical Trials

9.1 Introduction

Many developing nations, like India, are is in the state of a major shift in demographics towards elderly populations. As per the 1991 census, the population of the elderly in India was 57 million as compared with 20 million in 1951. Since then there has been a sharp increase in the number of elderly persons between 1991 and 2001 and it has been projected that by the year 2050, the number of elderly people would rise to about 324 million [1]. India, having the highest elderly population by 2025, has thus acquired the label of "an ageing nation" with 7.7 % of its population being more than 60 years. The increase in life expectancy of Indians has a direct and amplifying effect on the prevalence of edentulism. With the increase in age, there is a greater risk of hip fractures, knee fractures or other life-threatening diseases, like cardiac problems. Apart from the use of relevant drugs, hip/knee replacements require the use of biomedical devices, which have to undergo clinical trials.

It needs to be categorically mentioned that the clinical trial is the final end-point experiment for any biomedical device prototype with the aim of ensuring the performance validation in patients.

Typically, clinical trials are sponsored by a governmental organization or a pharmaceutical, biotechnology or medical device company and are very common in drug or new medical devices. These experiments generate data on safety and efficacy [1]. They are conducted only after they have received health authority/ethics committee approval in the country where the trials are to be conducted. The responsible institutional/national authorities are responsible for vetting the risk/benefit ratio of the trial and it needs to be understood that their approval does

B. Basu and S. Ghosh, *Biomaterials for Musculoskeletal Regeneration*,
Indian Institute of Metals Series, DOI 10.1007/978-981-10-3017-8_9

not mean that the therapy is 'safe' or effective, but only that the trial may be conducted. In the United States, all clinical trials submitted to the FDA as part of a drug approval process are independently assessed by clinical experts within the Food and Drug Administration, including inspections of primary data collection at selected clinical trial sites.

In the context of prosthodontics, the edentulism has also an indirect effect on the oral health related quality of life. The Quality of life is improved with good oral health that allows one to speak, chew, taste, smile, live without pain or discomfort and interact socially without embarrassment. The current status of the Indian population reveals that over half of the population has experienced tooth loss. The Level of complete edentulism in the general population varies from 14–16 % mostly in the 60-year age group. Studies have reported a higher rate of edentulism in both the urban and rural elderly populations with around 91.2 % in the elderly above the age of 65 years [2]. Considering the goals above, affordable treatment to the existing edentulous population, both in the urban and rural sectors, should be emphasized. Since dental equipment and machinery, as also dental materials, are quite costly, only a few can afford dental treatment and it is imperative that cost should be contained and reduced by developing alternative indigenous material. This has motivated a large number of researchers to develop dental implants, and their clinical efficacy is to be established using clinical trials.

9.2 Some Concepts and Definitions

9.2.1 Clinical Trial

A Clinical trial is an experimental research protocol of an authorized intervention carried out on human volunteers for evaluating the performance of a new biomedical device, method of treatment, drugs or vaccines, etc. in comparison with an existing treatment option or a placebo (no intervention).

As far as the terminology is concerned, animal experiments carried out to assess *in vivo* biocompatibility are termed pre-clinical trials. Similarly, human participants or human subjects taking part in experiments or investigation in clinical settings are known as clinical trials.

9.2.2 Clinical Trial Protocol

A clinical trial protocol is a documentation ordained by a committee of biomedical researchers/clinicians for defining and regulating the conduct of the trial. The study investigators must, in word and spirit, truly abide by the clinical trial protocol. The background, hypothesis, rationale, goals, design, experimental methods, suitable number of participants for each phase and the overall management of the trial is summarized in the trial protocol.

9.2.3 Good Clinical Practice (GCP)

GCP is a worldwide standard of clinical ethics in science, especially for organizing, implementing, data recording and reporting experimental trials that involve human volunteers. For any clinical trial, GCP guidelines also define the roles of clinical trial sponsors, research investigators, and monitors. Concurrence with good clinical practice (GCP) assures that the safety of the participants is paramount and that the clinical trial has credibility in data recording and analysis. The GCP is an internationally accepted quality standard conceived by the International Council for Harmonisation of Technical Requirements for Pharmaceuticals for Human Use (ICH), in order to expedite drug trials by the mutual acceptance of clinical data generated in the European Union, Japan, the United States and many other countries. A related set of rules for the clinical trial of biomedical devices is the international standard ISO 14155, which is accepted in the European Union as a harmonised standard. The ICH-GCP for drugs/pharmaceuticals is globally adopted, while the ISO 14155 is yet to be universally accepted. The general principles of ICH-GCP and ISO 14155 are the same. These regulatory standards for clinical trials are known as ICH-GCP or ISO-GCP guidelines. Their focus is different i.e., drugs versus medical devices.

9.2.4 Good Laboratory Practice (GLP)

GLP was originally introduced by the Food and Drugs administration (FDA) as a formal regulation to monitor the working of research laboratories in the United States. GLP is a set of principles that provides guidelines for laboratory studies with regard to planning, conducting, recording, reporting and archiving experimental results. GLP mainly pertains to non-clinical research studies. As a part of GLP, all the materials/reagents and instruments used for the study are quality checked and validated. GLP ensures infraudulent/unbiased conduct of experiments, data acquisition, analysis and reporting of the studies in a truly legitimate manner such that the conclusions of the studies can be utilized for the future design of experiments, be they sub-clinical or clinical. GLP accentuates the reliability of the laboratory generated experimental toxicity (acute and chronic) data, which forms the basis for the approval of similar studies in human volunteers or clinical trials. GLP is often confused with lab safety such as using lab coats, glasses, and gloves, which is a misnomer.

9.2.5 Klimisch Score

With reference to a GLP study, the reliability of evidence in general and test reports, publications or experimental data of a study, in particular, are indicated by the Klimisch score. This score is assigned a value as follows: 1—reliable without restrictions, 2—reliable with restrictions, 3—not reliable and 4—not assignable. Studies conducted in concurrence with GLP are assigned a Klimisch score of 1 and this is preferred by regulatory bodies. In the absence of GLP for a particular endpoint, a score of 2 is also acceptable.

CONSORT: The Consolidated Standards of Reporting Trials (CONSORT) provides guidelines as how to conduct clinical trials and develop protocols. A number of biomedical journal editors as well as funding agencies prior to funding any clinical trial always desire the investigators to follow CONSORT guidelines. In particular, such guidelines include evidence-based recommendations for conducting and reporting randomized trials. The goal of CONSORT is to develop a new yardstick for evaluating the quality of randomized controlled trial reports (RCT). As per the guidelines, the conduct of equivalence or non-inferiority design as well as the inclusion of a criterion for engaging human volunteers needs to be followed. An outcome of CONSORT is the CONSORT statement, which is a 25 item checklist concerning study details such as how the trial was organized, conducted and interpreted. More information is available at the website, www.consort-statement.org.

9.3 Some Illustrative Examples to Substantiate the Need of Clinical Trials

It is well known in the biomedical community that clinical trials are to be designed with respect to a specific disease model. This is in line with a similar rationale that is commonly followed while assessing the biocompatibility of implantable biomaterials.

To substantiate this, the cell lines or the animal model to be used for *in vitro/in vivo* biocompatibility assessment particularly depend on the end-point application of a given biomaterial.

In the following, a few examples will be presented for researchers to realise how a clinical trial could be planned for some biomaterials, which are proven to be biocompatible *in vivo*.

9.3.1 Osteoporosis Treatment

Osteoporosis, a commonly reported disease among post-menopausal women and an ageing population, is characterized by an imbalance in the functionality of osteoblast (bone forming) and osteoclast (bone resorption) cells, leading to the disruption of bone remodelling. This consequently makes the bone porous and prone to fracture. The prevention of such bone resorption is a primary requirement for the maintenance of the implant in elderly patients, who otherwise have to undergo revision surgery.

For the treatment of osteoporosis, the widely-used SrR (Strontium ranelate) drug, combines two atoms of stable strontium with the organic moiety (ranelic acid). The biophysical role of SrR is reported to be through the strontium cations themselves, which have been shown to work both *in vivo* and *in vitro* by stimulating osteoblasts to make new bone and preventing osteoclasts from resorbing bone, leading to the restoration of the normal bone remodelling balance [1]. Motivated by the recent use of strontium ranelate (SrR) under the trade name, Protelos, for the treatment of osteoporosis, a number of research groups have investigated the biocompatibility of Sr-containing glass-ceramics and bioactive glasses [2–5]. For example, the *in vivo* biocompatibility and bone healing of strontium (Sr)-stabilized bulk glass ceramics with the nominal composition of $4.5SiO_2$–$3Al_2O_3$–$1.5P_2O_5$–$3SrO$–$2SrF_2$ during implantation of up to 26 weeks in a rabbit (animal) model has been reported recently [References]. The overall biocompatibility assessment using fluorescence microscopy, histological analysis and micro-computed tomography unequivocally confirmed the long term implant stability as well as the osteoconductive property of 100 % Sr-substituted glass ceramics, which is comparable to that of a known bioactive implant. The rationale for pursuing research on Sr-stabilised glass ceramics is that Sr, like Ca, can stimulate the Ca–sensing receptors at the cell membrane of osteoblast cells. These, in turn, influence RANKL production and reduce the osteoblast proliferation or can even cause apoptosis. The enhanced proliferation and differentiation of osteoblasts in the presence of Sr is believed to increase neobone formation and this is the result of the stimulation of homeostatic local hypercalciuria around the implant [6–9].

In the context of the discussion in this chapter, it should be evident that Sr-stabilised biomaterials in osteoporotic human patients is warranted to realize clinical efficacy. In such clinical trials, the control subjects can receive the fore mentioned Strontium ranelate drug at clinician-prescribed doses for a planned treatment time. Bone-healing can be monitored and compared with reference to test subjects, who received the strontium-stabilised biomaterial devices. This is one of the examples where the clinical efficacy of using biomaterials vis-à-vis the currently used drug can be established.

9.3.2 *Clinical Trials on New Hip Implants*

There has been a consistent effort to develop biomaterials with better combinations of mechanical/tribological properties and uncompromised biocompatibility for articulating surfaces like hip joints, knee joints, elbow joints etc. For example, some Indian researchers have recently developed ZrO_2-toughened—Al_2O_3 femoral ball heads and acetabular sockets for total hip joint replacement applications. It can be perceived that these new device prototypes could have a different set of performance-limiting properties in comparison to commercially available (Depuy-Stryker etc.) similar biomedical devices of comparable compositions. The difference can be due to the sequence of processing/manufacturing stages adopted. In order to establish the clinical efficacy of lab developed prototypes, clinical trials are therefore warranted with two groups of human subjects, where one group receives the newly developed devices and the other group receives commercially available biomedical devices.

The long term survival and performance of human patients can be used as a reflection of the clinical efficacy of such methods.

9.3.3 *Prosthodontic Treatment*

A dental implant is an artificial tooth root made of a natural and/synthetic biomaterial. This biomaterial implant is used to support a dental prosthesis (implant-supported crown, bridge or denture), that resembles natural teeth (see Fig. 9.1). Implant prostheses offer a more predictable and fixed treatment option than traditional restorations. It is worthwhile to emphasize here that the number of implants inserted by dentists across the globe is increasing [1]. Therefore, the future trend is towards improving dental health care services promoting healthy aging with research and development on smart biomaterials and better designs of dental implants.

There are around 220 FDA approved implant systems globally. Most implant systems are from North America (USA, Canada), Europe (Netherlands, Italy, France, Switzerland, Asia (Israel, Korea) and South America (Brazil). The cost of the medical devices imported from international countries to India include a 10 % duty rate and CESS (3 % duty) with an additional countervailing duty [9]. An international dental implant per se costs a minimum of Rs. 4000–8000, while the implant system charges vary according to the manufacturer with a minimum of Rs.10,000 with abutment. An Indian implant system made of commercially pure titanium was developed by the Defense Research and Development Organisation (DRDO) in 2004. The Wane Core implant system made up of titanium alloys is also

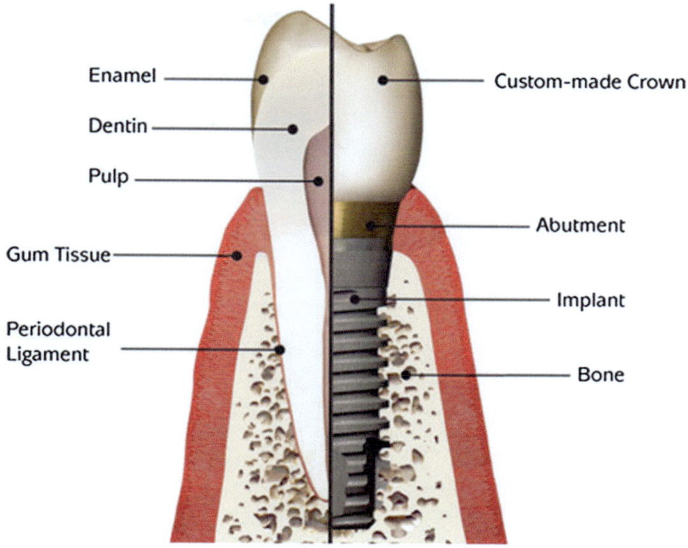

Fig. 9.1 Schematic of screw typed tooth implant showing the abutment, implant, and periodontal ligament and bone near to the implant

currently in the market. Titanium alloys are, at present, majorly used as dental implant biomaterial.

Implant macrodesign and surface topography are the critical factors responsible for the osseointegration of an artificial tooth implant. Various designs have been proposed for the root such as screw, cylindrical and hollow basket geometries (see Fig. 9.2). Hollow basket geometries (Fig. 9.2a) have become obsolete, but screws (Fig. 9.2b) and cylinders (Fig. 9.2c) are now synonymous with threaded and non-threaded implants, respectively. Both threaded and non-threaded implants are manufactured as straight, tapered, conical, ovoid and trapezoidal shapes [10].

It needs to be categorically mentioned here that if the implant design is altered for the same biomaterial, clinical trials are to be conducted. With reference to the above, if Ti_6Al_4V based threaded dental implants are investigated at a lab and commercially available non-threaded implants are routinely used with human patients, then clinical trials with a threaded implant are to be performed in human subjects and the results are to be compared with the performance of non-threaded implants in human subjects.

In view of the lower complexity in terms of the safety aspects associated with dental implants when compared to orthopaedic implants or cardiovascular implants, the clinical trials with dental materials are reported more in the literature [10–13].

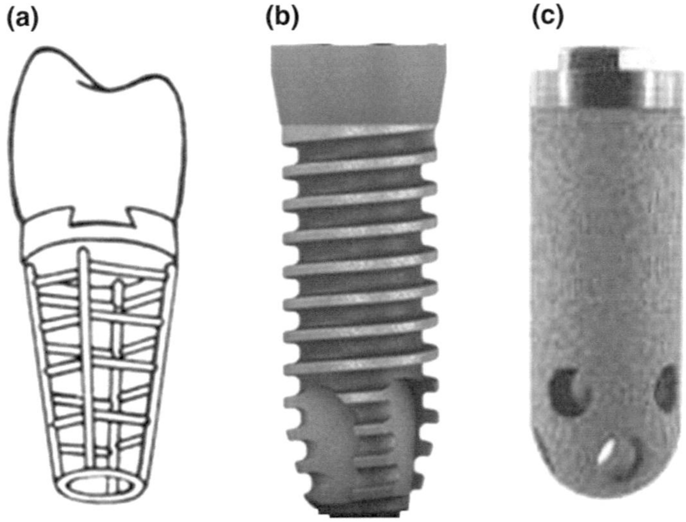

Fig. 9.2 Various types of geometry for tooth implant **a** hollow basket **b** screw and non-threaded (**c**)

9.4 Types of Clinical Trials and Randomised Control Trial

Figure 9.3 presents a summary of the various types of clinical trials. In a clinical observational study, the investigators observe the subjects and measure their outcomes. The researchers do not actively manage the study. In an interventional study, the investigators give the research subjects (human patients) a particular medicine or other intervention. Usually, they compare the treated subjects to subjects who receive no treatment or standard treatment. Then, the researchers measure how the subjects' health changes over the treatment time or thereafter.

Another way of classifying trials is by their purpose. The U.S. National Institute of Health (NIH) organizes trials into different types. For example, screening trials test the best way to detect certain diseases or health conditions. Diagnostic trials are conducted to find better tests or procedures for diagnosing a particular disease or condition. Treatment trials test experimental treatments, new combinations of drugs, or new approaches to surgery or radiation therapy.

There are various ways of conducting clinical trials but the method having the highest quality of evidence the Randomized controlled trial (RCT). It is widely considered a gold standard design of conducting clinical trials. RCTs have three important features.

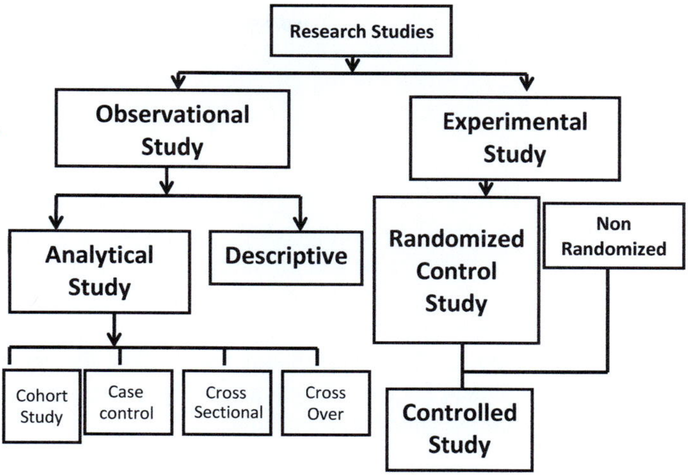

Fig. 9.3 Summary of various types of study related to clinical trials

9.4.1 Randomization

Randomization is a process to ensure that each participant has the same chance of being assigned to either intervention or control.

This is in view of the fact that the procedure selection bias has been documented to have a very strong effect on outcome variables. Each patient should have an equal chance to be given either an established drug or a new drug in the market. It should not be based on doctor choice, patient choice, or any other condition. It is done by using a computer generated number, and by utilizing the services of the nurse. Therefore, randomization is effective in reducing bias, because it guarantees that treatment assignment will not be based on the patient's prognostic factors. If randomisation is adopted, the investigators cannot favor one treatment group over another by assigning patients with better prognoses to them, either knowingly or unknowingly.

9.4.2 Allocation Concealment

The word 'concealment' means to keep something secret. Therefore, allocation concealment refers to the technique used to implement the sequence.

The allocation concealment mechanism aims to prevent participants and recruiters from knowing the study group to which the next participant will be assigned.

Allocation concealment is possible with all types of trials, including un-blinded trials, and is therefore universally recommended. To substantiate, a patient is allocated a new drug by a nurse (a sealed envelope inside which A or B was mentioned but not the drug name). The patient takes this envelope and goes to the pharmacy and shows the envelope which will be opened by the pharmacist and is hand over Drug A or Drug B. This will ensure that neither the nurse nor the pharmacist knows which patient is taking which drug—the color, shape, and outer covering of the drug should be same for both drugs.

9.4.3 Blinding

The word ‘blinding’ means masking.

The purpose of blinding is to reduce biases in the clinical trial experiment.

The use of each blinded group eliminates a different source of bias. It is most useful when there is a subjective component to treatment or evaluation. Again, this is hiding which patient has taken which type of drug from the investigator who will measure for example, reduction in pain after giving drug A or drug B. For example, the investigator would not know which patient has received drug A or B, as he would measure the outcome of drug intervention in terms of pain reduction. Thus, this aspect would make the RCT ‘investigator blinded’ or ‘outcome accessor blinded’. A single blind protocol requires that the experimenter does not know the identity of the samples or animals during the testing and calculations. It is appropriate when no human subjects are involved. A double blind protocol comes into play when human subjects are tested and requires insuring that neither the experimenter nor experimental subjects have a knowledge of the identity of the treatment or the results until after the experiment is complete.

With reference to drug trials, after the collection of how much reduction of pain occurs in the two groups (established drug or new drug), the data will be given to a statistician without providing any indication about the origin of the data. This is called statistician-blinded.

The international committee of medical journal editor (ICMJE) recommendations, which is followed by most peer-reviewed journals in the medical sciences, warrants the use of CONSORT (Consolidated Standards of Reporting Trials) (http://www.consort-statement.org/) guidelines of reporting Randomized controlled trials. It is a 25-item checklist, which is to be submitted mandatorily with article submissions to the Journal.

Depending on the product type/the stage of development, the investigators initially enroll volunteers and/or human patients into small pilot studies before conducting a clinical trial at a single research center or multiple centers in one country, or in multiple countries.

The biggest barrier to completing the studies is the shortage of people who take part. All drug and many device trials target a subset of the population, meaning not everyone can participate. Some drug trials require patients to have unusual combinations of disease characteristics. It is a challenge to find the appropriate patients and obtain their consent, especially when they may receive no direct benefit (because they are not paid, the study drug is not yet proven to work, or the patient may receive a placebo).

9.5 Clinical Trial Phases

Prior to the start of a clinical trial study on a biomedical device/implant, preclinical data with reference to *in vitro* experimental results as well as *in vivo* biocompatibility results (small/medium/large animal implantation experiments) are to be collected and appropriately assessed. This stage however does not require Ethical committee approval or a registration number.

A clinical trial normally comprises four phases and these phases are described with reference to the drug trial.

> The distinction between the different phases of the clinical trial is essentially based on the number of human subjects investigated at each phase and the number of subjects progressively increases at each phase to validate the clinical trial outcome in more populations with the endpoint being to ensure the safety and efficacy of a specific disease treatment in human patients.

It can be recalled here that an *in vivo* biocompatibility assessment on an implantable biomaterial is also carried out first on a small animal model, a medium-sized animal and finally on a large animal model with the rationale of validating the osseointegration or compatibility with osseous structures in a more and more complex biochemical environment *in vivo*.

Phase I pertains to human pharmacology, wherein the objective is to determine the maximum dosage tolerance and study the adverse reactions that can be elicited by the drug formulation in both healthy male and female volunteers. Pharmacokinetic parameters such as drug absorption, distribution in different tissues of the body, metabolism and excretion are ascertained to support the dosage regimen by testing in different age groups. Also, the pharmocodynamics of the formulation by determining drug concentrations in the blood at different time points after administration are evaluated in patients with the target disease. Phase I or the limited clinical trial can be conducted at a single center and can be carried out with 20–80 patients.

Phase II relates to therapeutic exploratory trials for further assessment of the safety and effective dosage in patients with mild and severe disease. The drug dosage tested here is below the highest dose in phase I and is also examined in concomitant medication for other ailments. As a norm, 100–300 patients are studied per dose and the trial is conducted in 3–4 centers.

Phase III of the therapeutic confirmatory trials is a multi-centered clinical trial in a large number of male and female patients (1000–3000). At this stage, the efficacy of the new drug formulation is compared with an existing standard drug or a placebo in case of the unavailability of standard drugs with the aim of assessing the performance of the new drug against the existing medication for the particular disease or ailment in question. The drug is approved for marketing and sale off the counter, after the successful completion of the phase III trial.

After the dissemination of the drug to the market in phase III, the final **phase IV** comprises the surveillance of drug performance and gathering information on the risks and/or benefits upon prolonged usage. This data must be correlated with the toxicity data in animals generated prior to the start of the clinical trial. Such data are necessary for optimizing the drug usage by the licensing authority. A drug may be withdrawn from the market and stopped from further production with immediate effect by the drug license regulatory bodies under instances of greater toxic effects than benefits reported after long term use.

Similar to drug trials, a host of biomedical devices need an evaluation of device performance in animals and humans. A medical device is defined as an inert diagnostic or therapeutic object that does not achieve any of its intended purposes by chemical action within or on the body. By virtue of risk evaluation, medical devices can be categorized into non-critical devices (e.g. thermometers, blood pressure apparatuses) and critical devices (e.g. pace makers, implants and catheters). It is now being widely realized that data on the efficacy of devices, and adverse reactions, if any, must be documented before pre-market certification. Implantable devices such as orthopedic implants (hip/knee joints), cardiac pacemakers, artificial heart valves and cardiovascular stents require a systematic pre-clinical and clinical investigation prior to their certification and marketing. Currently, most of the devices imported by India are approved by the Federal Drug Administration (FDA) of the United States of America. However, the need for developing indigenous biomedical technology with the goal of reducing production costs has spurred the setting up of the Indian Medical Device Regulatory Authority (IMDRA) by the Government of India. It is, however, important to make a clear distinction between devices and drugs. For clinical trials of devices, phase I is non-existent as devices cannot be tested on healthy volunteers. Hence, they are tested directly in patients in phase II. The risk assessment of critical devices used continuously, such as intraocular lenses, need greater attention compared to those used intermittently, like contact lenses. Such intra body devices need prolonged testing in phase III and phase IV. Overall, a rigorous investigation of biomedical implants and device components must be undertaken at the lab, followed by animal experiments before proceeding to clinical trial in humans for device certification.

Although the term "clinical trials" is most commonly associated with the large, randomized studies typical of Phase 3, many clinical trials are small. They may be "sponsored" by single researchers or a small group of researchers, and are designed to test simple questions. In the field of rare diseases, sometimes the number of patients is the limiting factor for the size of a clinical trial.

One aspect that needs some attention prior to the start of clinical trials on biomedical implants is sterilization. The sterilization of implants can be carried out using three modes. Also, many sterilization methods are taken from food processing industries. For most metallic or ceramic implants, steam sterilization is carried out in an autoclave at 120 °C under steam pressure. This is the most widely used. Another method can be chemical sterilization using Ethylene oxide (ETO). The third mode is Gamma ray sterilization, which is often used for polymeric implants. Typically, the Gamma chamber has a Co^{60} source, which works both at low and high doses. A typical gamma dose of 10 kGy is used for food processing, while 25 kGy is used for the regular disinfection of metal/ceramic/polymer implants. Depending on the dose rate of the instrument, the developed implants will be irradiated for a given timeframe. Since the gamma dosage can influence the material properties, the researchers have to investigate the phase assemblage (XRD/FTIR/Raman), microstructure (SEM) and mechanical properties (strength/hardness) to understand the effect of Gamma irradiation on a potential decrease in properties. It may be worthwhile to mention that the steam autoclave is by far the most widely used in clinical scenarios (hospitals). Depending on its material composition, the sensitivity of an implant towards any of the afore-mentioned sterilization modes differs and accordingly, the sterilization is to be conducted.

9.6 Ethical Issues and Approval Processes

The basis for conducting biomedical device and drug trials involving human volunteers must, in word and spirit, be in accordance with the fundamental principles of human ethics. These include respect for the participants, benefits to human healthcare, harmless or minimal risk to the participants, justified compensation under extreme circumstances and many more specific to the study in question. Prior to conducting any clinical trial, the goals of the study are to be explicitly stated and the proposal must be well supported and substantiated with laboratory and animal data proving the rationale for studies involving human subjects. In addition to the scientific aspects, the experiments must be performed in concurrence with the country-specific ethical and legal requirements of the study protocol. It is the primary duty of the principal investigator of the clinical trial to educate the participants and inform them of the potential risks/dangers and/or benefits of enrolling in the study. Before the commencement of the study, it is mandatory to obtain the informed consent of the participants or guardians for minors by furnishing important study information and details to them. The disclosures should include the purpose and duration of the study, procedures, risks, benefits, compensation policy, permission to store biological samples and volunteer data, free medical treatments in case of unforeseeable accidents arising from the study protocol, and withdrawal from the study at any stage as deemed fit by the participant. This information must be made available in language that can be easily understood by the local population, where the study is being conducted. All the official paperwork related to conducting the trial must be submitted and approved by the Institutional ethical committee and Drug Controller General (India) (DCGI), if the trial is conducted in India or by a similar competent authority in another country. The researchers conducting the study must be well qualified and competent to undertake an unbiased/fair the study, analysis and publication of the results.

The following points, therefore, need to be critically considered prior to the start of any clinical trial:

i. First of all, the necessary permission to use any new drug for human use needs to be taken from the Drug Controller General of India, New Delhi. The necessary paper work is generally filed by the manufacturing company, but can be done by any individual as well. The Investigator should enclose the institutional ethical committee approval with the application or, if already taken, then attach the approval letter. There is no need to take permission from the DCGI, if the drug/device is already in the market.
ii. After taking permission from the DCGI, the investigator has to take approval from the Institutional ethical committee of the institute/hospital where the clinical trial will be conducted. If the clinical trial is multi-centric, then all the centres have to take ethical approval separately. All the details about the clinical trial will be given to each patient in the written format and in their local language. If the patient agrees, then the patient has to sign a consent form. The following points need to be considered prior to the start of any clinical trial:

    a. The Clinical trial will not harm (even if chances of harm are present then they will not be more than routine harm which may occur in normal circumstances) the patient.
    b. The involvement of the patients will be completely voluntary and a patient will have the freedom to ask any question. Also, the patient can leave the clinical trial at any point of time without giving any reason.

iii. The next step is to the submit necessary paperwork for the inclusion of the study in the clinical trial registry of India (www.ctri.nic.in) or the WHO clinical trial registry (http://www.who.int/ictrp/en/). The Investigator can fill the clinical trial details online in only one database and it will automatically be shared between the above two databases. It is mandatory to register in the clinical trial registry, before enrolling the first patient. The various criteria to be considered while enrolling the patient typically include the type of drug, dose, investigations, duration of the clinical trial, place of trial, investigators contact details, ethical committee approval letter and other details. The clinical trial registry of India is managed by the Indian Council of Medical Research, New Delhi.

iv. After taking DCGI approval (if needed), Institutional ethical committee approval, (mandatory in any clinical trial), registering in the clinical trial registry of India (mandatory in any clinical trial), the first patient can be enrolled.

v. It is the duty of the principal investigator (PI) to register for the clinical trial. The Prime sponsor can register if so agreed in the Memorandum of Understanding. In a multi-centric or multi-sponsored trial, the lead PI or lead sponsor should register the trial. In a multi-national trial, the Indian PI should register the trial in CTRI, mentioning the relevant registration numbers in the registries of other countries.

vi. All volunteers being considered for a trial are required to undertake a medical screening. Requirements differ for different trials, but typically volunteers will have the following tests in a clinical laboratory:

(a) Measurement of the electrical activity of the heart (ECG)
(b) Measurement of blood pressure, heart rate and temperature
(c) Blood sampling
(d) Urine sampling
(e) Weight and height measurement.

9.7 Clinical Trial Protocol

Beginning in the 1980s, the harmonization of clinical trial protocols was shown as feasible across the countries of the European Union. At the same time, coordination between Europe, Japan and the United States led to a joint regulatory-industry initiative on international harmonization named after 1990 as the International Conference on Harmonisation of Technical Requirements for Registration of Pharmaceuticals for Human Use (ICH).

The clinical trial protocol contains a precise study plan to assure the safety and health of the trial subjects and to provide an exact template for the trial to be conducted by investigators. The format and content of clinical trial protocols sponsored by pharmaceutical, biotechnology or medical device companies in the United States, European Union, or Japan have been standardized to follow Good Clinical Practice guidance issued by the International Conference on Harmonization of Technical Requirements for Registration of Pharmaceuticals for Human Use (ICH). Regulatory authorities in Canada and Australia also follow ICH guidelines.

9.8 Illustrative Example—Design of Limited Clinical Trial Experiments in Prosthodontics

In any RCT on dental implants, the age of the patients can vary between 15 and 50 years, and no bias should be exercised concerning the sex (male/female) of the patient. Also, while selecting human subjects, the principal investigator of the clinical trial should ensure that the patient should not have any prior systemic diseases, and that at least one or two teeth bilaterally in the maxilla or mandible are absent in those test subjects. As far as the number of human subjects is concerned, 46 patients can be taken up to obtain statistically relevant evidence considering the sample size for a power of 90. It is worthwhile to reiterate here that adequate sample size is one of the important criteria in any RCT. As far as the study design is concerned, a parallel or split mouth design can be adopted. More importantly, appropriate randomisation, and allocation concealment is to be ensured. As discussed earlier, some human subjects can decide not to undergo the clinical study within a few weeks from the start. The principal investigator needs to follow up such drop out cases for any clinical evidence. In the context of the study design above, the dental implant supported with bone cement/screw for a single posterior tooth needs to be retained for at least a year. One important piece of clinical evidence for the RCT described above study is the measurement of adequate bone height/width in 4 months post-extraction healing time. The expected outcome measures include the following aspects,

(a) Patient centred outcomes: Oral health quality of life/Patient satisfaction
(b) Implant supported restoration: Longevity of restoration/technical complication
(c) Peri-implant health: Marginal bone level and probing depth of soft tissue
(d) Osseointegration analysis: Cone beam computed tomography (CBCT) analysis to quantify hard tissue stability/loss around the crest at various timepoints post-implantation
(e) Primary Stability analysis: At the surgical placement, this aspect for the threaded implant (host bone engagement) to be assessed using a Resonance frequency analyser (RFA)

Based on the following analysis, the outcome measures are to be validated and it is expected the design described above of the clinical trial study will provide meaningful results to assess the efficacy of a dental implant in prosthodontics treatment.

9.9 Clinical Trials—Case Studies

In the preceding sections, the relevance, principles as well as methodology to design a clinical study are described together with a substantial explanation of the ethical issues involved. Given such a background, some published clinical study results are discussed in this section so that the reader can develop an understanding of how to design a clinical study as well as how to assess the results of such a study to obtain clinically meaningful results.

Often, a clinical study is also termed a 'First in Man' study. It may be worthwhile to mention that any new biomaterial/biomedical devices may require limited single-centric and, subsequently, multi-centric trials to establish their clinical efficacy in a larger patient population.

It is important to reiterate that more the following case studies should, however, not be blindly followed to design future clinical studies, and are to be considered only as guidelines.

9.9.1 *Hydroxyapatite Based Burr-Hole Buttons for Cranioplasty Applications*

A burr-hole in the frontal region for a neurosurgical intervention often leaves an aesthetically inappropriate indentation in the treated craniofacial region of a patient in their post-operative phase. The burr-hole buttons made of metallic biomaterials transmit heat leading to chronic pain, while acrylic substances lead to irritation by transmitting the exothermic heat of polymerisation. Therefore, cranial neurosurgical procedures utilizing burr-holes of metallic/polymeric materials lead to the development of cosmetically unacceptable puckered scars on the scalp over burr-hole sites.

From this perspective, Easwar et al. [14], reported a study involving the use of custom-made hydroxyapatite-based ceramic burr-hole buttons to address the neurosurgical issues above. After getting the necessary ethical approval, a team of clinicians and biomaterial scientists used a specially designed bilayer HA button with one half being porous and the other half being dense (see Fig. 9.4). In particular, sixty-five HA buttons were implanted in 22 patients, who participated in this clinical study on a voluntary basis and also underwent cranial neurosurgical procedures. The cosmetic outcome (absence of puckered scars over burr hole sites, absence of allergic reactions or infections associated with the implant) as well as the radiological outcome with X-rays at specified intervals were assessed with particular reference to the interference in the postoperative neurological imaging due to the implants. The implants persisted as radio-dense opacities on the skull X-rays of recipients for up to 2 years (see Fig. 9.5). The bi-layer porous-dense HA buttons were useful in preventing cosmetic defects over burr-hole sites on the scalp after cranial neurosurgical procedures.

Concerning the clinical study, an informed consent explaining the cosmetic benefits and the possible risks, including allergic reactions and infections, was

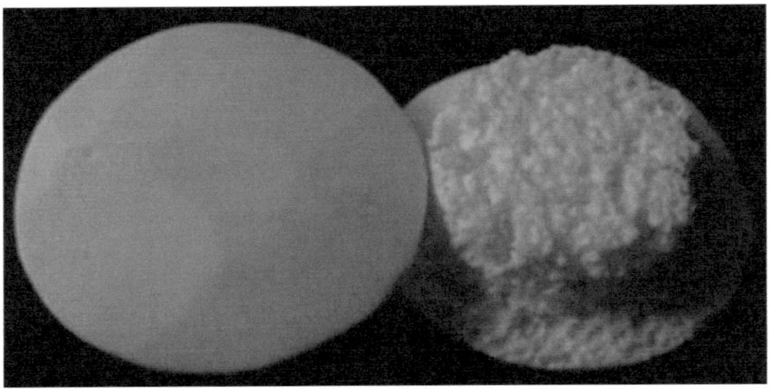

Fig. 9.4 Burr-hole button with two variants of hydroxyapatite components—a dense cap and a porous column [14]

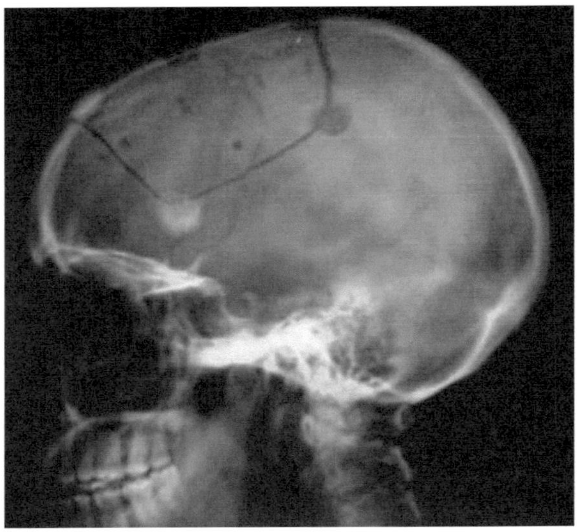

Fig. 9.5 X-ray skull showing burr-hole buttons in the frontal region in a human patient, post-surgery [14]

obtained from a group of patients in the age group of 14–60 years prior to implantation. The implantation was carried out on a patient undergoing craniotomy with one or more burr-holes located at cosmetically significant sites. The exclusion criterion was based on patients with adjuvant radiotherapy and/or chemotherapy, with short life expectancy, significant hepatic or renal diseases and those with compromised immunity. The burr hole buttons of various sizes were used during the surgery, and the one which fitted snugly into the burr-hole, was chosen for implantation. The buttons were implanted at the end of the surgery after the replacement of the free bone flap and the scalp was closed over it. No attempt was made to fix the button at the site of implantation. Radiological evaluations with X-rays were made in all patients at 6 months, 12 months, 18 months and at 2 years after implantation. Twelve patients (55 %) were available for a radiological follow-up of over 18 months, three patients (14 %) completed 2 years and seven (32 %) 6 months. Twenty-one patients (95 %) underwent a CT scan of the brain in the immediate postoperative period, while one patient had an MRI imaging for the same. Digital Subtraction Angiography was performed in another patient to confirm the adequacy of a surgical clipping of a cerebral aneurysm.

Summarising, the clinical study described above established that the two-component burr-hole button, composed of dense and porous hydroxyapatite, is safe and cosmetically effective in preventing puckered scars on the scalp over burr-hole sites on the cranium. Though the implants were still evident on X-ray radiographs of the skull after 2 years, they did not interfere in post-operative neurological imaging with CT, MR imaging or digital subtraction angiography.

9.9.2 *Degradable Hydroxyapatite-Bioactive Glass Ceramic for Iliac Crest Implant*

The reconstruction of bone defects in orthopedic surgery is an important application area, where bioceramics are considered a potential solution (see Fig. 9.6a). The following clinical study, conducted by Acharya et al. [15], prospectively validated the hypothesis that the iliac crest donor site morbidity is, perhaps, a structural issue, and the use of a hydroxyapatic-bioglass composite block to repair and reconstruct the iliac crest defect site can reduce its incidence. It may be worthwhile to mention that an ideal reconstructive material should integrate with the surrounding bone (bioactive and biodegradable), have optimum mechanical strength, and be safe, economical, and easily available.

After the necessary ethical committee approval and the obtaining of informed consent, 28 patients voluntarily underwent tricortical iliac crest graft harvesting and 26 (17 male and 9 female patients with age ranging from 18 to 57 years) were available for follow-ups up to the end of a year from surgery. The anterior iliac crest was reconstructed in 20 patients and the posterior crest in 6 patients. The Chitra-HABG based ceramic graft used in this study is a composite of hydroxyapatite (HAP) and a bioactive glass ceramic in the ratio of 80:20. A conventional sintering process is used to manufacture the HAP and the bioactive glass ceramic is prepared by a sol-gel method involving a low temperature processing route. The composite is 70 % porous by volume, and the pore size varies from 100 to 200 mm (see Fig. 9.6b). The trapezoidal-shaped ceramic blocks are implanted in the surgical space around the iliac crest of human patients (who voluntarily participated in the study), as shown in Fig. 9.6c.

The study had 2 outcomes the first was donor site morbidity: pain, wound hematoma, infection, iliac fracture, visceral prolapse, cosmetic deformity, and instability (defined as discomfort during ambulation or turning in bed). Pain was assessed by a simple scoring system as—severe (intolerable) pain, moderate pain

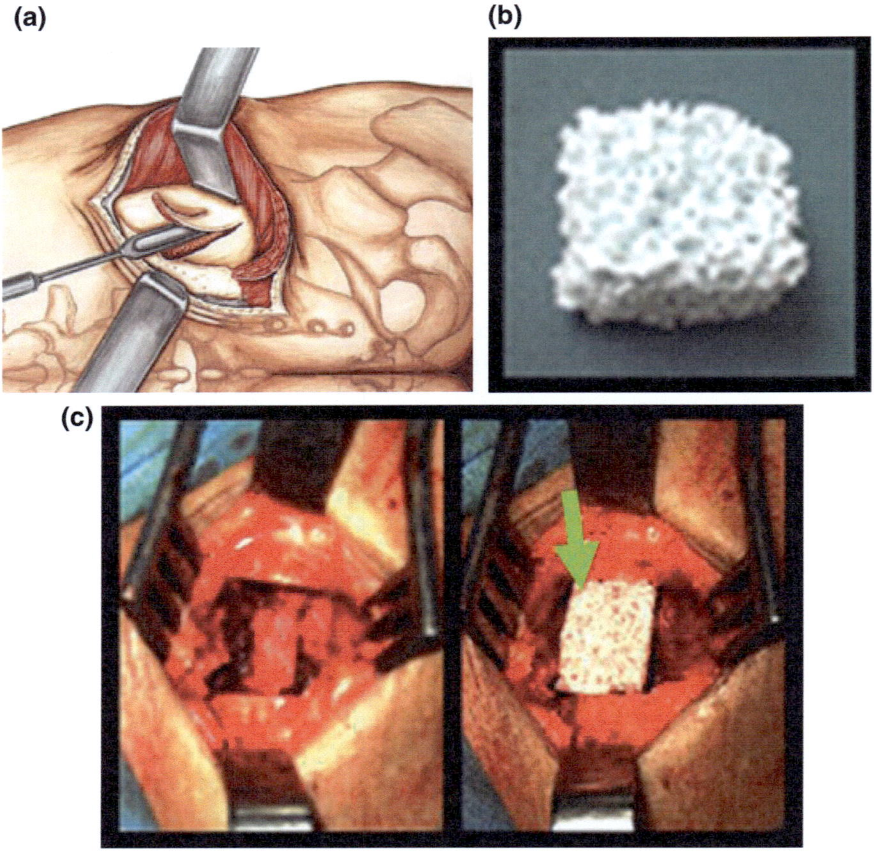

Fig. 9.6 Schematic of clinical site showing the place of surgery at iliac crest (**a**), Chitra-HABG block to be implanted in human patients (**b**) and the placement of the ceramic block at posterior iliac crest (**c**) [15]

(requiring analgesics), mild pain (tolerable without analgesics), and no pain. The second outcome measure was a radiologic assessment for ceramic incorporation, dissolution, fragmentation, and migration. Ceramic incorporation was defined as "complete" if there was a gradual blurring of the sharp margins of the block, loss of radio-opacity, and the establishment of trabeculae across the block, and "partial" if they were present, but only on one surface of contact. Ceramic dissolution meant the disappearance of the block with no visible bony replacement, whereas partial dissolution meant thinning or a reduction in the size of the block with or without incorporation. Fragmentation was defined as breakage with dispersion into the surrounding tissue and migration as displacement from the implanted site. At the end of 1 year from surgery, 25 of the 26 patients (96.15 %) had no pain at the donor site, which had been reconstructed. Radiologic evaluation showed that in 21 cases the ceramic incorporation was complete (see Fig. 9.7), partial in 3, and absent in 2.

A partial dissolution of ceramic was noticed in 3 patients and migration in 1. The clinical study described above established that the Chitra-HABG block is an excellent material for the reconstruction of the iliac crest defect, as it is incorporated into the surrounding bone with no adverse effects.

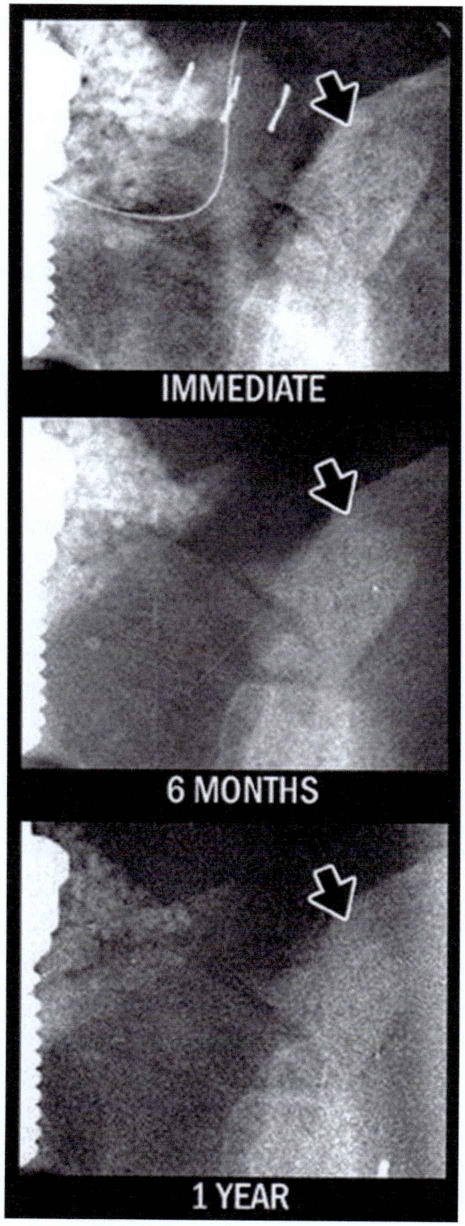

Fig. 9.7 X-ray radiograph of showing progressive incorporation of HABG ceramic block in posterior iliac crest.mg TID for 3 days and 0.2 % chlorhexidine mouthwash were prescribed postoperatively [15]

9.9.3 *Calcium Phosphate Cement for the Treatment of Human Periodontal Intraosseous Defects*

In osseous grafting, and tissue-regenerating scaffolds, either natural bone (autografts, allografts and bone derivatives) or synthetic osteoconductive materials (alloplasts) are provided at the defect site. They integrate with the host bone and participate in the remodeling of the defect. The technique of guided tissue regeneration (GTR) employs barrier materials to allow selective cellular repopulation on the root surface, to retard the apical migration of epithelium and to exclude gingival connective tissue from the healing wound. To this end, Rajesh et al. [16] reported a clinical study to establish the efficacy of a formulation of Calcium phosphate cements (CPC) with proven biocompatibility, mouldability and osteoconductivity in healing human periodontal intraosseous defects in comparison with hydroxyapatite ceramic granules. A notable feature of CPC is osteotransductivity (i.e., resorption of the cement graft in tune with new bone formation), which is absent in other alloplast materials.

After obtaining the necessary clearance from the ethical committee and informed consent forms, a total of 60 patients of both sexes in the age range of 20–45 years having moderate to severe periodontitis with angular osseous defects (two-wall or three-wall defects identified through probing) were involved in the clinical study. The exclusion criteria were the presence of any systemic illness, prior antibiotic therapy for any illness during the past 2 weeks and any known allergic reactions to drugs. The selected patients were given instructions on self-performed oral hygiene techniques. The progress was assessed at 3, 6, 9 and 12 month observation intervals through soft tissue parameters (probing depth, attachment level and gingival recession), which were appended by postoperative radiographs and surgical re-entry in selected cases. A check-up for oral hygiene was done after 2 weeks and only patients who maintained optimal oral hygiene were recruited for the study. Detailed medical and dental histories were obtained from each patient and a complete clinical examination was performed while recruiting.

In the patient group receiving CPC, the calcium phosphate cement prepared in putty form was placed in the osseous defects to the level of the respective crest (Fig. 9.8a) and similar grafting was done in the HAG test group, by mixing the hydroxyapatite granules with saline (Fig. 9.8b). In the control group, no implantation was done after the open flap debridement. Antibiotics (doxycycline, 100 mg BID for the first day followed by 100 mg OD for 5 days), anti-inflammatory analgesic (ibuprofen 400) were given.

Some clinically relevant measurements were recorded preoperatively (after the hygienic phase) and at the review visits (3, 6, 9 and 11 months postoperatively) to calculate the soft tissue parameters. The pocket depth reduction was determined by subtracting the measurement obtained at each observation period from the preoperative values of probing depth. Gingival recession was found by taking the

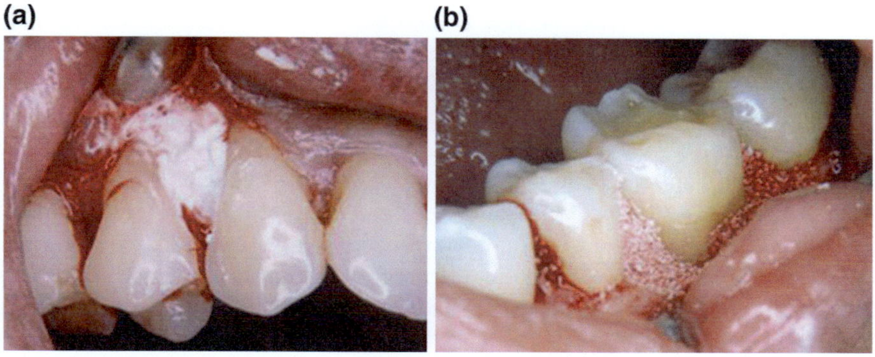

Fig. 9.8 A close-up view of the patient's teeth repaired with calcium phosphate cement grafts (**a**) and with the hydroxyapatite ceramic granules (**b**) during a clinical trial study on Indian patients [16]

difference in the gingival margin levels from the reference point at each time period. The clinical attachment level was determined as the depth to the base of the pocket from the cemento-enamel junction and the attachment gain was calculated by subtracting the value obtained at each time period from the preoperative values of the attachment level. There was no gingival inflammation or soft tissue reactions adjacent to the grafted materials, in any of the cases in this study. Also, the radiographs taken in the follow-up visits did not show any signs of resorption of root or bone in any of the cases, when compared with the preoperative radiographs. Irrespective of the radioopacity of the graft materials, bone regeneration could be identified through trabecular continuity, progressively at the postoperative periods. The clinical study described above also indicates the potential use of CPC as a graft material without the use of a GTR barrier graft for periodontal regeneration.

9.9.4 Treating Tibial Plateau Fractures with Synthetic Porous Hydroxyapatite Granules

In an early study, Menon and Varma [17] investigated the potential to use synthetic porous HAP as a cancellous bone substitute to act as a space filler and also as a scaffold to facilitate new bone growth, while providing structural support in metaphyseal fractures. A total of 58 patients underwent the operative procedure during the study period of which 28 consecutive patients with a minimum of 12-month follow-ups were included in the clinical trial. They were between the ages of 20–50 years. There were seven females and 21 males. The exclusion criterion involved the immuno-compromised patients like those with malignancies, hepatic and renal diseases, HIV infection, but other co-morbidities like Diabetes and Hypertension were included. Chitra HAP porous granules of the size 3–6 mm with a pore size of 200–300 microns and a grain size less than 4 microns and a pore volume of 70 % and 10–15 ml graft was used in each case. The granules were introduced in small quantities into the funnel and pushed into the metaphyseal defect created by the elevation of an articular depression.

All fractures were treated by a standard surgical protocol. Under appropriate anesthetic conditions and under a tourniquet, the fracture was stabilised and aligned with a femoral distractor. A 6-mm incision was made on the medial tibial meta-diaphyseal junction about 5 cm below the tuberosity level. A similar window was created in the bone with a 6-mm drill and through this a 4-mm punch was introduced under an image intensifier and the resultant defect filled with Chitra porous HAP granules in measured quantities (volume) inserted through a 5-mm auroscope funnel and impacted well. The plater slab and the sutures were removed after 14 days and a snug fitting femoral cast brace was applied with a knee hinge. Active knee mobilization was encouraged. Partial weight bearing was allowed from 12 weeks onwards progressing to full weight bearing as tolerated.

The outcome measures included the radiological evaluation of fracture union to investigate the subsidence of the tibial articular surface and the morphological characteristics of the implanted materials. For all the 28 patients included in the trial, radiological evaluation was done at 6 weeks, 3 months, 6 months and 1 year. Standard antero-posterior and lateral films were obtained in the lying position at 1 m distance. All the X-rays were evaluated by an independent radiology consultant and analyzed with reference to the following parameters, (a) subsidence of the articular surface measured in millimeters and compared with the immediate post-operative films, (b) graft Dissolution in volume, (c) migration into the joint, (d) incorporation assessed by a clear appreciation of the borders of each granule, (e) Fragmentation of the granules, (f) Quality of the radio-opacity: increasing or decreasing radio opacity, (g) Changes in the surrounding bone (specifically, sclerotic reactions or lytic responses were looked for). All the patients had excellent post-operative recovery from the surgery. There was one case of superficial infection, which healed with parenteral antibiotic without adverse effects.

It was observed that the quality of the radio-opacity progressively decreased or it became less radio-dense. In the study group of 28 patients, none had any biochemical or structural changes due to the use of the HAP granules and the fractures united completely both clinically and radiologically in the normal timeframes for similar fractures. Histological specimens from a human spinal fusion mass using the Chitra HAP have also demonstrated a fibrous tissue-free bone-ceramic interface. Based on the radiological evidence presented here, it is apparent that the implant-bone interface becomes stable by about 6 months, post-operatively. Therefore, it was concluded that porous HAp is a safe and clinically useful bone graft substitute for cancellous bone augmentation in fractures.

9.10 Closure

Clinical trials are only a small part of the research that goes into developing a new treatment. Potential drugs, for example, first have to be discovered, purified, characterized, and tested in labs (in cell and animal studies) before ever undergoing clinical trials. In all, about 1,000 potential drugs are tested before just one reaches the point of being tested in a clinical trial. For example, a new cancer drug has, on average, 6 years of research behind it before it even makes it to clinical trials. But the major holdup in making new cancer drugs available is the time it takes to complete clinical trials themselves.

This chapter has provided some fundamental elements to understand the need and importance of clinical trials in biomaterials research. One point that needs to be mentioned is that even if one new biomaterial is found to have excellent biocompatibility, both *in vitro* and *in vivo* (up to a large animal model), the same material based device prototype may not produce clinically acceptable outcomes measures in clinical trials. Therefore, any new biomaterial must undergo clinical trials before commercialisation. Also, while *in vitro/in vivo* biocompatibility assessment is routinely carried out using test samples of simple geometry, the clinical trials must be conducted with a biomedical device assembly based on that specific biomaterial in human patients. Most importantly, all the relevant ethical issues as well as the necessary approvals at the institute at the national/country level need to be sought before conducting any clinical trial.

After describing how to design the clinical trial study as well as the outcome measures in the field of prosthodontics, the chapter also discusses the background as well as the illustrative results of the clinical trial studies on ceramic based materials from the published literature. It is expected that the discussion in this chapter will provide a better platform for active researchers to perceive the importance of conducting clinical trials in an ethically recommended manner as well as to realize how different the overall clinical study design is when compared to research on biomaterials involving *in vitro* study alone. It also needs to be emphasized that the sensitivity and ethical issues to be considered while conducting clinical trials are at the highest level and can never be compromised in the research space. Also, the manner in which the clinical study results are analysed and the use of related equipment/analysis schemes is similar to some extent to pre-clinical studies, but very different from *in vitro* studies or studies related to processing or the physical property evaluation of biomaterials.

References

1. Gentleman E, Fredholm YC, Jell G, Lotfibakhshaiesh N, O'Donnell MD, Hill RG, et al. The effects of strontium-substituted bioactive glasses on osteoblasts and osteoclasts *in vitro*. Biomaterials. 2010;31:3949–56.

2. Ravi ND, Balu R, Sampath Kumar TS. Strontium-substituted calcium deficient hydroxyapatite nanoparticles: synthesis, characterization, and antibacterial properties. J Am Ceram Soc. 2012;95(9):2700–8.
3. Goel A, Rajagopal RR, Ferreira JM. Influence of strontium on structure, sintering and biodegradation behaviour of CaO–MgO–SrO–SiO_2–P_2O_5–CaF_2 glasses. Acta Biomater. 2011;7(11):4071–80.
4. Vestermark MT, Hauge EM, Soballe K, Bechtold JE, Jakobsen T, Baas J. Strontium doping of bone graft extender. Acta Orthop. 2011;82(5):614–21.
5. Hamdy NA. Strontium ranelate improves bone microarchitecture in osteoporosis. Rheumatology. 2009;48:9–13.
6. Brennan TC, Rybchyn MS, Green W, Atwa S, Conigrave AD, Mason RS. Osteoblasts play key roles in the mechanisms of action of strontium ranelate. Br J Pharmacol. 2009;157: 1291–300.
7. Buehler J, Chappuis P, Saffar JL, Tsouderos Y, Vignery A. Strontium ranelate inhibits bone resorption while maintaining bone formation in alveolar bone in monkeys (Macaca fascicularis). Bone. 2001;29:176–9.
8. Canalis E, Hott M, Deloffre P, Tsouderos Y, Marie PJ. The divalent strontium salt S12911 enhances bone cell replication and bone formation in vitro. Bone. 1996;18:517–23.
9. Fromigue O, Hay E, Barbara A, Petrel C, Traiffort E, Ruat M, et al. Calcium sensing receptor-dependent and receptor-independent activation of osteoblast replication and survival by strontium ranelate. J Cell Mol Med. 2009;13:2189–99.
10. Singh PK, Singh BP, Alvi HA, et al. Effect of various treatment modalities in sleep bruxism. J Prosthet Dent. 2015;. doi:10.1016/j.prosdent.2015.02.025.
11. Mall P, Chand P, Siddarth R, Rao J, Singh BP et al. Effectiveness of positioning stents in radiation induced xerostomia having tongue carcinoma: a randomized controlled trial. Int J Prosthodont. (accepted).
12. Talwar N, Chand P, Singh BP et al. Evaluation of the efficacy of a prosthodontic stent in determining the position of dental implants. J Prosthodont. 2012;21(1):42–7.
13. Agrawal KK, Tripathi A, Chand P, Singh RD, Rao J, Singh BP. A study to evaluate the effect of oral stereognosis in acceptance of fixed prosthesis. Indian J Dental Res. 2011;22(4):611.
14. Easwer HV, Rajeev A, Varma HK, Vijayan S, Bhattacharya RN. Cosmetic and radiological outcome following the use of synthetic hydroxyapatite porous-dense bilayer burr-hole buttons. Acta Neurochir. 2006;149:481–6.
15. Narayana K. Acharya, Chinnappa V. Mahajan, Renjit J. Kumar, Hari Krishna Varma, and Venugopal K. Menon. Can iliac crest reconstruction reduce donor site morbidity? A study using degradable hydroxyapatite-bioactive glass ceramic composite. J Spinal Disorder Technol. 2010;23:266–71.
16. Rajesh JB, Nandakumar K, Varma HK, Komath M. Calcium phosphate cement as a "barrier-graft" for the treatment of human periodontal intraosseous defects. Indian J Dental Res. 2009;20(4):471–9.
17. Menon KV, Varma HK. Radiological outcome of tibial plateau fractures treated with percutaneously introduced synthetic porous hydroxyapatite granules. Eur J Orthop Surg Traumatol. 2005;15:205–13.

Chapter 10
Case Study: Development of Constructs for Maxillofacial Reconstruction

10.1 Introduction

Craniofacial defects, particularly in the maxilla and mandible regions resulting from periodontal diseases, oncological resection or other age related factors offer a major orthopaedic challenge [1]. A significant population in the world suffers from congenital facial deformities. Overwhelming number of traffic accidents and urban violence are also causing increasing cases of maxillofacial trauma. Current repair strategies include the implantation of cortico-cancellous bone graft harvested from anterior or posterior iliac crest or allografts depending upon whether the bone graft substitute has been harvested from the recipient's body or from another individual. Although the use of grafts (autografts/allografts) is considered the current 'gold standards', the technique does not go without complications [2]. It is even clear now that the autogenous grafts, often isolated from iliac crest, fail to reproduce the complex geometries of the facial contour [3]. Alternatively, several synthetic materials have been exploited in repairing maxillofacial injuries including metals [4], polymers [5], bioceramics [6], biocomposites [7] and bioactive glasses [8]. However, inspite of some promising data obtained in preclinical evaluation, successful clinical outcomes are limited.

One major drawback in planning maxillofacial reconstruction using either autologous tissue or synthetic scaffolds is the inability to precisely reproduce complex facial contours. Even though this method results in recovery of deformed facial appearances, its biological responses may not be akin to the native lost bone. Modelling of the facial anatomy is highly complex owing to the sharp esthetics, as well as requirement for complex load bearing during daily functions, such as chewing foods, talking, jaw movement. Moreover, the presence of the temporomandibular joint, one of the most active joints of the skeleton, consisting of the head of mandible and the articular disc as a part of the segmental mandible region aggravates the biomechanical complexity of the facial skeleton. At the same time, mandibulectomy also demands resection of soft tissues, such as nerves, blood

B. Basu and S. Ghosh, *Biomaterials for Musculoskeletal Regeneration*,
Indian Institute of Metals Series, DOI 10.1007/978-981-10-3017-8_10

vessels, gingiva, for which a soft and flexible scaffold or prosthesis is required. Therefore maxillofacial reconstructive surgery demands custom-made tissue replacements that can properly fit the deformed structures and regenerate the lost tissue of individual patients. Pure titanium and its alloys are most commonly used metallic biomaterials to repair craniofacial defects, but these prostheses often fail due to wear or corrosion and eventually mechanical fracture [9].

Therefore an ideal strategy to precisely reconstruct large and complex maxillofacial deformities would be the ability to fabricate desired 3D printed prototypes of the lost tissue to produce an artificial bone graft or soft tissue substitute that exactly fits the dimensions of the defect site [10, 11]. However, fabrication of such 3D printed prototypic templates that can exactly mimic the deformed facial anatomy is still in its infancy. Research in that direction is ongoing for quite some time. The first application of this computer-aided (CAD) technology for bone repair was reported as early as 1985, when the Sirona CEREC (Chairside Economical Restorations of Esthetic Ceramics), combined digital teeth scanning with CAD models of prosthetic crowns was developed by researchers of University of Zurich, Switzerland [12]. Polymethylmethacrylate (PMMA) was a popular material for developing customized skull plates for over a decade [13]. But it has now slowly been replaced by other improved materials such as Polyether ether ketone having more relevant properties for bone substitution.

Several applications of 3D printed prototypes are currently utilized:

(a) Maxillofacial models are being made by 3D printing techniques the fabrication of skull models using patient-specific imaging data from MRI/MicroCT for preoperative surgical planning and communication between surgeons and patients, surgical training and teaching purposes (Fig. 10.1). In addition to the orthognathic surgery planning, models can be used to reconstruct the templates to determine the size and shape of the bone grafts. It is also possible to print soft

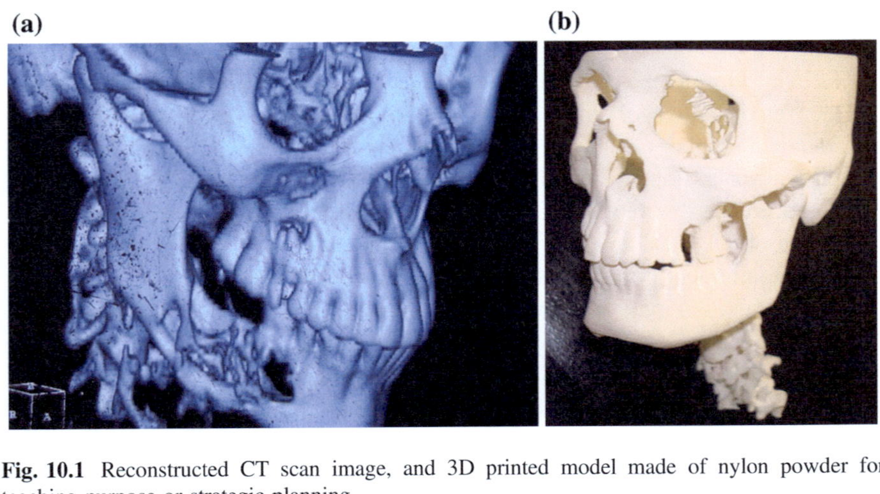

Fig. 10.1 Reconstructed CT scan image, and 3D printed model made of nylon powder for teaching purpose or strategic planning

tissue models and mark/highlight the tumours and other anatomical structures of interest, for example nerves, arteries or veins in the models [14, 15].
(b) manufacturing of inert and customised reconstruction plates for denture or implants, thus offer improved passive fitting [16],
(c) production of surgical instruments and special equipment associated with the operations,
(d) fabrication of postoperative guides and long-term supports [5].

Therefore, the idea of utilizing clinical imaging data of the patient to create arbitrary 3D structures to fit complex defects does not look far fetched with this technique today.

An early method of 3D printing was the use of inkjet/binder system which utilizes a powder bed (which was usually Tricalcium phosphate in the case of scaffolds of bone reconstruction) which is contacted by a binder solution (citric acid, phosphoric acid etc.) dispensed using a nozzle to prepare 3D shapes and architectures. Presently several additive manufacturing technologies being used for fabrication of 3D printed patient-specific implants for cranial reconstruction, for example fused deposition modeling, stereolithography, selective laser sintering, selective laser melting and electron beam melting. 3D printing techniques are advantageous to stereolithography due to faster printing, more precision, accuracy and lower cost. In laser-based techniques a laser beam is used to sinter the particles together; resulting in resolution in the range of several hundred microns depending upon the diameter of the laser used. While the system had the potential to enable the construction of large scaffolds which could be put to use for clinical applications, it lacked the precision of printing smaller feature size for efficiently reproducing the small structures. In this context of finer resolution, an improved strategy based on extrusion-based is 3D Printing called 'Direct-write assembly' which employs a computer-controlled platform which controls movement of a pattern-generating device (e.g., ink deposition micronozzle) in a layer-by-layer sequential manner, to fabricate complex 3D architectures and composition. The working principle involves extrusion of ink under constant pressure through a fine cylindrical nozzle with micro-meter scale inner diameter to create a filamentary element [17, 18], avoiding need for laser or chemically reactive binder. The technique drastically opened up new strategy to fabricate scaffolds having microperiodic biomimetic structures with predefined internal pore architecture (i.e. shape, size, orientation, geometry, interconnections) [19, 20]. This design process will undoubtedly hold the capability to efficiently regenerate the complicated maxillofacial tissues by fabricating scaffolds matching the complex 3D structures. But along with the 3D architecture, printing the exact compositional pattern of the native tissue along with surface chemistry for promoting cellular interactions and molecular signaling are some challenges that need to be addressed.

In these case studies, our main emphasis will be on the recently developed textile braided and 3D Printed scaffolds within the domain of maxillofacial reconstruction. In order to develop a deeper mechanistic insight about 3D printing-based maxillofacial reconstruction; we would briefly recapitulate the complex anatomy of the maxillofacial region and the commonly associated disorders that have developed the quest for the fabrication of improved orthopedic materials.

10.2 Clinical Perspective

Maxillofacial reconstruction refers to the wide range of procedures employed to repair and rebuild the soft and hard tissues in the maxillofacial region, comprising of a complex array of nerves, blood vessels, bone and cartilage [21, 22]. Scoliosis of the mandibular arch, missing alveolar bone in cleft palates, fractures of orbital floor and rim, osteoradionecrosis, ameloblastoma, intraosseous mucoepidermoid carcinoma, extremely deficient chin are some common maxillofacial disorders demanding immediate resurrection procedures [23]. In addition, osteosarcoma, defect in oral-facial soft tissues such as cleft palate and other intraoral soft tissue lesions that are also very important in shaping the facial contour demand a reliable reconstruction strategy [24].

For small mandibular defects of less than 3–6 cm, nonvascularized corticocancellous grafts harvested from the iliac crest are commonly used [25]. Larger defects (>6 cm wide) are often treated with vascularized fibula grafts [26, 27] which offer the advantage of incorporating other components such as skin, muscle and bicortical bone, if required. While vascularised tissue flaps are the current gold standards in treating maxillofacial defects, the technique is not without complications [21, 28]. Extended hospitalisation, limited availability of autologous bone, substantial donor site morbidity resulting in numbness and limited motion, infection, malformation and loss of function are some of the drawbacks associated with vascularised tissue flaps that render high inconvenience to the patients [23, 29–31]. Another alternative, distraction osteogenesis, is a technique increasingly being used as endogenous bone tissue engineering in cases such as mandible lengthening, also renders severe complications in more than 35 % cases including pin-tract infections, scarring of the defect site, device failure, and failure to form a bony union [32, 33]. Then there are biodegradable Lactosorb implants comprising of poly (L-lactic acid) and polyglycolic acid which usually resorb within a year of implantation for facilitating congruent bone ingrowth [34]. Though they serve excellent purpose for soft bones such as in the case of infants or in the orbit, they lack adequate compressive properties and stability for repairing adult bones. Therefore there is an urgent need to explore improved methods for maxillofacial reconstruction in clinical setup. In some cases, clinicians have replaced all or most parts of the mandible using 3D printed titanium prosthetics. However, the long-term effects of such implants directly placed under the gingivae risk dehiscence.

As next alternative, three dimensional scaffolds have been used, that can to some extent regenerate bone tissue and can be excellent alternative or adjuncts to bone grafting. But the conventional scaffolds, developed from polymers or hydroxyapatite (HA) etc., often fail to recapitulate its native structural and biological complexity required for maxillofacial reconstruction. To address the above issues, researchers are on a quest for fabricating advanced materials and methods for recapitulating the microenvironment of the complex craniofacial anatomy.

Calcium phosphate ceramics including beta-tricalcium phosphate (β-TCP) and HA have obtained clinical acceptance in maxillofacial reconstruction due to their comparable chemical composition and compressive strength to the human bone tissue [35, 36]. However a major drawback with β-TCP scaffolds is the high rate of resorption which weakens the mechanical properties of the scaffolds to support osseointegration [37]. HA scaffolds have been commonly used as fillers in bone defects due to their excellent osteoconductivity. However, a major drawback associated with commercial HA scaffolds is that they are manufactured in generic dimensions based on the shape and size of an average patient with a 'one shape fits all' strategy. Therefore the surgeon may need to spend considerable time in manual pre-shaping of constructs to fit the patient's defect site, a method prone to human errors and therefore might lead to fatigue fractures [38]. Besides geometry, internal parameters such as pore size, pore shape, surface area to volume ratio, pore interconnectivity, porosity, pore arrangement and size distribution are of paramount importance for developing an efficient vascular supply to establish functionality of the neo-osseous tissue; a major problem associated with repairing large size defects. Conventional scaffold fabrication techniques are unable to precisely control these internal dimensions. Besides they offer other limitations such as brittleness of the materials, inadequate control over the material's resorption properties *in vivo*, etc. Therefore, a major challenge for maxillofacial scaffolds is to fabricate 3D materials that are defect-specific as well as patient-specific and can perfectly fit into the facial defects, which appear more delicate and complicated than the defects of appendicular skeleton [10, 17, 18].

Taking into consideration the above the limitations offered by the current clinical treatments, the following requirements explain the properties of an 'ideal implant' that can be readily used by the maxillofacial surgeons: (i) the scaffold should possess a 3D architecture, exactly mimicking the deformed facial contour; this demands for a custom-made, on demand fabrication strategy, (ii) an interpenetrating network of pores and interconnects with small feature size for enhanced cellular attachment, infiltration and nutrient diffusion. (iii) The scaffold should also possess adequate mechanical properties comparable to the intended site of implantation. For human mandibular implantation in the cortical bone region, the anisotropic elastic moduli was found to be in the range of 11000–30000 MPa while the values ranged from 120 to 450 MPa in trabecular region for mandibular condyle and 114–910 MPa from the mid-line to ramas [39]. Moreover, in the craniofacial anatomy, it is well demonstrated that the mechanotransduction exerted by the four muscles of mastication are very well defined within the spatiotemporal pattern and therefore the shape and alignment of the reconstructed bones immensely affect the pattern of load bearing within the region. (iv) The scaffold be able to resorb in the body overtime into non-toxic byproducts. (v) should be easy to manufacture, cost effective and reproducible into any desired shape and size. Therefore the engineering process for designing scaffolds for maxillofacial reconstruction should provide a precise control over the scaffold's external shape as well as the internal architecture. Apart from the bony structure, the craniofacial region is also made up of an interconnected network of soft tissues such as muscles, cartilage and microvasculature. Therefore, for reconstructing the structural and functional aesthetics of the maxillofacial region, it is of utmost importance that this soft tissue regeneration occurs in tandem with the bone. This presses the need for an implant containing appropriate geometric features that can facilitate the growth of both soft and hard tissues simultaneously to restore the lost tissue function close to the native bone.

10.3 Textile Braided Structure

Textile technology offers versatility for the development of 3D spatial structures with tailor-made properties in the order of micro- and macro meters useful for application in medical sector [40]. Some critical factors that determine the ultimate fate of such fibre-based scaffolds include the type of material (silk, collagen, chitosan), yarn structure and fabric manufacturing technique used (woven, knitted, braided), complex mechanical properties. By using a suitable combination of the above factors, innovative scaffolds/implants can be fabricated demonstrating properties similar to the native target tissue in order to match the ideal scaffold criteria. Silk fibres are extensively used for various tissue regeneration strategies due to tailorable biodegradability [41], mechanical properties [42], cell compatibility and tolerable immunogenecity in situ [43].

10.3.1 Preparation and Characterization of Silk Braids

Braiding is a technique that offers dimensional stability (especially torsional) and tailor-made mechanical properties to the scaffold structure as a result of the specific orientation and alignment of the criss-cross network of yarns [44]. Braided prosthesis can be prepared by intertwining two sets of yarns with each other at a specifically determined alignment angle to the axis of the fabric, so that each crosses alternatively over and under another set of yarn. Braided textile fabrics can be prepared in various forms such as flat braid, circular braid or triaxial braids. Flat braids are consist of simple criss-crossed oriented yarns. Circular braids are prepared by interlacing diagonally aligned three or more yarns by braiding over a mandrel. In a circular or tubular braid yarns are intertwined in a way that a hollow core is formed. Biaxial braids consist of two sets of braided strands. Whereas the triaxial braids has an additional set of yarns or braided strands aligned in the direction of braid axis. Biaxial braid can offer higher breaking extension. The triaxial braids can offer high rigidity and bending strength. In general mechanical characteristics of braided fabrics are governed by the braid angle (orientation of fibers or yarn in a specific angle) and pick spacing.

Due to the above mentioned advantages several studies applied braided textile for muscle or ligament TE [45]. However, recently their potential in bone tissue engineering applications is also surfacing [44]. Textile braids can be promising candidates for bone tissue engineering as apart from the exceptional mechanical properties, they offer specific orientation of the yarns that can help in precisely regulating the orientation and crystallisation of hydroxyapatite deposition on the surface of the fibre. Hence the function of the silk braided structure in a way would resemble collagen type I [7], another fibrous protein, which forms an important element of the native bone matrix of native bone and therefore holds promising potential as osteogenic construct.

10.3.2 Case Study Using Silk Braids

Tooth extraction is a common procedure to resolve issues related to complicated periodontal disease or unrestorable caries. After tooth extraction, the alveolar bone progressively undergoes atrophy. At the same time, after removal of bundle bone, the surrounding tissues, such as alveolar bone, alveolar ridge as well as soft tissue undergo dimensional changes [46]. In order to avoid this modification of alveolar ridge, several preservation techniques are used, such as the filling of the socket by using autografts, allografts, xenografts or synthetic filling materials [47]. Compared to natural healing without any filling material, xenografts were found to provide best support and restoration of alveolar ridge, followed by allografts and synthetic materials [48, 49]. Clinicians are still looking for a better material for alveolar ridge preservation.

The present case study focused on exploring application of braided silk fabrics for socket seal surgery after extraction of maxillary anterior tooth using braided silk textile biomaterial. The interdental papillae and buccal gingiva were undermined using a tunneling approach and silk braids were used to augment the socket under general anesthesia (Fig. 10.2). Patients were asked to quantify level of pain (Visual Analog Scale) [50]. Anatomic features were evaluated at various time point. Tables 10.1 and 10.2 depict the findings, which showed that silk braids helped to augment the socket and preserve the alveolar ridge, and putatively helped in regeneration of the tissue with minor inflammation initially, which subsided with time.

Therefore, the reported technique seems appropriate for preservation of the alveolar ridge using silk braids in functionally and esthetically relevant areas.

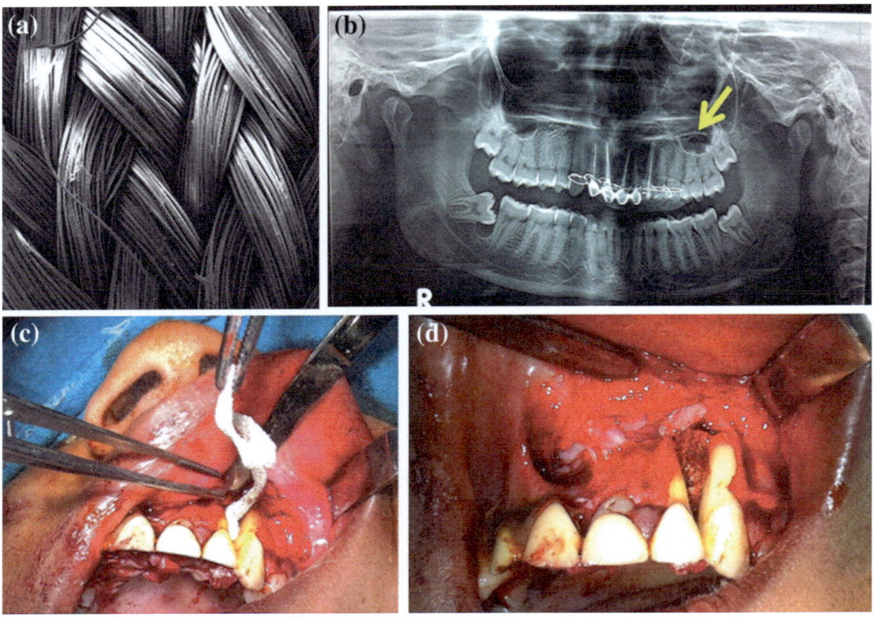

Fig. 10.2 Application of silk braid: **a** SEM image of silk braids, **b** *Dental* X-Ray—Orthopantomogram) *arrow* is indicating the defect site, **c**, **d** Intra-operative images for implantation of silk braided fabric

Table 10.1 Pre-operative and post-operative clinical parameters for patient A

	Pre-operative	1st day	1st week	4th week
Pain (Visual Analog Scale 1-10)	0	5	0	0
Swelling	nil	Moderate	No	No
Ginival color	Normal	Normal	Normal	Normal
Infection	Not evident	Not evident	Not evident	Not evident
Graft Exposure	-	No	No	No
Paraesthesia	No	No	No	No

Table 10.2 Pre-operative and post-operative clinical parameters for patient B

	Pre-operative	1st day	1st week	4th week
Pain (Visual Analog Scale 1-10)	0	3	0	0
Swelling	Mild	Moderate	No	No
Ginival color	Normal	Normal	Normal	Normal
Infection	Not evident	Not evident	Not evident	Not evident
Graft Exposure	-	No	No	No
Paraesthesia	No	No	No	No

10.4 3D Printed Hydroxyapatite

For maxillofacial tissue engineering, HA provides a good substrate due to its superior load bearing properties and enhanced osteoconductivity. In order to produce customised scaffolds for complex defects such as in the case of maxillofacial deformations, HA scaffolds could be printed using advanced rapid prototyping using a whole array of approaches including stereolithography, fused deposition modelling, selective laser sintering and 3D printing. For instance, Gittard et al. fabricated scaphoid and lunate bone prostheses from a photoreactive acrylate polymer by direct light projection stereolithography approach in a layered structure with build layer corresponding to 50 μm [51]. The bone prostheses were created using 3D models of patient's micro-CT data. However, most commercial SLA machines produce structures with minimum feature sizes of 70 μm, though a few structures with 1 μm thick feature size have been reported [52–54]. The laser beam spot size and powder bonding properties make it impossible to build feature of <400 μm in size [52]. Medical grade poly (e-caprolactone) scaffolds using fused deposition modelling approach were used as whole burr plugs in cranioplasty in more than 200 clinical cases [55]. A major limitation of these general prototyping techniques is the inability to build structures with small feature sizes [51, 55]. Since the typical size of a biological cell is only about 20–30 μm, they demonstrate a more flattened morphology when cultured on filaments exceeding this size range. Therefore it is important to fabricate constructs with small feature sizes (say 5–10 μm i.e. smaller than the size of a typical cell) which would force the cells to form aggregates and would more closely mimic the *in vivo* physiological matrix for cellular adhesion and differentiation.

Further due to its colloidal nature, direct-write printing of HA offers a tremendous control over the structural feature size, micro- as well as macroporosity and surface roughness to fabricate custom-made materials that can be used for specific applications in bone reconstruction [56]. However, each variable in the fabricating procedure such as ink composition (water or alcohol based inks), viscoelasticity, high colloid volume fraction (to reduce drying-induced shrinkage of scaffold), pH and sintering temperature can have a significant effect on the geometry, surface chemistry and mechanical strength of the printed scaffold [20, 57]. Colloidal HA inks for Direct-write must meet following important criteria- the ink should display shear thinning behavior, so that it can pass through the micronozzle easily without choking. After printing elastic modulus should be higher than viscous modulus so that it can solidify as soon as possible. Microperiodic 3D ceramic structures have successfully developed with nanoscale feature size with rheological behaviour optimized by keeping a lower colloid volume fraction. Michna et al. [20] printed a concentrated colloidal ink of HA by tailoring its viscoelasticity and showed that 45 vol% solids loading and a high elastic shear modulus of 10^5 Pa of HA is required to directly print self-supporting HA microfilaments. Then Simon et al. [56] fabricated cylindrical, multilayered HA scaffolds from concentrated suspensions (49 vol%) possessing 3D periodic arrays of quadrant architectures (consisting of macro-pores

& micro-pores) which showed promising osseo-integrative potential in preclinical small animal bone defect remodelling. Thus prior to patterning through direct-write assembly, optimising the ink development and corresponding ink rheology are pre-requisites to tailor viscosity and elasticity of the ink.

Direct-write assembly enables the fabrication of three dimensional complex patterned structures in a layer-wise deposition strategy, without the requirement for expensive tooling, laser, dies, sterolithography-based masks or nano-engineering fabrication devices. Hence this technique is advantageous over others in terms of flexibility, versatility, precision by robotic control and low cost.

Furthermore, composites of HA such as silk-HA of appropriate printable viscosity (10^4 Pa s), which is higher than its individual components (i.e. silk and HA) were also printed [58]. The silk-HA ink was readily extruded through a 200 μm micro nozzle, showing 10^5 plateau shear elastic modulus under low shear stress. Hence concentrated ink of ceramic materials can be directly deposited and air drying increases the strength to prevent structural deformity of the printed microfilament.

The above studies clearly suggest that exploring ink formulations that permit an easy control over the above listed parameters will offer an advantage in the fabrication of custom-made, on demand tissue replacement scaffolds for reconstructing complex facial contours.

10.5 Characterization

10.5.1 in vitro Cell Culture

Biomaterial properties play an indispensable role in precisely regulating the corresponding biological interactions including cell attachment, alignment, proliferation, differentiation, extracellular matrix formation and, ultimately, tissue regeneration. Therefore extensive preclinical testing is currently underway to screen HA patterns with enhanced biological responses for validating their clinical usage in skeletal reconstruction. Besides chemical composition, the physical properties of HA scaffolds such as 3D architecture, porosity and pore size distribution, surface roughness have profoundly affected their resultant in situ interactions with the host [59–61]. For instance, honeycomb-like HA scaffolds demonstrated significantly increased osteogenesis over particles or 3D porous blocks [62].

One important consideration while designing HA scaffolds is that they should facilitate vascular infiltration in tandem with bone formation. A common reason for the failure of most HA implants in large maxillofacial lesions is the formation of a necrotic core. As the maxillofacial region comprises of extensive microvasculature [63], the repair mechanism should include an implant that guides blood vessel infiltration concomitantly with osseous tissue formation. In an attempt to achieve this, Sun et al. [58] fabricated a 3D printed direct write silk/HA composite with pore sizes in the range of 200–750 μm to investigate the effect of pore geometries on osteogenesis and angiogenesis. *in vitro* investigations showed extensive vascular network formation induced by co-seeding human mammary microvascular endothelial cells and mesenchymal stem cells. However, pore size was seen to have a significant impact on the extent of vascularisation within the construct. When a co-culture of mesenchymal stem cells and microvascular endothelial cells were seeded on these silk/HA scaffolds with varied pore sizes, intricate networks of extracellular matrix and vasculature were revealed within 10 days of culture, observed only in pores with an average diameter of 400 μm [58]. We already know that angiogenesis and osteogenesis are symbiotic processes in bone tissue engineering [64]. Therefore scaffolds with an appropriate pore size are a must for facilitating capillary ingrowth, nutrient diffusion and tissue infiltration to promote osseointegration of the scaffold with the host bone.

Facial trauma, ablative surgery or neoplastic malignancies in the maxilloficial region leads to major bone defects which destroy functional as well as aesthetic aspects. Such bone defects require immediate reconstruction to reinstate function of the jaws, talking, breathing, swallowing of foods as well as aesthetic outcome. Maxillofacial reconstruction surgery involving composite microvascular flaps is challenging, time consuming, needs significant effort to intraoperatively optimize the configuration and symmetry of the facial skeleton. Moreover, need of simultaneous muscle attachment, innervations and microvasculature make it even more challenging. Fabrication of patient-specific 3D printed construct will enable the surgeons to manufacture the implants before surgery, and hence will solve above mentioned problems.

10.5.2 *in vivo* Osseointegration

The challenge in the management of large ameloblastoma of the mandible is as follows: (a) to excise the tumor completely in order to make the patient tumor free, (b) to prevent recurrence of cancer, (c) to provide the best reconstruction for aesthetics and daily function of mandible in terms of talking, chewing and swallowing of food etc. In addition to promoting bone growth to achieve adequate anatomical reconstruction of the maxillofacial region, simultaneous muscle attachment, innervations and microvasculature should take place to allow complete functional and structural restoration of the area. In order to assess the clinical outcomes of the 3D printed HA in maxillofacial reconstruction by Direct-write assembly, we recently performed a collaboration study with the surgeons at Lady Harding Medical College, New Delhi.

Treatment options for ameloblastoma vary from conservative to wide bone radical resections, with or without the need for reconstruction [65]. While autologous bone transplants are associated with several drawbacks in terms of limited availability, extended time of surgery, donor site morbidity, pain, infection, excessive blood loss, and high costs which offer severe inconvenience to patients. Alternatively, use of particulate materials such as hydroxyaptite, ostacalcium phosphate, β-tricalcium phosphate has been clinically accepted owing to their similar stochiometry to native bone and minimal batch-to-batch variations [66]. However, the conventional fabrication strategies used for the production of these synthetic scaffolds fail to recapitulate the precise structural and functional aspect of native bone. Commercially used methods for reconstruction of mandibular bone utilize titanium or dacron-urethane trays which often encounter difficulties when attempting to match with the requirements of facial contour, jaw relationship and condylar position as the trays are prefabricated [67]. Moreover, use of metal or dacron-coated polyurethane trays, which include wound dehiscence also face serious complications such as tray exposure and post-operative infection leading to partial or total loss of graft often demanding a revision surgery. Additionally, higher modulus of elasticity of the metal tray compared to bone result in fast resorption due to stress shielding [68]. Hence, 3D printing via direct-writing technique has fascinating prospect for fabricating custom made tissue analogs by sequentially depositing appropriate biodegradable and biocompatible materials specifically at the desired location, in a layer-by-layer fashion, to fabricate patient- as well as tissue-specific constructs. The progenitor cells from neighboring tissues can attach, proliferate and integrate onto the implanted 3D printed construct and eventually form neo-tissues at the repair site, thereby replacing the printed construct.

In the following study, a 31 year old lady reported about swelling and mild pain of the right side of her face which was gradually increasing since past 1 month. She had a history of surgical curettage of a lesion involving the mandible, and a diagnosis of Plexiform ameloblastoma of mandible a few years back. Preliminary investigation revealed that the lesion had recurred. The swelling was firm to hard in consistency and non tender in nature. An OPG revealed a multilocular radiolucency with respect to 44, 45, 46, region involving mandibular lower border and extending up to ramus sparing the condyle. Incisional biopsy was done under local anesthesia, whose report confirmed Plexiform ameloblastoma. The CT scan demonstrated that the swelling as $4 \times 3.3 \times 2.5\ cm^3$ multi-lobular cystic lesion, showing cortical thinning, buccal more than lingual. As the patient was young, the challenge was to preserve her facial contours while conducting radical excision and immediate reconstruction. A submandibular incision along with intraoral incision was conducted and the lesion was resected maintaining a 1 cm of safety margin.

An anatomically shaped patient-tissue-specific tray was fabricated with 3D printed hydroxyapatite plates fused with bone cement (Simplex® P Bone Cement, Stryker, USA) and was sutured onto to the Titanium reconstruction plate (Fig. 10.3). Corticocancellous bone graft harvested from iliac crest and filled within the 3D printed tray. There was rapid postoperative wound healing. There was no infection or symptoms of rejection of implanted HAP tray. The patient was

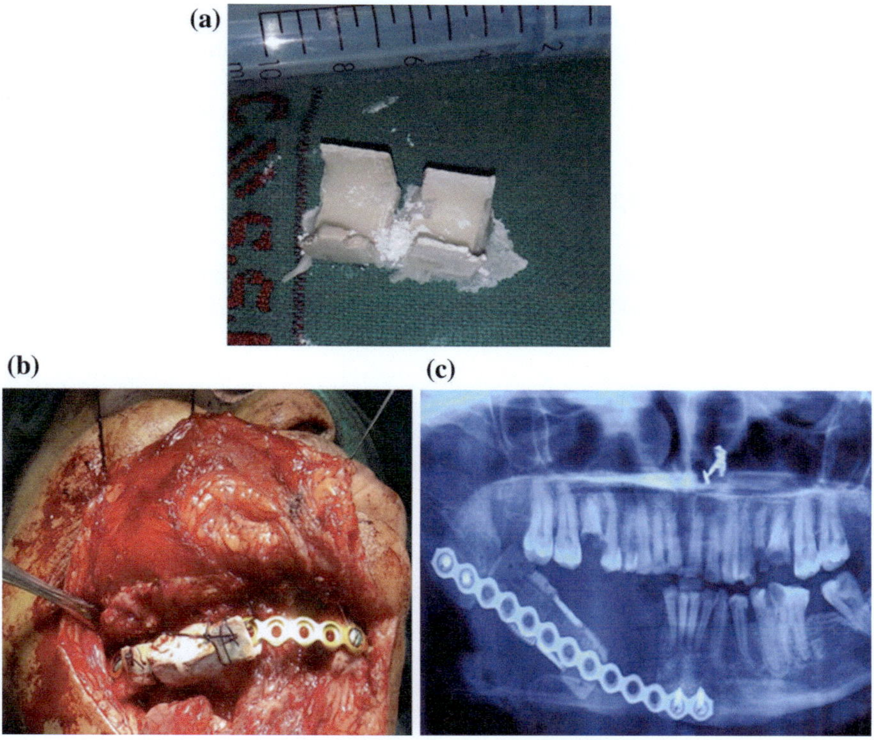

Fig. 10.3 **a** 3D printed hydroxyapatite tray, **b** Per-operative photo showing custom made tray in situ, **c** Post operative OPG x-ray image of mandible after 6 months

followed up for 1 year with no recurrence, and continuity of mandible bone was maintained. However there was a mild dehiscence of plate intraorally (a known complication which occurs in 5–10 % patients), which was initially managed conservatively and later after 3 weeks primary closure was done. Taken together, innovative clinical use of 3D hydroxyapatite tray showed promises for mandibular bone reconstruction for oral cancer patients.

This same procedure was repeated for a 29 years old patient suffering from ameloblastoma, and subsequent follow up showed absence of any wound break down intra or extra orally. Tissue morbidity was minimal in both the patients with normal gingival color. Clinical form, function and aesthetics of the patients were highly satisfactory. Continuity of mandible bone was preserved. Function of the mandible was more or less achievable, as far as daily activities like chewing, talking and swallowing are concerned. Both the patients were followed up with no tumor recurrence and are still on regular follow up for evaluating quantitative and qualitative bone formation and osseointegration.

Taken together, the present study provided a new and better modality of treating recurrent malignant tumor and providing satisfactory reconstruction in terms of both aesthetic and functional rehabilitation. The stem cells present in the bone marrow chips placed in the 3D printed tray would presumably help in bone regeneration as well as minimize inflammatory response.

10.6 Future Perspective

There is immense scope of improvement on the current treatment methods for the reconstruction of maxillofacial disorders or defects that can offer a more patient-specific as well as patient-friendly strategy. While the boost in 3D printing technology will ensure more precisely fabricated architectures using improvised material compositions that closely mimic the mechanical and biochemical aspect of craniofacial anatomy to facilitate more congruent tissue ingrowth (both hard and soft tissue) within the defect region. In parallel, researchers are currently investigating *ex vivo* methods to make the surface of these 3D printed constructs more conducive for cell and subsequently tissue attachment and integration. For instance, conjugation of engineered protein coatings on the surface of such orthopaedic implants (69) will help stabilize the initial cellular response of these implants in situ immediately post-implantation; an ultimate factor which eventually determines the fate of such constructs. Moreover, early attachment and infiltration of osteogenic and angiogenic precursor cells via integrin associated cascades triggered by the surface coatings (such as RGD motifs) will help in guiding specific cellular signalling pathways for the development of more specialized cellular differentiation and tissue regeneration within the construct well integrated with the host tissue.

References

1. Goiato MC, Santos MR, Pesqueira AA, Moreno A, dos Santos DM, Haddad MF. Prototyping for surgical and prosthetic treatment. J Craniofac Surg. 2011;22(3):914–7.
2. Yadav SK and Shrestha S, Current and Advancing Concepts in Pedicled Flaps, Old and New, for Oral and Maxillofacial Reconstruction. Medical and Clinical reviews 2016 Mar; 2 (18).
3. Rogers GF, Greene AK. Autogenous bone graft: basic science and clinical implications. J Craniofac Surg. 2012;23(1):323–7.
4. Schubert W, Gear AJ, Lee C, Hilger PA, Haus E, Migliori MR, Mann DA, Benjamin CI. Incorporation of titanium mesh in orbital and midface reconstruction. Plast Reconstr Surg. 2002 Sep 15;110(4):1022–30; discussion 1031-2.
5. Conejero JA, Lee JA, Parrett BM, Terry M, Wear-Maggitti K, Grant RT, Breitbart AS. Repair of Palatal Bone Defects Using Osteogenically Differentiated Fat-Derived Stem Cells. Plast Reconstr Surg. 2006;117(3):857–63.
6. Adamopoulos O, Papadopoulos T. Nanostructured bioceramics for maxillofacial applications. J Mater Sci Mater Med. 2007;18(8):1587–97.
7. Marelli B, Ghezzi CE, Alessandrino A, Barralet JE, Freddi G, Nazhat SN. Silk fibroin derived polypeptide-induced biomineralization of collagen. Biomaterials. 2012;33(1):102–8.
8. Gosain AK. Bioactive glass for bone replacement in craniomaxillofacial reconstruction. Plast Reconstr Surg. 2004;114(2):590–3.

9. De Water VR, Rubio EJ, Schouten JW, Koudstaal MJ. Deformation of a Titanium Calvarial Implant following Trauma: A Case Report. Craniomaxillofac Trauma Reconstr. 2016;9 (2):158–61.
10. Arora A, Datarkar AN, Borle RM, Rai A, Adwani DG. Custom-Made Implant for Maxillofacial Defects Using Rapid Prototype Models. J Oral Maxillofac Surg. 2013;71(2): e104–10.
11. Choi JW, Kim N. Clinical Application of Three-Dimensional Printing Technology in Craniofacial Plastic Surgery. Arch Plast Surg. 2015;42(4):513.
12. Mörmann W, Brandestini M, Ferru A, Lutz F, Krejci I. Marginal adaptation of adhesive porcelain inlays in vitro. Schweiz Monatsschr Zahnmed (1984). 1985 Dec;95(12):1118–29.
13. Pochon JP, Klöti J. Cranioplasty for acquired skull defects in children–a comparison between autologous material and methylmethacrylate 1974-1990. Eur J Pediatr Surg. 1991;1(4): 199–201.
14. Christensen AM, Humphries SM. Role of rapid digital manufacture in planning and implementation of complex medical treatments. Advanced Manufacturing Technology for Medical Applications in Advanced Manufacturing Technology for Medical Applications: Reverse Engineering, Software Conversion and Rapid Prototyping (ed I. Gibson), John Wiley & Sons, Ltd, Chichester, UK. doi:10.1002/0470033983.ch2.
15. Ritacco LE, Di Lella F, Mancino A. Gonzalez Bernaldo de Quiros F, Boccio C, Milano FE. 3D Printed Models and Navigation for Skull Base Surgery: Case Report and Virtual Validation. Stud Health Technol Inform. 2015;216:1025.
16. Williams RJ, Bibb R, Eggbeer D, Collis J. Use of CAD/CAM technology to fabricate a removable partial denture framework. J Prosthet Dent. 2006;96(2):96–9.
17. Ghosh S, Parker ST, Wang X, Kaplan DL, Lewis JA. Direct-Write Assembly of Micro-Periodic Silk Fibroin Scaffolds for Tissue Engineering Applications. Adv Funct Mater. 2008;18(13):1883–9.
18. Lewis JA. Direct Ink Writing of 3D Functional Materials. Adv Funct Mater. 2006; 16(17):2193–204.
19. Das S, Pati F, Chameettachal S, Pahwa S, Ray AR, Dhara S, Ghosh S. Enhanced redifferentiation of chondrocytes on microperiodic silk-gelatin scaffolds: toward tailor-made tissue engineering. Biomacromolecules. 2013;14(2):311–21.
20. Michna S, Wu W, Lewis JA. Concentrated hydroxyapatite inks for direct-write assembly of 3-D periodic scaffolds. Biomaterials. 2005;26(28):5632–9.
21. Fearraigh PO. The Modern Methods in the Surgical Reconstruction and Rehabilitation of the Orofacial Region. A Review of the Literature. TSMJ. 2008;9:44–7.
22. Tobita M, Mizuno H. Oral and Maxillofacial Tissue Engineering with Adipose- Derived Stem Cells. Regenerative Medicine and Tissue Engineering.2013. Prof. Jose A. Andrades (Ed.), ISBN: 978-953-51-1108-5, InTech, doi:10.5772/55899.
23. Tardy ME, Kastenbauer ER. Head and neck surgery. 1995. Vol. I.2nd rev. ed. New York, NY: Thieme Medical Publishers, Inc.
24. Kadam D, Pillai V, Bhandary S, Hukkeri RY, Kadam M. Indian Facial contour deformity correction with microvascular flaps based on the 3-dimentional template and facial moulage. Indian J Plast Surg. 2013;46(3):521–8.
25. Pogrel MA, Podlesh S, Anthony JP, Alexander J. A comparison of vascularized and nonvascularized bone grafts for reconstruction of mandibular continuity defects. J Oral Maxillofac Surg. 1997;55(11):1200–6.
26. Genden E, Haughey BH. Mandibular reconstruction by vascularised free tissue transfer. Am J Otolaryngol. 1996 Jul–Aug;17(4):219–27.
27. Hidalgo DA. Fibula free flap: a new method of mandible reconstruction. Plast Reconstr Surg. 1989;84(1):71–9.
28. Mehta RP, Deschler DG. Mandibular reconstruction in 2004: an analysis of different techniques. Curr. Opin. Otolaryngol. Head Neck Surg. 2004;12:288–93.
29. Wolff J, Sándor GK, Miettinen A, Tuovinen VJ, Mannerström B, Patrikoski M, Miettinen S. 2013. GMP-level adipose stem cells combined with computer aided manufacturing to

reconstruct mandibular ameloblastoma resection defects: Experience with three cases. Ann Maxillofac Surg. 2013 Jul–Dec; 3(2): 114–125.
30. Ling XF, Peng X. Review What is the price to pay for a free fibula flap? A systematic review of donor-site morbidity following free fibula flap surgery. Plast Reconstr Surg. 2012; 129(3):657–74.
31. Alsberg E, Hill EE, Mooney DJ. Craniofacial tissue engineering. Crit Rev Oral Biol Med. 2001;12(1):64–75.
32. McCarthy JG, Schreiber J, Karp N, Thorne CH, Grayson BH. Lengthening the human mandible by gradual distraction. Plast Reconstr Surg. 1992 Jan;89(1):1–8; discussion 9-10.
33. Mofid MM, Manson PN, Robertson BC, Tufaro AP, Elias JJ, Vander Kolk CA. Craniofacial distraction osteogenesis: a review of 3278 cases. Plast Reconstr Surg. 2001 Oct;108(5): 1103–14; discussion 1115-7.
34. Wiltfang J, Merten HA, Schultze-Mosgau S, Schrell U, Wénzel D, Kessler P. Biodegradable miniplates (Lacto Sorb): Long-term results in infant minipigs and clinical results. J Craniofac Surg. 2000 May;11(3):239–43; discussion 244-5.
35. Moreira-Gonzalez A, Jackson IT, Miyawaki T, Barakat K, DiNick V. Clinical outcome in cranioplasty: critical review in long-term follow-up. J J Craniofac Surg. 2003;14(2):144–53.
36. Burstein FD1, Williams JK, Hudgins R, Boydston W, Reisner A, Stevenson K, Cohen S. Hydroxyapatite cement in craniofacial reconstruction: experience in 150 patients. Plast Reconstr Surg. 2006 Aug;118(2):484–9.
37. Tanaka T, Kumagae Y, Saito M, Chazono M, Komaki H, Kikuchi T, Kitasato S, Marumo K. J. Bone formation and resorption in patients after implantation of beta-tricalcium phosphate blocks with 60 % and 75 % porosity in opening-wedge high tibial osteotomy. J Biomed Mater Res B Appl Biomater. 2008 Aug;86(2):453–9.
38. Kokemueller H, Spalthoff S, Nolff M, Tavassol F, Essig H, Stuehmer C, Bormann KH, Rücker M, Gellrich NC. Prefabrication of vascularized bioartificial bone grafts in vivo for segmental mandibular reconstruction: Experimental pilot study in sheep and first clinical application. Int J Oral Maxillofac Surg. 2010;39(4):379–87.
39. Hollister SJ, Lin CY, Saito E, Lin CY, Schek RD, Taboas JM, Williams JM, Partee B, Flanagan CL, Diggs A, Wilke EN, Van Lenthe GH, Müller R, Wirtz T, Das S, Feinberg SE, Krebsbach PH. Engineering craniofacial scaffolds. Orthod Craniofac Res. 2005;8(3):162–73.
40. Aibibu D, Hild M, Wöltje M, Cherif C. Textile cell-free scaffolds for in situ tissue engineering applications. J Mater Sci Mater Med. 2016;27(3):63.
41. Altman GH, Diaz F, Jakuba C, Calabro T, Horan RL, Chen J, et al. Silk-based biomaterials. Biomaterials. 2003;24(3):401–16.
42. Midha S, Murab S, Ghosh S. Osteogenic signaling on silk-based matrices. Biomaterials. 2016;97:133–53.
43. Bhattacharjee M, Schultz-Thater E, Trella E, Miot S, Das S, Loparic M, Ray AR, Martin I, Spagnoli GC, Ghosh S. The role of 3D structure and protein conformation on the innate and adaptive immune responses to silk-based biomaterials. Biomaterials. 2013;34(33):8161–71.
44. Midha S, Tripathi R, Geng H, Lee P, Ghosh S. Elucidation of differential mineralization on native and regenerated silk matrices. Mater Sci Eng, C. 2016;68:663–74.
45. Leong NL, Petrigliano FA, McAllister DR. Current tissue engineering strategies in anterior cruciate ligament reconstruction. J Biomed Mater Res A. 2014;102(5):1614–24.
46. Araújo MG, Silva CO, Misawa M, Sukekava F. Alveolar socket healing: what can we learn? Periodontol 2000. 2015 Jun;68(1):122–34.
47. Tan WL, Wong TL, Wong MC, Lang NP. A systematic review of post-extractional alveolar hard and soft tissue dimensional changes in humans, Clin Oral Implants Res. 2012 Feb;23 Suppl 5:1–21.Jambhekar S, Kernen F, Bidra AS. Clinical and histologic outcomes of socket grafting after flapless tooth extraction: a systematic review of randomized controlled clinical trials, J Prosthet Dent. 2015 May;113(5):371-82.
48. Stein JM, Hammächer C. Postextraction Socket Seal Surgery with an Epithelized Connective Tissue Graft Using a Subpapillar Tunneling Procedure. Int J Periodontics Restorative Dent. 2015 Nov-Dec;35(6):877–84.

49. Hawker GA, Mian S, Kendzerska T, French M. Measures of adult pain: Visual Analog Scale for Pain (VAS Pain), Numeric Rating Scale for Pain (NRS Pain), McGill Pain Questionnaire (MPQ), Short-Form McGill Pain Questionnaire (SF-MPQ), Chronic Pain Grade Scale (CPGS), Short Form-36 Bodily Pain Scale (SF-36 BPS), and Measure of Intermittent and Constant Osteoarthritis Pain (ICOAP). Arthritis Care Res (Hoboken). 2011;63(Suppl 11): S240–52.
50. Gittard SD, Narayan RJ, Lusk J, Morel P, Stockmans F, Ramsey M, Laverde C, Phillips J, Monteiro-Riviere NA, Ovsianikov A, Chichkov BN. Rapid prototyping of scaphoid and lunate bones. Biotechnol J. 2009;4(1):129–34.
51. Yang S, Leong K, Du Z, Chua C. The design of scaffolds for use in tissue engineering. Part 2. Rapid prototyping techniques. Tissue Eng. 2002;8(1):1–11.
52. Gittard SD, Narayan R. Laser direct writing of micro- and nano-scale medical devices.
53. Expert Rev Med Devices. 2010 May; 7(3): 343–356.
54. Zhang X, Jiang XN, Sun C. Micro-stereolithography of polymeric and ceramic microstructures. Sens Actuator A. 1999;77(2):149–56.
55. Hutmacher DW, Cool S. Concepts of scaffold-based tissue engineering–the rationale to use solid free-form fabrication techniques. J Cell Mol Med. 2007 Jul-Aug;11(4):654–69.
56. Simon JL, Michna S, Lewis JA, Rekow ED, Thompson VP, Smay JE, Yampolsky A, Parsons JR, Ricci JL. In vivo bone response to 3D periodic hydroxyapatite scaffolds assembled by direct ink writing. J Biomed Mater Res A. 2007;83(3):747–58.
57. Smay JE, Cesarano J III, Lewis JA. Colloidal Inks for Directed Assembly of 3-D Periodic Structures. Materials Science and Engineering Department, University of Illinois at Urbana-Champaign, Urbana, Illinois 61801, and Sandia National Laboratories, Albuquerque, New Mexico 87106. Langmuir. 2002;18(14):5429–37.
58. Sun L, Parker ST, Syoji D, Wang X, Lewis JA, Kaplan DL. Direct-write assembly of 3D silk/hydroxyapatite scaffolds for bone co-cultures. Adv Healthc Mater. 2012;1(6):729–35.
59. Deligianni DD, Katsala ND, Koutsoukos PG, Missirlis YF. Effect of surface roughness of hydroxyapatite on human bone marrow cell adhesion, proliferation, differentiation and detachment strength. Biomaterials. 2001;22(1):87–96.
60. Specchia N, Pagnotta A, Cappella M. Effect of hydroxyapatite porosity on growth and differentiation of human osteoblast-like cells. Journal of Materials Science. 2002;37(3):577–84.
61. Rosa AL, Beloti MM, Noort RV, Hatton PV. Surface topography of hydroxyapatite affects ROS17/2.8 cells response. Pesqui. Odontol. Bras. 2002;16(3):209–15.
62. Fröhlich M, Grayson WL, Wan LQ, Marolt D, Drobnic M, Vunjak-Novakovic G. Tissue engineered bone grafts: biological requirements, tissue culture and clinical relevance. Curr Stem Cell Res Ther. 2008; 3(4):254–64.Miller RJ, Edwards WC, Boudet C, Cohen JH. The Mandibular Symphysis in 3D Revised Maxillofacial Anatomy: 2009. 1(2):1-7.
63. Kanczler JM, Oreffo RO. Eur Cell Mater. Osteogenesis and angiogenesis: the potential for engineering bone. Eur Cell Mater. 2008;15:100–14.
64. Montoro JR, Tavares MG, Melo DH, Franco Rde L, Mello-Filho FV, Xavier SP, Trivellato AE, Lucas AS. Mandibular ameloblastoma treated by bone resection and immediate reconstruction. Braz J Otorhinolaryngol. 2008;74(1):155–7.
65. Kawai T, Echigo S, Matsui K, Tanuma Y, Takahashi T, Suzuki O, Kamakura S. First Clinical Application of Octacalcium Phosphate Collagen Composite in Human Bone Defect. Tissue Eng Part A. 2014;20(7–8):1336–41.
66. Samman N, Luk WK, Chow TW, Cheung LK, Tideman H. Clark RK Custom-made titanium mandibular reconstruction tray. Aust Dent J. 1999;44(3):195–9.
67. Tonino A, Davidson CL, Klopper PJ, Linelau LA. Protection from stress in bone and its effects. J Bone Joint Surg Br. 1976;58(1):107–13.
68. Raphel J, Karlsson J, Galli S, Wennerberg A, Lindsay C, Haugh MG, Pajarinen J, Goodman SB, Jimbo R, Andersson M, Heilshorn SC. Engineered protein coatings to improve the osseointegration of dental and orthopaedic implants. Biomaterials. 2016;83:269–82.

Chapter 11
Assessment of Technology and Manufacturing Readiness Levels

11.1 Introduction

Traditionally, research in academia has been largely restricted to implant/scaffold development using the conventional manufacturing route. Also, a set of traditional material testing methods has been used to determine the physical properties of such materials. In the context of biomaterials, a test of *in vitro* and *in vivo* biocompatibility needs to be conducted and this is an important step in biomaterials' research. As discussed in this chapter, the above will establish the 'proof-of-concept' behind the development of a new biomaterial or a new manufacturing strategy. Translational research involving the 'bench to bedside' concept, however, requires one to adopt a series of higher level testing to take a given technology to a more mature stage. From the perspective above, this chapter introduces the reader to the concept of TRLs and MRLs.

11.2 Concept of Technology Readiness Level (TRL)

11.2.1 Various TRL Stages

> Technology Readiness Levels (TRLs) are a systemic qualitative assessment system used to estimate the maturity level of a particular technology, and to compare the maturity between different types of technology.

Each technological project is evaluated during a Technology Readiness Assessment (TRA) that examines technology requirements, program concepts, and demonstrated technology capabilities against the parameters for each technology level. Originally, TRLs with only seven levels were developed by NASA in the

B. Basu and S. Ghosh, *Biomaterials for Musculoskeletal Regeneration*,
Indian Institute of Metals Series, DOI 10.1007/978-981-10-3017-8_11

1980s. These seven levels were later expanded to nine. Based on the progress of the project, a TRL rating is assigned. TRL 1 has the lowest, while TRL 9 has the highest maturity. The regulatory authorities in each country consider the recommended TRLs, when assessing the program risk for any technology or product development.

> It is, therefore, worthwhile to recognise that in order to study product validation, it is important to involve an entrepreneur or device manufacturing companies particularly to assess Technology Readiness Levels (TRLs) and to explore the possibility of technology transfer to biomedical companies.

Such involvement would involve an extensive examination of the existing product technologies and competition before planning the exact research targets. Since the R&D agenda is strongly dictated by marketable product features/properties, market research inputs, competitive business scenarios and IP potentials, the final specifications of the product towards market relevance need to be assessed. With reference to Fig. 11.1, a brief discussion of each technology readiness level has been provided below.

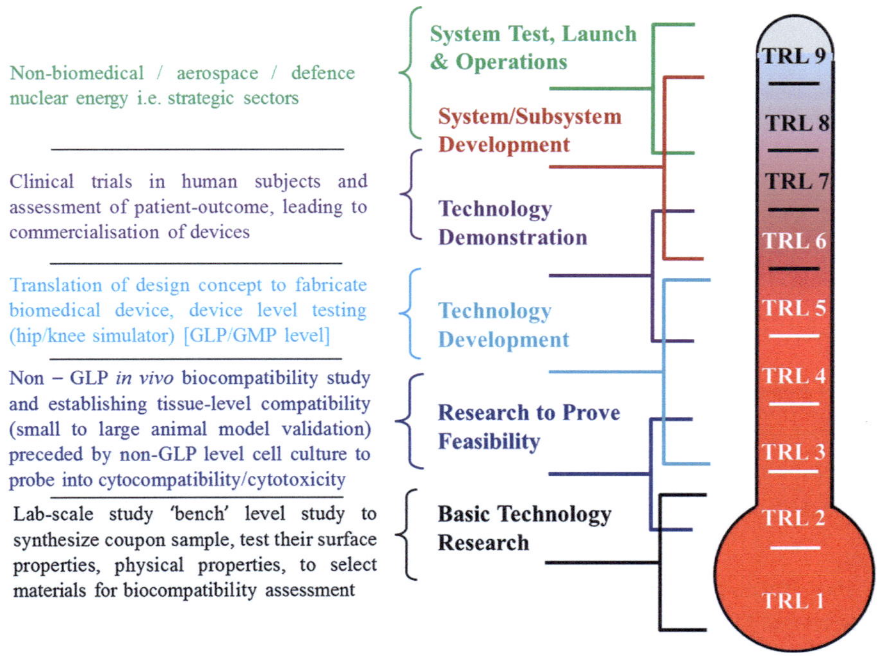

Fig. 11.1 Schematic and conceptual description of various stages of Technology Readiness assessment using TRL levels (adapted from Ref. [1])

TRL 1—Review of scientific knowledge base (basic research): At this stage, experimental results or scientific findings are reviewed and assessed as a foundation for characterizing new technologies. This is the lowest "level" of technology maturation, with the assessment of scientific research to be translated into applied research and development. The assessment should be done in the light of already published research or other references to understand the principles or novelty of the process or material properties.

TRL 2—Technology concept and/or application formulated (applied research): This level can generate research ideas, hypotheses, and experimental designs for addressing related scientific issues. Computer simulations or other virtual platforms to test hypotheses can be used with the aim of confirming the potential of a material or a process to satisfy a technology. The step up from TRL 1 to TRL 2 moves ideas from basic to applied research. Once the basic principles are observed, practical applications can be identified at the next level of maturity. The applications may still be speculative, and there may be no proof or detailed analysis to support the assumptions. Most of the work at this level may be analytical or paper studies with an emphasis on understanding the science better. Experimental work can be designed at this stage to corroborate the basic scientific observations made during TRL 1.

TRL 3—Establishing proof-of-concept: At this stage of technology maturation, the early stage of product development is pursued with data collection and related analyses to the test hypotheses. Some illustrative examples include research on device components, or processes, that are not yet integrated. In particular, both analytical and lab-based studies should constitute "proof-of-concept" validation of the applications/concepts formulated at TRL 2. In some cases, the modeling and simulation may be used to complement physical experiments.

TRL 4—Laboratory environment testing/validation of component/process. TRLs 4–6 represent the bridge from scientific research to engineering i.e. from development to demonstration. TRL 4 is the first step in determining whether an individual component will work while being part of a system. This level of maturity is characterized by basic technological components, which are integrated to establish that the components will work together to achieve concept-enabling levels of performance. The related information to assess at TRL 4 includes the results of integrated experiments as well as the estimation of how the experimental components/test results differ from the expected system performance. The goal of TRL 4 should be to narrow down possible options to integrate devices.

> At TRL 4, non-GLP laboratory research is conducted to refine the hypothesis and to identify the relevant parametric data required for technological assessment in a rigorous experimental design.

In particular, the candidate device is evaluated in a laboratory and/or animal models to identify and assess potential safety problems, adverse events, and side effects. The methodologies to be used during non-clinical and clinical studies in evaluating candidate devices are identified. The design history file, design review, and a master device record, if required, are initiated to support PMA (Pre-Market approval).

In the context of tissue engineering research on scaffolds, the following experiments are recommended to be conducted as part of TRL 4 evaluation: (a) *in vitro* cell compatibility including differentiation, (b) *in vivo* tissue compatibility, (c) *in vitro* and *in vivo* mechanical strength, (d) *in vivo* tolerance of allogenic/xenogenic cells and (e) *in vitro* shear stress simulation of a 3D construct.

Summarizing, the biological performance validation in the lab/animals for safety/side effects/adverse events for biomedical devices in TRL 4 would involve the following stages:

(a) rigorous experiments with varying critical parameters.
(b) initial specification of the device and subsystems, based on failure limits.
(c) procedures for nonclinical and clinical studies for device evaluation.
(d) designing of history files and a master device record for a regulatory pathway.
(e) device evaluation in the lab and/or animals for safety, adverse events and side effects.

TRL 5—Component, system and/or process validation: Once the proof-of-concept and the safety of devices are demonstrated in defined cell/animal models, the technology can be moved forward to TRL-5. In TRL 5, the biomedical device performance is compared to existing modalities, and indications for use and equivalency are demonstrated in model systems. Examples include devices tested through simulation in tissue or organ models, or animal models, if required. All component suppliers are identified and qualified and vendors for critical components are audited for cGMP (certified Good Manufacturing Protocol)/QSR (Quality System Regulation) compliance. Component tests, component drawings, design history files, design reviews, and any master device records are to be verified with the endpoint objective to draft a product development plan. For Class III devices, the IDE (Investigational Device Exemption) is submitted to the CDSCO (central drugs standard control organization) for the necessary approval. In a nutshell, GLP compliance needs to be established at TRL 5. Following success in the review of IDE by CDSCO, clinical investigations can be initiated in suitably designed clinical trials, which require the strictest regulatory approval. At this stage, the basic technological elements must be integrated with reasonably realistic supporting elements so that the total applications (component-level, sub-system level, or system-level) can be tested in a 'simulated' or somewhat realistic environment similar to the final application in almost all respects.

The major difference between TRL 4 and 5 is the increase in the fidelity of the system to the actual application. At the end of TRL 5, the system tested is almost prototypical.

In summary, the validation for equivalency in lab simulation/tissue/organ/ animals involved in TRL 5 should ensure the following aspects,

(a) device to be compared to the benchmark and equivalency to be demonstrated.
(b) all component vendors are identified and audited for cgmp/qsr compliance.
(c) component tests, design history files and master device records verified.
(d) product development plan drafted.
(e) closer interaction with regulatory body for proposed regulatory regime (e.g. regulatory pathway deliberation with CDSCO).
(f) planning of clinical trial study and seeking necessary clearance from ethical bodies.

TRL 6—System/subsystem model or prototype demonstration in a relevant operational environment: TRL 6 begins with the true engineering development of the technology as an operational system. The major difference between TRL 5 and TRL 6 is the step up from the laboratory scale to the engineering scale and the determination of scaling factors that will enable the design of the final system. A model or prototype of the technology system or subsystem may be demonstrated, that can simulate and validate all system specifications within a test house, test track or similar operational environment. The engineering pilot scale demonstration should be capable of performing all the functions that will be required of a full manufacturing system. The refinement of the cost model is expected at this stage, and, simultaneously, efforts should be made to the reduce engineering risk. In the context of biomedical applications the following points are to be strictly followed:

(a) Clinical trials to demonstrate the safety of candidate Class III medical device in relevant clinical settings.
(b) Class IIa medical devices final prototypes to be demonstrated for substantial equivalency in actual clinical settings.
(c) Component tests, component drawings, design history files, design reviews and any master device records to be updated and verified.
(d) Production technology demonstrated in production-scale cGMP plant qualification. It is important to mention that TRL 6 is normally the last stage of assessment before the technology is transferred to the biomedical device manufacturing company. The following are to be additionally followed in the case of mission-oriented projects for strategic sectors (defence, space, atomic energy, etc.)

TRL 7—Demonstration of Integrated Pilot System: TRL 7 would normally be performed in cases where the technology and/or subsystem application is mission critical and at relatively high risk. The driving force for achieving this level of maturity is to ensure system engineering and confidence development. At TRL 7, the final design should be almost complete. The goal of this stage is to further reduce any engineering and manufacturing risk.

TRL 8—System Incorporated in Commercial Design: At this stage, the technology with validated performance is proven to work in its final form and under expected conditions. In almost all cases, TRL 8 represents the end of true system development. This might include the integration of new technology into an existing system. True manufacturing costs will be determined together with product performance.

TRL 9—System proven and ready for full commercial deployment: By definition, all technologies being applied in actual systems have to go through TRL 9. The actual technology system has been qualified through operational experience and it is in its final form, being operated under the full range of operating real-world conditions. Some related examples include steady state 24/7 manufacturing costs, yield, and output projections.

In summary, although TRL 1–9 stages are strictly followed in technology development in various strategic sectors, like defense, aerospace applications, the translational research on biomaterials and tissue engineering towards technology maturation as well as subsequent commercialization is possible once a biomedical device passes through various TRL stages up to TRL 6.

11.3 Concept of Manufacturing Readiness Level (MRL)

Similar to TRL, Manufacturing Readiness Levels (MRLs) are used to assess the maturity of a given technology from a manufacturing perspective.

Manufacturing readiness begins, continues during the development of systems, and continues even after a system has been in the field for a number of years. The purpose of MRLs is to provide decision makers with a common understanding of the relative maturity of the stages of manufacturing technologies, products, and processes. Manufacturing readiness and technology readiness go hand-in-hand. Ideally, a technology or process matures adequately in conjunction with Technology Readiness Levels (TRLs).

In 2004, the definition of MRL was created by the Department of Defence (DOD) under the auspices of the Joint Defense Manufacturing Technology Panel (JDMTP). Their goal was to develop a maturity model along the lines of the Technology Readiness Level (TRL) to serve a similar purpose, which is to provide a widely-accepted common language and measurement scale to enhance communication on the subject of manufacturing risk and readiness.

11.4 Various MRL Levels

As summarized in Fig. 11.2, the scales of MRL consist of up to 10 levels. Each level characterises the progress in the development of a technology, from the idea (level 1) to the full development of the product in multiple marketplaces and multiple applications (level 10). In the following, a set of MRL descriptions of the criteria necessary to define each level and the associated acquisition phase are briefly mentioned.

MRL 1—Basic manufacturing implications identified: This is the lowest level of manufacturing readiness in which basic research expands scientific principles that may have manufacturing implications. The focus is on a high-level assessment of manufacturing opportunities and to address manufacturing shortfalls.

MRL 2—Manufacturing concepts identified: This level is characterized by an analysis of material and process approaches. New manufacturing concepts and the feasibility of manufacturing along with processes are determined.

MRL 3—Manufacturing proof-of-concept developed: This level is characterized when a manufacturing proof-of-concept has been developed. Experimental hardware or processes have been created in a laboratory environment, but are not yet integrated due to limited functionality. Further evaluation and functionality are required for the materials and/or processes that have been characterized for manufacturability and availability. Initial manufacturing cost projections should be made at this stage.

MRL 4—Capability to produce the technology in a laboratory environment: At this level of manufacturing readiness, the capability exists to produce the technology in a laboratory or prototype environment with the technologies to be matured to at least TRL 4. Key design performance parameters in manufacturing technology development need to be identified with the assessments of design

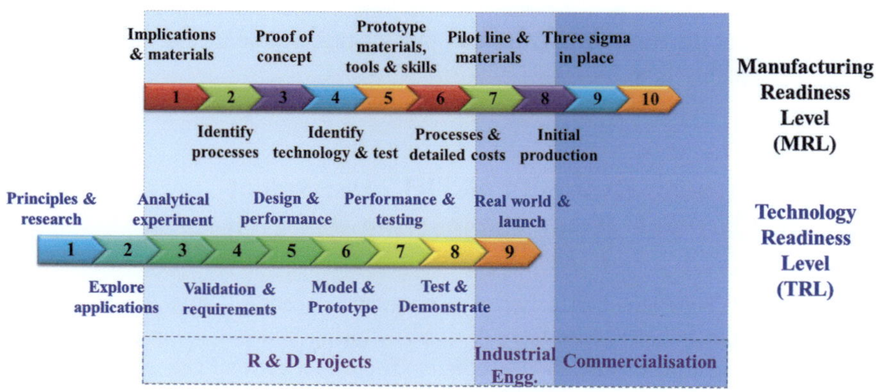

Fig. 11.2 Relationship of Technology readiness level (TRL) to manufacturing readiness level (MRL) (Adapted from Ref. [2])

concepts. This level indicates that the technologies are ready for the Technology Development Phase of acquisition.

MRL 5—Capability to produce prototype components in a production relevant environment: This level of maturity is assigned when the capability exists to produce prototype components in a production relevant environment. The prototype materials, tooling and test equipment, as well as personnel skills can be demonstrated with components in a production-relevant environment. Many manufacturing processes and procedures and productivity assessments of key technology and components can still be in the development stage of MRL 5. Technologies should have matured to at least TRL 5.

MRL 6—Capability to produce a prototype system or subsystem in a production relevant environment: At this stage, the majority of manufacturing processes are to be defined and characterized. Technologies should have matured to at least TRL 6. Cost estimates need to be allocated and approved as viable.

MRL 7—Capability to produce systems, subsystems, or components in a production representative environment: At MRL 7, material specifications are approved and materials are available to meet the planned pilot line schedule. Manufacturing processes and procedures have been demonstrated in a production representative environment.

MRL 8—(Pilot line capability demonstrated; Ready to begin Low Rate Initial Production) and **MRL 9** (Low rate production demonstrated; Capability in place to begin full scale production): At MRL 8 and MRL 9, the system or component is expected to be in small scale initial production. The related technologies should have matured to TRL 9. This level of readiness is normally associated with readiness for entry into full scale production. Now, all systems engineering/design requirements should have been met such that there are minimal system changes.

MRL 10—Full-rate production demonstration: This level of manufacturing is normally associated with the production or sustainment phases of the acquisition life cycle. Engineering/design changes are generally limited to quality and cost improvements. Systems, components or items are in full-scale production and meet all engineering, performance, quality and reliability requirements. It is expected that at the end of MRL 10, the manufacturing process capability is at the appropriate quality level.

11.5 Illustrative Examples of TRL Assessment in Cartilage Tissue Engineering Research

Clinically, cartilage repair is a two-step surgical procedure, often involving significant risk and patient morbidity. A manufacturing process to prepare artificial cartilages can lead to off-the-shelf products. There is a general acceptance that the next generation of cartilage repair therapy will combine allogenic and/or autologous

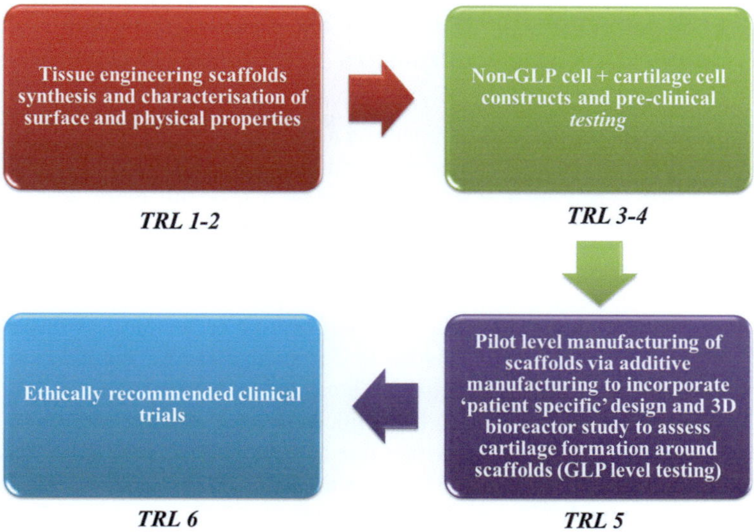

Fig. 11.3 Various TRL level to be assigned to different stages of research to develop natural bone-mimicking biomaterials

chondrocyte transplantation (ACT) in a biodegradable scaffold for direct implantation into the defect via a minimally invasive surgical procedure. For chondral/osteochondral lesions, cell-based therapies such as autologous chondrocyte transplantation (ACT) and matrix associated ACT are established clinical procedures. A prospective tissue-engineered treatment would arguably prevent the progression of cartilage damage and would significantly reduce the total knee or hip replacement. In Fig. 11.3, the author's own understanding as to how various TRLs can be assigned to research on cartilage tissue engineering is presented. The first thing should be the selection of appropriate scaffold materials, which can either support the growth/proliferation of chondrocyte cells or guide the stem cell differentiation to chondrocytes. Once the scaffold surface and physical properties are established, the technology is at TRLs 1–2. Subsequently, the non-GLP level cell culture (2D/conventional) and pre-clinical study with appropriate animal models are conducted to take the research to TRLs 3–4. Now, the technology is ready to mature to TRL 5, if the pilot level manufacturing of patient-specific cartilages is accomplished together with 3D cell bioreactor studies to assess cartilage formation. Based on the extensive test results generated, the ethical committee approval can be sought prior to conducting clinical trials in human patients. The success of a single/multi-centric clinical trial can take the technology to the last level of maturity to TRL 6. Now, the technology of cartilage repair is deemed ready for commercialization.

11.6 TRL Concept Illustration in Developing Natural Bone-Mimicking Composites

It can be reiterated here that natural bone is a nanocomposite of collagen I, a fibrous protein and hydroxyapatite, a mineral. Load-bearing bones have a porous core (cancellous) and dense shell (cortical) structures. Currently, there is no process technology to achieve porous hierarchical bone-mimicking composite structures from the nanometer to the macro-scale dimensions.

The synthesis of 10 nm–100 μm HA powders and patient-specific bone implant dimensions demand the use of additive manufacturing techniques, as discussed in one of the preceding chapters in this book. The motivation for the innovative and rapid manufacturing of realistic artificial bone tissue for orthopaedic surgery is basically need driven. The market for artificial bone is one of the largest, and current alternatives are inadequate either due to poor performance, high cost or in their in ability to address complex clinical situations.

In Fig. 11.4, the author's personal understanding of TRL assessment in the case of bone-mimicking scaffold development is summarized. At the initial stage, HA nanoplatelets of various sizes (10–100 nm) are to be synthesized at the lab scale and collagen-HA based nanocomposites can be synthesized using conventional manufacturing approaches. Once the compressive strength or other basic mechanical properties together with 2D/3D microstructural characteristics (e.g. pore architecture) are properly analysed, the technology is set to be at TRLs 1–2 and is ready to enter into the next level of maturity. As part of the biocompatibility assessment, either the osteoblast cell growth or the osteogenesis of stem cells is to be assessed

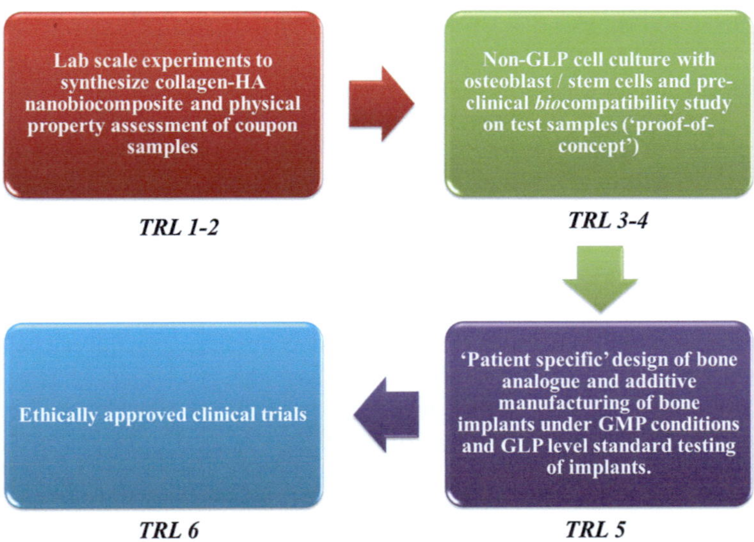

Fig. 11.4 Summary of TRL assessment of cartilage repair/replacement technology

before conducting a pre-clinical study in animal models (e.g. rabbit). Once the tissue level short/long term compatibility is established, the technology can be considered to pass through TRLs 3–4. During this stage, various ISO recommended biocompatibility testing (ISO 1993) with animal studies in rabbits and in higher animals (sheep/dogs) are to be conducted. At the next level, when additive manufacturing can be adopted to allow patient-specific bone implant fabrication under GMP conditions and together with GLP level testing, the technology is set to pass through TRL-5. Here again, bio-additive manufacturing can be adopted with the delivery of osteoblast/stem cells in a cell-interaction binder while fabricating a collagen-HA nanobiocomposite at physiologically relevant manufacturing conditions. Such a process can be assessed further for prototyping (MRL 3–5). Also, ISO31486 (quality management system) and ISO14971 (risk management) are to be carefully assessed.

Like before, the final stage of technology maturity can be realized once the ethically approved clinical trials in human patients are conducted and a satisfactory outcome is recorded. Now, the technology is at TRL 6 and is ready to be transferred to a biomedical company for commercialization.

11.7 Closure

The discussion in this chapter can be summarised with reference to a general understanding of TRL/MRl assessment (see Fig. 11.5). In the lab research, the process parameters are to be optimised in an effort to obtain clinically relevant

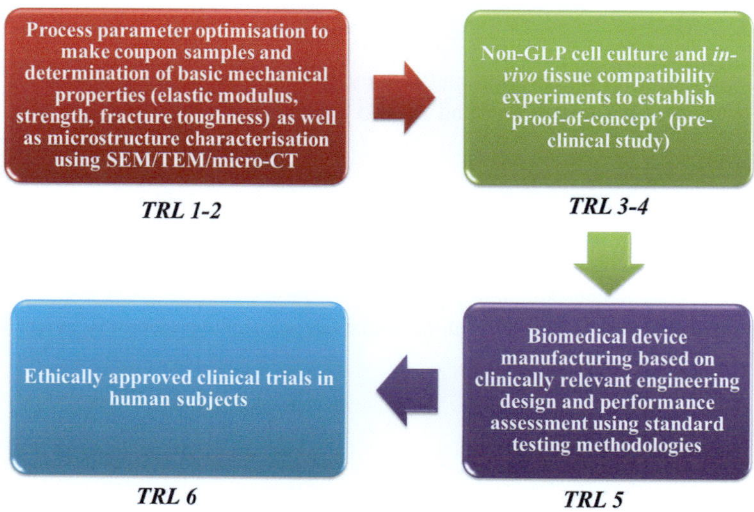

Fig. 11.5 Overall understanding of TRL assessment in establishing biomedical device/implant development

scaffold/biomaterial properties (physical/mechanical). At this stage, MRLs 1–3 are established with 'proof-of-concept' being proven, while characterizing the optimally processed material. Also, the microstructure at different length scale can be sufficiently analysed to realise bone-mimicking structures in the context of targeted orthopedic applications. Once the above research results with test samples are being analysed, TRLs 1–2 are established. At the next level, non-GLP level cell culture and animal studies are to be conducted. Here, the application-specific cell line and disease specific animal model are to be used. Given that the material has acceptable biocompatability at the non-GLP level, the technology is now at TRLs 3–4. At a still higher level, the engineering design of patient-specific implants or the design of biomedical devices with a range of sizes are to be conceptualised, and MRL assessment of device/implant prototypes is to be pursued. The MRL assessment at this stage is more challenging than at an earlier level and various stages of device manufacturing are often accompanied by close dimensional tolerance or device-specific property measurements at the GLP-level. Once the prototype fabrication is established, the concept is considered to pass through MRLs 3–5. Also, when the GLP-level device testing, i.e. hip/knee simulator for THR/TKR, has been successfully conducted to assess the biomechanical performance, the overall technology is said to reach the end of TRL 5. The next level of TRL would involve clinical trials following ethically approved protocols. Once the outcome measures are clinically acceptable, TRL 6 is established and the technology is ready to be transferred to a biomedical company.

> Prior to large-scale industrial production, higher MRL levels i.e. MRL 6 or above are to be assessed depending on regulatory approval to ensure the strictest control over device performance in human healthcare.

The discussion in this section and particularly in the last two sections are based on the author's own perception and should be used with due diligence.

References

1. http://as.nasa.gov/aboutus/trl-introduction.html.
2. https://www.ucl.ac.uk/electrochemical-innovation-lab/people/phd-students/jon_morgado.

Chapter 12
A Way Forward

12.1 Patient-Specific Implants

Traditionally, the biomedical devices are available in different sizes or shapes suited to human population for a given country. Each and every patient has a unique requirement of target bone replacement, especially with respect to mechanical and biocompatibility properties. Therefore, a combination of various porous designs composed of different materials (titanium/hydroxyapatite and neovascularization-promoting polymer materials) can be manufactured by additive manufacturing technologies, and their mechanical properties can be measured to obtain a property database of medical materials. This is in line with the existing need to develop implants for those patients, whose bone condition at any location has deteriorated so much that it cannot bear the load, or to respond to any other cause generating any vulnerable situation in the human body due to disease or trauma. However, the existing off-the-shelf implants used as remedial measures can take care of the problem only partially. Through developing a production method and business plan to commercialise patient-specific customized implants, it is expected that better performance with precise adaptation to the region of surgery, reduced procedural times, better aesthetics and most importantly greater integration with the injury site will be obtained.

To this end, engineering tools like the finite element (FE) analysis method can be adopted in order to predict the mechanical properties desired for a particular patient. Several parameters will be taken as the input with the most important being the patient's bone condition and the patient's physical parameters. Such an approach to the manufacturing of bone replacement is not yet reported in the literature. In the last decade, the use of this mathematical approach has enhanced our understanding of bone fracture remodelling and most importantly enabled us to understand the relationships between load transfer and optimized tissue integration or load redistribution after insertion of orthopedic implants. Coupled with the advent of more advanced imaging modalities, development of models that closely represent the

B. Basu and S. Ghosh, *Biomaterials for Musculoskeletal Regeneration*,
Indian Institute of Metals Series, DOI 10.1007/978-981-10-3017-8_12

biomechanics of individual patients is now possible. However, improvements in accuracy, reliability, and ease of use of the applied FE modelling method are needed before clinical adoption of patient-specific FE models. Therefore, one important aspect that concerns the development of proper patient-specific models is the incorporation of correct material properties after image segmentation and meshing. This step requires an estimation of patient-specific parameters from the CT/MRI scan of patients.

The idea of patient-specific implants for musculoskeletal reconstruction is rapidly gaining acceptance amongst surgeons and changing the way surgeons plan and execute their operations. Patient-specific customized implants have shown a better performance than their generic counterparts worldwide for maxillofacial operations owing to the precise adaptation to the region of surgery, reduced procedural times, better aesthetics, fast activation of regeneration related to signaling mechanisms and most importantly greater integration with the surrounding tissues at the defect. However, this paradigm has not been adopted in clinics in many developing nations, to a large extent.

A part of the problem lies in the fact that the development of such products and processes essentially requires interdisciplinary efforts drawing on the expertise of diverse areas. For example, advances in radiology along with image processing, analysis and patient-specific model development which are now possible should be combined with 3D printing and other additive manufacturing (AM) technologies. Integration of computer-aided design tools with computer-aided manufacturing especially AM [e.g.3D printing (3DP)] tools holds promise in the context of successful production of mass customized implants. A suitable business model is also required to be framed for mass customization models.

Such manufacturing approaches are expected to benefit economically marginalized members of society, who are afflicted with rare disorders impairing mobility and need customized assisting devices or prosthetics but are not served by industries owing to lack of business incentives. They are often provided with off-the-shelf implants and devices and hence subjected to significant discomfort. This discomfort is not always expressed, because of social pressures on them, and hence serving this marginalized population would be beneficial for social diversity.

The direct production of orthopedic implants for bone replacement application by additive manufacturing will eliminate the constraints of shape, size, curvature, complex internal geometries and mechanical properties, enabling them to better conform to the physical and mechanical requirements in the region of implantation, hence providing improved clinical outcomes. In one of the research publications, the appropriate handling of computer generated design files prior to the start of the AM process has been advised [1]. However, several fundamental challenges remain to be addressed before the clinical adaptation of AM technologies. Developing more reliable algorithms allowing precise feature extraction from medical imaging modalities like CT and MRI, specifically for medical AM, is an objective yet to be achieved. Though finite element methods are often used for design of implants for pre-operative planning, such methods are not yet adopted in clinical practice for patient bone condition specific design of implants. This will significantly help in optimizing treatment and predicting the prognosis of therapies, as well as in the intelligent design of clinician training. Patient-specific modeling allows measurement for a patient which otherwise cannot be recorded, and equips physicians with new data or new interpretation of the previously existing data.

12.2 Design Related Challenges for Porous Scaffolds

A number of biomedical applications demand the use of porous scaffolds, which can be fabricated using a host of AM techniques. One of the great advantages of AM-based processes is the ability to control external as well as internal morphology easily during scaffold/implant generation. External morphology (size and shape) can and should be adapted to the dimensions of the patient-specific defect. For this, 3D data from the patient is needed, which can be provided mostly as a CT or MRI dataset. Several software solutions for translating these datasets into command files for the RP machines have been developed, which is why this problem can be regarded as solved. In contrast, defining the internal morphology of (porous) scaffolds and implants still needs individual optimization and cannot be fully automated.

Despite the wider use of porous scaffolds, their design and geometry optimization for successful integration in tissue engineering requires further developments to identify optimal design. This has motivated researchers to develop new designs for orthopedic/dental applications, for example, optimal bone and cartilage regeneration. These aspects included criteria in geometry selection for the 3D porous scaffold library, and minimum ratio between pore size and layer thickness, which can be used in the design process. Although the requirement for fabrication of materials with gradient in porosity has been well perceived, there have not been enough efforts to create various porous architectures with complex pore size distribution.

The fabrication of scaffolds with complex porous architectures therefore represents a major challenge in the field of tissue engineering. The pore size and shape should be specified for different biomedical applications. For example, for hard tissue application, the pore size should be optimized for enhanced bone growth without causing the mechanical properties to deteriorate. Although a large number of studies have demonstrated the possibility of constructing tissue-engineered bone repairing scaffolds with pore size distributions using rapid prototyping techniques, the design aspects of mimicking complex pore shape as well as spatial distribution of pore sizes of natural hard tissue remain still unexplored. Computer simulation can be effective in optimizing pore characteristics with a size and mechanical requirement constraint. This should be the starting point for the design of a 3D powder-printed scaffold.

The mathematical modeling of porosity in bone implants may be prospective in the context of creating multi-scale pore sizes in the same layer of a particular scaffold segment. In addition, complex pore shapes, depending on various anatomical sites can also be generated in theoretical modeling. The mathematical modeling of scaffold pore designs becomes necessary when the complex positions, orientations, shapes and sizes of the pores in natural bone cannot be mimicked appropriately in CAD built scaffold architectures or using commercial 3D design software.

When designing 3D scaffolds, the mechano-biological aspects need to be considered. Computer simulations to predict tissue formation use mechano-regulation algorithms to formulate how mechanical forces modulate tissue differentiation and bone remodeling. Numerical simulations of bone growth into scaffolds based on mechano-biological models are however relatively new. For example, Sanz-Herrera et al. [2] used a two-dimensional (2D) finite element model to analyze bone growth within a scaffold implanted in the femoral condyle of a rabbit. Similarly, Adachi et al. attempted to model bone regeneration in a unit cell of a three-dimensional scaffold microstructure. The fabrication of the designed porous scaffolds in line with the mechano-biological property requirements demands the use of an advanced fabrication technique, which can translate design into product.

12.3 Challenges in Mechanical Property Evaluation

Although routine mechanical property measurements for porous scaffolds similar to those carried out for non-porous scaffolds or implants are being reported in the literature, this author believes that a different experimental strategy needs to be adopted, particularly to capture the effect of porous architecture on the mechanical properties. For example, in situ tensile or compression set-up can be used in combination with micro-CT to perform a series of thoughtful experiments with varying crosshead speed to probe into the influence of strain rate on stress–strain response of porous scaffold. In addition, micro-CT images of the scaffold at various timescales during deformation can be acquired in order to interrupt and analyze the progressive morphological changes of a 3D scaffold under given loading.

The failure of scaffolds under impact loading also represents another potential research area. The impact loading can be simulated using high strain rate mechanical testing and this can be conducted using a specialized experimental set-up like the Split-Hopkinson Pressure Bar test (SHPB) [3, 4]. The SHPB test for ceramics scaffolds needs to be conducted under compression mode; using a high-speed camera, the collapse of the porous architecture can be recorded at various timescales to understand the failure mechanism. At present, such experiments for 3D scaffolds with potential biomedical application are not yet reported in open literature. Here again, one can vary the strain rate and thereby can capture strain rate dependent dynamic stress response for ceramic or metallic scaffolds. Taken together, it needs to be highlighted that the strain rate dependent stress–strain response for porous scaffolds must be given due importance in the case of a load bearing scaffold.

One of the major concerns for a ceramic based scaffold is the strength reliability, which is quantified by the determination of the Weibull modulus (m). The measurement of 'm' requires a large number of samples to be tested in compression, shear or tensile mode. For example, Meininger et al. [5] recently reported the Weibull modulus of magnesium phosphate (MgP) based scaffolds to vary in the range of 5–8, which is comparable with the monolithic alumina ceramics. Prior to the clinical application of any bioceramics scaffold, it is imperative that the Weibull modulus of the scaffold be determined using either compression or bending mode.

Concerning the microstructure-property relationship, it is instructive to use different porous architecture for a given material so that the functional relationship between the architectural parameters (pore size, pore interconnection density) and elastic modulus or strength is established. Such a relationship can be subsequently used as a design guideline for scaffold manufacturing for any targeted application, which may experience a combination of mechanical properties. Although such functional relationship may exist in the case of non-porous materials, this is somehow not reported for a porous scaffold. It must however be recognized that such a relationship will be sensitive to material composition. As far as the failure mechanism is concerned it is important to establish a specific pore architecture which can facilitate progressive failure i.e. delayed failure once maximum strength is reached. The ideal scenario would be that under a load, the entire scaffold fails in a layer-by-layer manner, the way the scaffold was constructed during the additive manufacturing process.

It is believed that the above-mentioned experimental analysis approaches can be integrated to develop an overall understanding of the deformation and failure mechanism of different scaffolds under different loading conditions.

12.4 Additive Manufacturing Related Challenges

The fundamental concepts related to low temperature additive manufacturing of porous scaffolds are discussed extensively in a preceding chapter. In the context of powder-based 3DPP/3DPL of scaffolds, particle size and shape are equally important. While finer particle sizes with a larger interfacial area would have more surface interaction with a binder, the spherical-shaped powder would have a better flowability property. If powder size is too small, it would not be homogenously spread in thin layers, because of low powder flowability. Moreover, the spread roller may overpress the powder. On the other hand, if the powder size is too large, the structure would lack good strength and small parts would not be strong enough. The powder properties, such as particle size, flowability and roughness, have to be determined as key factors with predictive value for final 3DP outcome.

Another challenge lies in formulating the suitable binder and optimizing the binder saturation volume corresponding to a particular layer thickness to print a new material by 3DPP technique. The type and amount of the binder need to be optimized to fabricate a 3D printed scaffold. The use of a reactive binder (e.g. H_3PO_4) is restricted by the print-head of the 3DPP machine. In the current scenario, most (if not all) of the 3D printers used in biomedical applications have commercially patented print-head technology, which is optimized for regular ink printing on paper. A better solution is the development of suitable print-heads (piezoelectric/thermal DOD) such that they can print a chemically reactive binder. Hence, it remains challenging to find out the composition and the suitable amount of the binder for printing a homogeneous/gradient porous scaffold.

As far as other processing related challenges, (e.g. binder-material interaction), are concerned, the wetting of powder by the binder depends on many parameters such as the contact angle between binder and powder, the binder viscosity, the topography of the powder bed surface (depending on powder shape and size) and the chemical reactions occurring between binder and powder. Such interactions can lead to swelling or partial dissolution of the particles constituting the powder bed. The amount of binder solution absorbed and the volume distributed within the powder bed determine the resolution and mechanical properties. The binder spreading can be prevented by a reactivity that is too high, whereas a very low reactivity might favor intensive binder spreading. So, the timing and reactivity of the binder reaction are crucial and need to be optimized for the final printing accuracy, and for the consolidation of two consecutive layers.

The current methods available for depowdering, like brushing, air blowing and vibration, are not very helpful for complex or internal features. Moreover, in the case of small pores (<100 μm), pore-architecture and also scaffold design have to take depowdering into account. Indeed, minimum pore sizes produced by 3DP are currently ~100 μm, although the printing resolution with a 600 dpi print-head is in the range of approximately 40 μm. This is partially due to the previously mentioned spreading of binder in the powder bed, but removal of unbound powder (with particle size of 20–30 μm and irregular particle shapes blocking each other) from very small pores is another limiting parameter. Therefore, a new technique with better control needs to be discovered to completely depowder complex/gradient porous structure. This might be performed by simultaneously printing reactive chemicals (e.g. EDTA as Ca^{2+} complexing agent) into the pores, which would lead to a partial dissolution of the unbound calcium phosphate powder, reducing particle size in the pores, post-printing.

Concerning the sintering-based post-processing, phase transformation may take place at higher temperature during sintering; that would result in conversion of the scaffold material into an undesired different phase. Porosity can also be affected while the binder part is being removed. Hence, optimized post-processing heating is required to remove the excess binder as well as to develop materials with good strength property.

The development of novel bio-inks, suitable as pastes for 3D plotting (3DPL) as well as binders for 3D powder printing (3DPP) is considered as a new approach in fabricating bio-scaffolds in ambient temperature. Bio-inks for utilization in RP should ideally mimic the extracellular matrix (ECM) of the tissue to be fabricated. Unfortunately, the engineering properties of most ECM constituents and especially the ECM as a whole are extremely poor, which limits their application in biomaterial processing and additive manufacturing. The development of novel, ECM-based bio-inks with the desired composition is to be investigated in future to overcome the shortcomings. The provision for integration of live cells in the 3D plotting process with the bio-inks needs to be investigated. The state-of-the-art biodegradable biomaterials (synthetic and natural polymers) and polymer/calcium phosphate composites can be used in this aspect. Natural polymers (biopolymers) on the other hand are more cell interactive, but often inconsistent in properties like viscosity and mechanical strength owing to charge differences. In addition, impurities like endotoxins, polyphenols etc., may occur in medical grade polymers, leading to major limitations for large-scale medical use.

In the case of 3D plotting, the variability of pore size and morphology is limited compared to 3DPP. This is in view of the fact that the pores can only be defined by arranging strands with commonly round profile in different ways for the 3DPL route. Much work in this respect has been done with thermoplastic polymers using FDM, which can be translated to 3D plotting. Not much has been done concerning mathematical modeling of such morphologies and one can assume that novel designs will be developed, which can be reproduced by 3D plotting. Especially when different materials like stiff mineral phases and elastically compliant (bio) polymers are to be combined within one construct, mathematical optimization of the pore arrangement would be beneficial.

Similarly, additive manufacturing is introduced as an automatic process to create rapid prototypes and functional end-use parts, taking virtual designs from Computer Aided Design (CAD) software and transforming them into layer-wise cross-sections, until the model is complete. Today, AM is a growing market in every manufacturing sector, with a global additive manufacturing market of $1843.2 million in 2012, and is expected to grow at a CAGR of 13.5 % to reach $3471.9 million by 2017. The global demand for 3D printers and related materials and software is projected to rise 21 % per year to $5.0 billion in 2017. Prototyping continues to account for the majority of demand, but more rapid growth is now seen in production and consumer applications. The 3D printer is now increasingly being used to manufacture direct production parts and finished goods in a wide variety of applications. Recently, Wipro, an Indian multinational giant, has also ventured into this area with an investment of $1 million. It can therefore be concluded that patient-specific implant manufacturing using radiological data, cloud computing and data transmission, together with additive manufacturing using bio-functional and biological components, present an attractive techno-economic scenario to venture into in the near future.

In terms of the business potential, the cost-effectiveness of such approaches needs to be established by carefully estimating the cost structure or cost-benefit analysis. The patient-specific approach can provide the hospital with an additional 28 min of available operation theatre time per procedure, based on reduction in preparation and operative times compared to conventional methods, and an additional 67 min compared to computer navigation based on this model. This time saving is likely to give the health care system a greater economic impact along with implant-related cost savings. In India presently a total hip arthroplasty procedure costs approx. 5000USD. It is expected that the production efficiency due to additive manufacturing will take it downwards.

12.5 Challenges in Biomedical Device Development

In the last few decades, the healthcare system in many developing nations, such as India, China is gradually embracing the western procedures and protocols for diagnostics, treatment, and patient-care. As a result, the quality and efficacy of healthcare have significantly improved, but the cost of quality healthcare has escalated beyond the reach of the economically challenged population and has become a burden for most segments of the large population. The principal reason for this can be attributed to the high cost of imported biomedical devices.

As mentioned earlier in this book, the development of competitive indigenous biomedical devices is widely seen as an urgent need in many of the developing nations around the world. A silver lining in this dark cloud is the emerging pockets of innovation in academic institutions and industry. What is lacking is the supporting infrastructure to foster the development of biomedical devices. Bridging this gap and then extending help to researchers, inventors, and industry in the nation should be the focus. The creation of a comprehensive facility as well as of a research ecosystem that supports the development of biomedical devices from concepts to marketable products is essential to ensure the growth of this field of major societal relevance. A vision towards achieving such a goal is described below, so that a reader can realise the need to adopt a multidisciplinary approach in developing biomedical devices. The necessity for the next generation of biomedical researchers is also highlighted (Fig. 12.1).

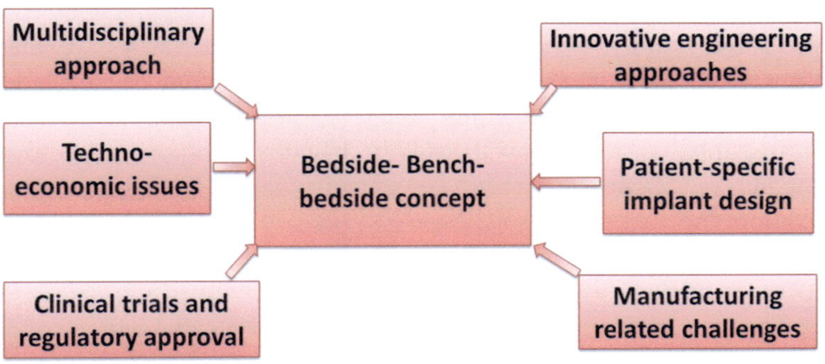

Fig. 12.1 Various issues restricting effective and smooth translation of the 'bedside-bench-bedside' concept to facilitate better human healthcare, as discussed in this chapter

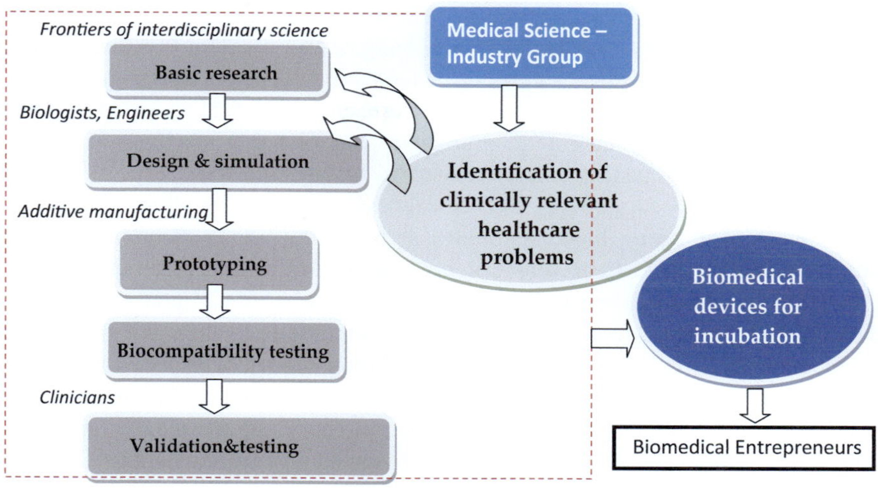

Fig. 12.2 Overview of the concept of translation of research to biomedical devices

As shown in Fig. 12.2, self-reliance in healthcare devices demands the availability of a formal Biomedical science-industry group comprising clinicians and biomedical industry professionals. This is deemed to be important in many nations, including India. In addition, it is equally important to bring together engineers, scientists (biologists in particular), clinicians and end-users or patients. Access to clinical advice and participation is critical for efficient development of biomedical devices. Therefore, clinicians should be hired as consultants to enable them to work at an academic institute. This Medical Science-Industry Group should be involved in all stages of the development of biomedical devices (see Fig. 12.2).

Facility for Biomedical Devices: Another important step would be to set up a central integrated facility for developing biomedical devices together with a computational platform. The human ecosystem around such a facility should make it possible to identify important and relevant problems and would provide guidance from the viewpoints of clinical departments and biomedical industry.

Such a facility should have a workshop with modern machines for prototyping biomedical devices and facility for biocompatibility testing. With all this, a researcher can move from the conceptual stage to the clinical validation stage, all in one facility, following the advice of clinicians and biomedical industry experts. Ample support in terms of the necessary basic research should also be available at the facility. As shown in Fig. 12.2, the facility could also act as an incubation centre for biomedical devices.

Prototyping of Biomedical Devices: Prototyping of biomedical devices during the development stage is another practical difficulty. The conventional workshops available in many engineering departments are clearly inadequate for this purpose. In fact, even established biomedical industries find prototyping prohibitively expensive because making a die or a mould for a device whose design is yet to be finalized and tested is indeed expensive. The use of the new class of 3D printers emerging from additive manufacturing can mitigate this problem to a large extent. As biomedical devices use a variety of materials including polymers, metals, and ceramics, an ideal facility should have a comprehensive fabrication workshop with an access to traditional and emerging additive manufacturing machines.

Biocompatibility testing: As compared to testing and validation of general materials science based products, biomedical devices have to pass through the extra requirement of ensuring biocompatibility. It is worthwhile to mention that as a proof-of-concept, *in vitro* testing can be used to screen the materials prior to pre-clinical testing (*in vivo*). The testing facility should have equipment for culture facilities for bacteria, cells, and tissues. It should also have equipment for haemocompatibility, toxicology, and histopathology along with essential microscopy and cell level analysis. Access to a good animal facility is also important. In addition to the equipment, the facility will have professionally managed well-trained technical staff who are expected to maintain the equipment and help the researchers.

Clinical trials and regulatory approval: In reference to the discussion in one of the chapters in this book, any biomedical device, once proven to be biocompatible pre-clinically either at implant or device level, needs to undergo clinical trials in human patients. This requires strict regulatory approval, apart from the human ethics committee approval and the funding to conduct clinical trials. This has often been the bottleneck in translational research on biomaterials, which also demands the involvement of busy clinicians. The sequence of the flow of concepts and research results in the translation of research concept to final biomedical device is shown in Fig. 12.3. Once a device is proven to provide clinically appropriate patient satisfaction, the technology for device manufacturing can be transferred to biomedical device companies. While the biomedical research can be very satisfying to a researcher because of its societal relevance, it is important to mention here that such research also actively involves people with widely differing expertise which demands effective integration.

Human Resource Training: The creation of a facility for biomedical devices, while helping research and development, will also help in training people. By way

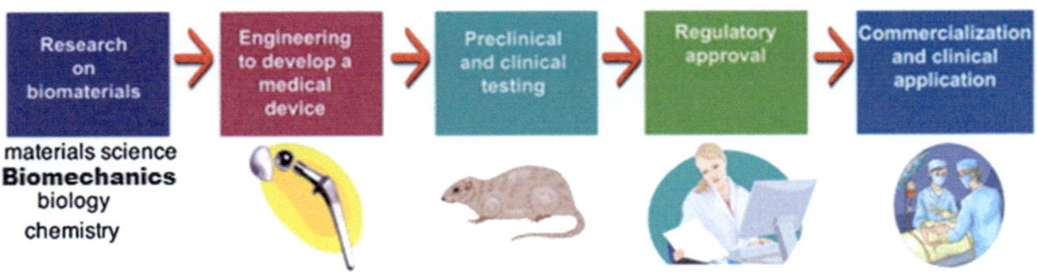

Fig. 12.3 The translational path from basic science, to a medical device, to clinical application. Although a rabbit animal model is shown under pre-clinical testing, larger animal models are to be used to validate small/medium animal models. Also, for representative purposes only, a total hip joint replacement device is shown

of symposia, summer schools, specialized programmes and internships for medical students and doctors, and familiarization workshops for undergraduates and high-school students, such a facility can play its part in human resource training.

References

1. Huotilainen E, Jaanimets R, Valášek J, Marcián P, Salmi M, Tuomi J, Mäkitie A, Wolff J. Inaccuracies in additive manufactured medical skull models caused by the DICOM to STL conversion process. J Cranio-Maxillofac Surg. 2014;42(5):e259–65.
2. Boss JH. Osseointegration. J Long Term Eff Med Impl. 1999;18:1–10.
3. Gorustovich AA, Steimetz T, Cabrini RL, Porto Lopez JM. Osteoconductivity of strontium-doped bioactive glass particles: a histomorphometric study in rats. J Biomed Mater Res A. 2010;92(1):232–7.
4. Gauthier O, Muller R, von Stechow D, Lamy B, Weiss P, Bouler JM, et al. In vivo bone regeneration with injectable calcium phosphate biomaterial: a three-dimensional micro-computed tomographic, biomechanical and SEM study. Biomaterials. 2005;26(27): 5444–53.
5. Meininger S, Mandal S, Kumar A, Groll J, Basu B. Uwe Gbureck; strength reliability and in vitro degradation on three-dimensional powder printed strontium-substituted magnesium phosphate scaffolds. Acta Biomaterialia. 2016;31:401–11.

Index

B. Basu and S. Ghosh, *Biomaterials for Musculoskeletal Regeneration*,
Indian Institute of Metals Series, DOI 10.1007/978-981-10-3017-8

MIX
Papier aus verantwortungsvollen Quellen
Paper from responsible sources
FSC® C105338

If you have any concerns about our products,
you can contact us on
ProductSafety@springernature.com

In case Publisher is established outside the EU,
the EU authorized representative is:
Springer Nature Customer Service Center GmbH
Europaplatz 3, 69115 Heidelberg, Germany

Printed by Libri Plureos GmbH
in Hamburg, Germany